MW01056919

Algebra

Pure and Applied
UNDERGRADUATE TEXTS · 11

Algebra

Mark R. Sepanski

American Mathematical Society
Providence, Rhode Island

2010 *Mathematics Subject Classification.* Primary 00–01;
Secondary 20–01, 12–01, 13–01, 16–01.

For additional information and updates on this book, visit
www.ams.org/bookpages/amstext-11

Library of Congress Cataloging-in-Publication Data

Sepanski, Mark R. (Mark Roger)
Algebra / Mark R. Sepanski.
p. cm. — (Pure and applied undergraduate texts ; vol. 11)
Includes index.
ISBN 978-0-8218-5294-1 (alk. paper)
1. Algebra—Textbooks. I. American Mathematical Society. II. Title.

QA152.3.S46 2010
512—dc22

2010022789

∞ The paper used in this book is acid-free and falls within the guidelines
established to ensure permanence and durability.
Visit the AMS home page at http://www.ams.org/

10 9 8 7 6 5 4 3 2 1 15 14 13 12 11 10

To Laura, Sarah, Ben, and Shannon
and also to Mom, Dad, Peter, Sheila, and Andy

Contents

Preface

Even to this day, I recall my first algebra course as wonderfully delightful and exciting. In my youthful pride, I also recall secretly fearing that when I told people I was studying algebra, they would think I was doing the type of mathematics taught to twelve-year-olds. After all, once you can solve $2x = 4$ and know how to use the quadratic equation, what else is there?

In fact, algebra is a large subfield of modern mathematics that has been intensely researched over the last few hundred years. It is very different from the type of mathematics that most students see during the bulk of their college years. Usually the majority of undergraduate classes are devoted to studying the subfield of mathematics known as analysis, which deals with limits, derivatives, integrals, and all sorts of related concepts.

In contrast, algebra is the study of mathematical systems that, in some sense, generalize the behavior of numbers. For example, the set of integers, the set of polynomials, and the set of $n \times n$ matrices are all very different types of objects. Nevertheless, it makes sense to "add" and "multiply" in each of these cases. Moreover in each of these settings, the formal behavior of addition and multiplication is strikingly similar. As a result, it is possible to abstract the properties of addition and multiplication in order to prove a single theorem applying simultaneously to each different case. This allows for a great deal of efficiency.

More importantly, these "algebraic" techniques frequently make it possible to tackle problems that are otherwise completely inaccessible. As a remarkable example, the ancient Greeks were interested in constructions using only a compass and straightedge. For instance, they could construct a regular 3-gon, 4-gon, 5-gon, 6-gon, 8-gon, and 10-gon, but they never succeeded in constructing a regular 7-gon, 9-gon, or 11-gon. For centuries afterward, geometers tried and failed to repair this gap. In fact, it took a few thousand years and the advent of algebraic techniques for people to figure out what was going on. In 1837, Pierre Wantzel succeeded in showing that the difficulty did not lie with a failure of human ingenuity, but rather that it was *impossible* to construct a regular 7-gon, 9-gon, or 11-gon (Exercise 4.179).

This book is intended as a first course in algebra. For the most part, the only formal prerequisite is the ability to add and multiply integers. In principal, my nine-year-old daughter is therefore more than prepared to study algebra and ought to have no problem jumping right in on page one. Excepting the rare genius, of course, this is a big fallacy. Generally speaking, algebra is considered one of the most difficult courses taken by an undergraduate mathematics major. While it is technically true that the greatest prerequisite for this material is familiarity with the integers, the true prerequisite is mathematical maturity. For many students, a "how to prove things" course is a good way to obtain this experience, though this is certainly not the only way. Besides maturity, it is also assumed that students are familiar with the basics of linear algebra such as matrix multiplication and that they know how to multiply complex numbers.

When I teach this course, I begin with the integers in order to start off gently and to lay a foundation for future work. The integers and modular arithmetic can be polished off fairly quickly and then I move on to group theory, ring theory, and finally field theory. In terms of big goals, I like to get my students all the way through Sylow's theorems and Galois theory. The text is therefore partly designed with efficiency in mind so that these milestones can be reached by the adventurous or the insane in one semester (perhaps omitting a few sections such as §2.5.3 of Chapter 1, §7.3 and §10 of Chapter 2, and §5.3 of Chapter 3). Of course there is much to be said for adopting a more leisurely pace and the text is equally appropriate for such a relaxed style spanning a year-long course. In fact, there are over 750 exercises in this book and spending plenty of class time on these problems is an excellent investment. In either case, I highly recommend regularly meeting with the students during or outside of class and having them present proofs and exercises on the board. In my experience, this is one of the best ways to help students absorb and master the material.

In terms of the arrangement of the book, Chapter 1 deals with arithmetic. Section 1 examines the integers, mainly presenting the division algorithm and the notion of divisors. This naturally leads to modular arithmetic in Section 2 where congruence classes and their structure are developed. As applications, the Chinese Remainder Theorem, Euler's Phi Function, Fermat's Little Theorem, and Public-Key Cryptography are studied. Finally, the definition of a general equivalence relation is given at this juncture since it fits nicely with the notion of congruence.

Spingboarding off modular arithmetic, Chapter 2 is devoted to the study of groups. Section 1 gives the basic definition and examples of groups. Both multiplicative and additive notation are simultaneously developed and examples are drawn from arithmetic, matrix groups, symmetric groups, and dihedral groups. Section 2 studies the most basic properties of groups, especially the notion of order. Section 3 examines how to construct new groups from already existing groups by means of subgroups or direct products. Section 4 develops the notion and basic properties of morphisms between groups. Section 5 is devoted to the study of quotients of groups. The definition of a coset is given in §5.1, Lagrange's Theorem is in §5.2, the notion of normality is in §5.3, the Correspondence Theorem is in §5.4, and the First Isomorphism Theorem is in §5.5. Section 6 presents the Fundamental Theorem of Finite Abelian Groups. Section 7 studies the symmetric group. Cayley's Theorem is in §7.1, the Cyclic Decomposition Theorem is in §7.2, partitions and the classification of conjugacy classes are in §7.3, and parity and the alternating group are in §7.4. Section 8 studies group actions and includes the Orbit Stabilizer Theorem, the Orbit Decomposition Theorem, Burnside's Counting Theorem, and the Class Equation. Section 9 contains Cauchy's Theorem and the three Sylow theorems. Section 10 studies simple groups and composition series and proves the Jordan-Hölder Theorem for finite groups.

Further generalizing modular arithmetic, Chapter 3 is devoted to the study of rings with a special emphasis on polynomial rings over fields. Because a number of early concepts in ring theory are completely analogous to ones in group theory with virtually identical proofs, the rate at which these early concepts are introduced is greatly accelerated. Section 1 presents the basic definition and properties of a ring. The definition is given in §1.1, examples drawn from arithmetic, matrix rings, and polynomial rings are in §1.2, special properties relating to units, zero divisors, and domains are in §1.3, subrings are in §1.4, and direct products are in §1.5. Section 2 is devoted to the study of morphisms and quotients. Morphisms are developed in §2.1, ideals are introduced in §2.2, quotients, the Correspondence Theorem, prime ideals, and maximal ideals are in §2.3, and the First Isomorphism Theorem is in §2.4. Section 3 begins the study of polynomial rings and roots. The Division Algorithm is in §3.1, roots and the fact that a finite subgroup of a field is cyclic are contained in §3.2, the Rational Root Test is in §3.3, and the Fundamental Theorem of Algebra is in §3.4. Section 4 studies the irreducibility of polynomials including Gauss's Lemma, Eisenstein's Criterion, and Modular Reduction. Section 5 develops the theory of unique factorization domains (UFD's). The definition and two equivalent conditions in terms of the ascending chain condition on principle ideals (ACCP), primes, and gcd's are given in §5.1, quotient fields are in §5.2, and the fact that a polynomial ring over a UFD is a UFD is in §5.3. Finally, Section 6 gives the basic theory of principle ideal domains (PID's) and Euclidean domains including Fermat's Theorem on the Sum of Two Squares and factorization of the Gaussian integers.

Making heavy use of the theory of polynomial rings, Chapter 4 is devoted to the study of fields. Section 1 gives the basic theory of finite and algebraic extensions. Section 2 develops the theory of splitting fields. Existence, isomorphic extension theorems, and normality are in §2.1 and the existence of algebraic closures is in §2.2. Section 3 gives the theory of finite fields. Section 4 contains Galois theory. The notion of the Galois group is in §4.1, separability is in §4.2, and the Fundamental Theorem of Galois Theory on intermediate fields is in §4.3. Section 5 gives proofs of some famous impossibilities involving ruler and compass constructions and factoring the quintic. Section 6 gives the theory of cyclotomic fields over $\mathbb{Q}$.

Throughout the text of the book, any text cross-reference to a section (subsection or subsubsection) refers to that section (subsection or subsubsection) in the current chapter unless otherwise specified.

There are a number of resources that had a powerful impact on this work and to which I am greatly indebted. I was first introduced to the subject in the summer of 1988 while reading *Abstract Algebra: A First Course* by L. Goldstein (Prentice-Hall, Inc., 1973). It was a remarkably enjoyable experience and has shaped what material I use and how I teach algebra to this day. More recently, I have used T. Hungerford's excellent book *Abstract Algebra: An Introduction* (Thomson Brooks/Cole, 2nd ed., 1996) when teaching algebra and the text has had a strong impact on my presentation of the subject. There are also two books that deserve special mention due to their outstanding quality: K. Nicholson's *Introduction to Abstract Algebra* (Wiley-Interscience, 3rd ed., 2007) and C. Lanski's *Concepts in Abstract Algebra* (Thomson Brooks/Cole, 2005). Both texts are exemplary introductions to algebra and have been extremely useful and influential to me. In addition, M. Artin's *Algebra* (Prentice Hall, Inc., 1991) and D. Dummit and R. Foote's *Abstract Algebra* (John Wiley & Sons, Inc., 3rd ed., 2004) are both excellent works that played a role in shaping certain sections of this book. Finally, the author is grateful to the Baylor Sabbatical Committee for its support during part of the preparation of this text and to M. Hunziker for help with a number of diagrams and pictures.

CHAPTER 1

Arithmetic

Mathematics begins with the integers. The branch of mathematics known as *algebra* devotes itself to studying mathematical systems that are, in some sense, generalizations of the integers or closely related gadgets. This chapter gathers together some of the important properties of the integers and modular arithmetic that will be used throughout the rest of the book.

1. Integers

1.1. Basic Properties. The set $\mathbb{Z}$ of *integers* is the well-loved set

$$\mathbb{Z} = \{\ldots, -4, -3, -2, -1, 0, 1, 2, 3, 4, \ldots\}$$

that most of us learned about in grade school. Instead of beating around the bush, we assume the reader is familiar with the elementary properties of addition and multiplication in $\mathbb{Z}$. At the risk of the reader rolling his or her eyes, we list some of the basic properties of $\mathbb{Z}$ since our later generalizations closely mimic them.

AXIOM 1 (Additive and Multiplicative Properties). Let $a, b, c \in \mathbb{Z}$.
The properties of addition are:

(1) *Associativity:* $(a + b) + c = a + (b + c)$.
(2) *Commutativity:* $a + b = b + a$.
(3) *Identity:* $0 + a = a$.
(4) *Inverses:* $a + (-a) = 0$.

The properties of multiplication are:

(1) *Associativity:* $(ab)\,c = a\,(bc)$.
(2) *Commutativity:* $ab = ba$.
(3) *Identity:* $1 \cdot a = a$.

There is a law relating addition and multiplication:

(1) *Distributivity:* $a\,(b + c) = ab + ac$.

There is a *zero divisor* law:

(1) $ab = 0$ if and only if $a = 0$ or $b = 0$.

We also assume that the reader is familiar with the standard elementary properties of inequalities, which we do not even bother to list. However, a related notion called the *Axiom of Well-Ordering* is worth mentioning. Depending on how one sets up the integers, the Axiom of Well-Ordering can be viewed as a straight-up axiom or as a corollary of the properties of positive integers, which we also do not bother to explicitly list. Instead, recall that a subset $S \subseteq \mathbb{Z}$ is *bounded above* if there exists $N \in \mathbb{Z}$ so that $a \leq N$ whenever $a \in S$. For example, the set $S = \{\ldots, -2, -1, 0, 1, 2, 3\}$ is bounded above (by, say, 3 or 4 or 5) while the set $E = \{\ldots, -2, 0, 2, 4, 6, \ldots\}$ of even integers is not bounded above. Notice the set

$S = \{\ldots, -2, -1, 0, 1, 2, 3\}$ has a maximum element, namely 3, while the set E of even integers has no largest element.

AXIOM 2 (Well-Ordering). If a nonempty set $S \subseteq \mathbb{Z}$ is bounded above, then S contains a unique largest element, i.e., there is a unique $m \in S$ so that $m \geq a$ for all $a \in S$.

In a similar fashion, a nonempty set $S \subseteq \mathbb{Z}$ that is *bounded below*, analogously defined, contains a unique smallest element. The reader should note that the Well-Ordering Axiom is a special property of $\mathbb{Z}$ that is not shared by the rational numbers, $\mathbb{Q}$, or by the real numbers, $\mathbb{R}$ (Exercise 1.1).

Finally, three important subsets of $\mathbb{Z}$ pop up frequently enough to deserve their own names. The set $\mathbb{N}$ of *natural numbers* or *positive integers* is

$$\mathbb{N} = \{1, 2, 3, 4, \ldots\},$$

the set $\mathbb{Z}_{\geq 0}$ of *nonnegative integers* is

$$\mathbb{Z}_{\geq 0} = \{0, 1, 2, 3, 4, \ldots\},$$

and the set $\mathbb{Z}^\times$ of *nonzero integers* is

$$\mathbb{Z}^\times = \mathbb{Z} \backslash \{0\} = \{n \in \mathbb{Z} \mid n \neq 0\}.$$

In general if A and B are sets, we use the notation $A \backslash B = \{a \in A \mid a \notin B\}$.

It is worth remarking that there are different conventions for the use of the symbol $\mathbb{N}$. Most often in North America, 0 is not viewed as a natural number. This is the convention we follow in this text. In Europe, however, the opposite viewpoint is often adopted.

1.2. Induction. Mathematical induction is a technique of proof that shares some similarity to the process of knocking over a whole chain of dominoes by simply knocking over the first one. It is a formal way of saying "and just keep doing the same thing".

THEOREM 1.1 (Mathematical Induction). *For each $n \in \mathbb{N}$ suppose there is a proposition P_n so that:*

(1) *The first statement P_1 is true.*

(2) *Whenever P_n is true, then P_{n+1} is also true.*

Then P_n is true for every $n \in \mathbb{N}$.

PROOF. Let $S = \{n \in \mathbb{N} \mid P_n \text{ is a true proposition}\}$. We desire to show $S = \mathbb{N}$, but hypothesis (1) only initially shows $1 \in S$. We can use hypothesis (2) with $n = 1$ to conclude that P_2 is also true, i.e., that $2 \in S$. Similarly since $2 \in S$, hypothesis (2) with $n = 2$ shows $3 \in S$ and so forth. The trick is to make the "and so forth" precise. While it is possible to create a careful proof out of this sort of reasoning, there is another approach.

By way of proof by contradiction, suppose for the moment that $S \neq \mathbb{N}$ so that $\mathbb{N} \backslash S$ is nonempty. Since $\mathbb{N} \backslash S$ is bounded below (by, say, 0), it has a minimal element $m \in \mathbb{N} \backslash S$. Since m is minimal in $\mathbb{N} \backslash S$, it follows that $m - 1 \in S$, i.e., that P_{m-1} is a true statement. However, hypothesis (2) with $n = m - 1$ then forces P_m to be true and thus $m \in S$. This contradicts the fact that $m \in \mathbb{N} \backslash S$ so that $m \notin S$. Therefore the supposition $S \neq \mathbb{N}$ cannot be true. As the only other possibility is that $S = \mathbb{N}$, the proof is finished. □

As a first example, consider the claim (known as *Gauss's Formula*) that

$$1 + 2 + 3 + \cdots + n = \frac{n(n+1)}{2}$$

for any $n \in \mathbb{N}$. First view the above statement as a family of propositions $P_1, P_2, \ldots, P_n, \ldots$ where

$$\begin{aligned} P_1 &: \ 1 = \frac{1(1+1)}{2}, \\ P_2 &: \ 1 + 2 = \frac{2(2+1)}{2}, \\ &\vdots \\ P_n &: \ 1 + 2 + 3 + \cdots + n = \frac{n(n+1)}{2}. \end{aligned}$$

The goal is to show each proposition P_n is true for any $n \in \mathbb{N}$. By inspection, P_1 is obviously true. The principle of mathematical induction allows us to finish the entire problem by proving P_{n+1} is true whenever P_n is true. Thus suppose P_n is true; i.e., suppose $1 + 2 + 3 + \cdots + n = \frac{n(n+1)}{2}$. The whole problem now reduces to verifying that P_{n+1} is true as well; i.e., we must show $1 + 2 + 3 + \cdots + n + (n+1) = \frac{(n+1)(n+2)}{2}$. This is now just simple algebra coupled with the *induction hypothesis* (the assumption that P_n is true):

$$\begin{aligned} 1 + 2 + 3 + \cdots + n + (n+1) &= (1 + 2 + 3 + \cdots + n) + (n+1) \\ &= \frac{n(n+1)}{2} + (n+1) \\ &= (n+1)\left(\frac{n}{2} + 1\right) \\ &= \frac{(n+1)(n+2)}{2} \end{aligned}$$

as desired.

As a fancier example, consider the claim

a set S containing n elements has 2^n subsets

where $n \in \mathbb{N}$. To prove this by induction, again view the above statement as a family of propositions, P_n. The first statement, P_1, says a set S with one element has 2^1 subsets. This is clearly true since if $S = \{a_1\}$, the only subsets are $\varnothing$ and $\{a_1\}$. Induction now allows us to establish the claim simply by proving P_{n+1} is true whenever P_n is true. Thus suppose P_n is true; i.e., suppose any set with n elements has 2^n subsets. We must show this forces P_{n+1} to be true as well; i.e., we must show any set S with $n+1$ elements has 2^{n+1} subsets.

To see this, write $S = \{a_1, \ldots, a_{n+1}\}$ so that S has $n+1$ elements and consider the subset $S_0 = \{a_1, \ldots, a_n\} \subseteq S$. Note that S_0 has n elements so that our induction hypothesis guarantees that S_0 has 2^n subsets. The trick here is to observe that subsets of S come in two flavors: a subset either contains a_{n+1} or it does not. The subsets that do not contain a_{n+1} are actually just the subsets of S_0. We already know there are 2^n of these. To obtain all the remaining subsets that contain a_{n+1}, simply take any subset of S_0 and take the union of it with $\{a_{n+1}\}$. This therefore gives another 2^n subsets of S. All together, this means S has $2^n + 2^n = 2 \cdot 2^n = 2^{n+1}$ subsets, just as we wanted.

In general, if S is a set, write $\mathcal{P}(S)$ for the set of all subsets of S. The set $\mathcal{P}(S)$ is called the *power set* of S. Recall for a finite set S that the *order of the set*, written $|S|$, is the number of elements of S. Our argument above shows that if $|S| = n$, then

$$|\mathcal{P}(S)| = 2^n. \tag{1.2}$$

As a final remark, it should be noted that while induction is often a powerful and efficient method of proof, it usually fails to provide a real explanation of "why" something is true. As a result, sometimes there are much more natural non-inductive proofs that are shorter or give more insight into why a theorem is true.

1.3. Division Algorithm. Going back to grade school, dredge up in your mind the days of long division. For instance, say we wanted to divide 14 by 5. In elementary school, we had to write the answer like this:

$$\begin{array}{rl} & \ \ 2 \quad R\,4. \\ 5 & \overline{)14} \end{array}$$

In words, 5 goes into 14 exactly 2 times with a remainder of 4. This can be written more concisely as

$$14 = 5 \cdot 2 + 4.$$

Of course the *remainder*, 4, has to be at least 0 and strictly less than 5 or else the *quotient*, 2, can be increased. The fact that this grade school procedure of long division always works provides us with one of the most important tools for the study of $\mathbb{Z}$.

THEOREM 1.3 (Division Algorithm). *Let $a \in \mathbb{Z}_{\geq 0}$ and $b \in \mathbb{N}$. There exist unique $q, r \in \mathbb{Z}_{\geq 0}$ so that:*

(1) $a = bq + r$.

(2) $0 \leq r < b$.

PROOF. Consider existence first. The idea for figuring out q is to subtract as many copies of b from a as possible without going below 0. To that end, let $S = \{x \in \mathbb{Z}_{\geq 0} \mid a - bx \geq 0\}$. Since $0 \in S$, S is not empty. Moreover, S is bounded above by $\frac{a}{b}$ (cf. Exercise 1.11). Therefore the Axiom of Well-Ordering shows that S has a maximal element which we call q. From there, let $r = a - bq$. Thus by construction, $a = bq + r$ and $r, q \geq 0$. It only remains to show $r < b$. By way of proof by contradiction, suppose $r \geq b$. Then, since $r = a - bq$, $a - bq \geq b$. Thus $a - b(q+1) \geq 0$ and so $q + 1 \in S$. This is a contradiction to the choice of q as the maximal element of S. Hence $r < b$ as desired and existence is settled.

To show uniqueness, suppose also that $a = bq' + r'$ with $0 \leq r' < b$ for some $q', r' \in \mathbb{Z}$. Subtracting $a = bq+r$ gives $0 = b(q-q')+(r-r')$ so that $r'-r = b(q-q')$. However the inequality $0 \leq r, r' < b$ shows $b > |r' - r|$ so that $b > |b(q - q')|$. Hence $1 > |q - q'|$. Note that the only integer strictly between -1 and 1 is 0. Thus $q - q' = 0$ which also shows $r' - r = 0$, i.e., $q = q'$ and $r' = r'$ as desired. □

Of course if $a = 14$ and $b = 5$, we have already noted that we can write $14 = 5 \cdot 2 + 4$ so that the q and r from the Division Algorithm are given by $q = 2$ and $r = 4$.

COROLLARY 1.4. *Let $a, b \in \mathbb{Z}$ with $b \neq 0$. There exist unique $q, r \in \mathbb{Z}$ so that:*
(1) $a = bq + r$.
(2) $0 \leq r < |b|$.

The proof of Corollary 1.4 is left to Exercise 1.9.

1.4. Divisors.

DEFINITION 1.5. Let $a, b, p \in \mathbb{Z}$.
(1) For $a \neq 0$, we say *a divides b* or that a is a *divisor* of b, written $a \mid b$, if

$$b = ka$$

for some $k \in \mathbb{Z}$. In other words, $a \mid b$ if the remainder of division of b by a is 0.
(2) For $p \neq 0, \pm 1$, we say p is *prime* if its only divisors are $\pm 1, \pm p$.

For instance, the set of divisors of 12 is $\pm\{1, 2, 3, 4, 6, 12\}$ while the set of divisors of 7 is $\pm\{1, 7\}$. In particular, 12 is not prime while 7 is prime.

Notice that $+1$ and -1 obviously satisfy the condition of having no other divisors besides themselves and ± 1. Nevertheless, we explicitly kick them out of club prime. The rationale for this exclusion stems from a technicality related to the uniqueness of factoring integers as a product of primes (cf. Theorem 1.14). For instance, the number 12 can be essentially uniquely factored as $12 = 2^2 \cdot 3$. However, if we were to count the number 1 as being prime, we would lose the uniqueness of the factorization by writing things like $12 = 1 \cdot 2^2 \cdot 3$, $12 = 1^2 \cdot 2^2 \cdot 3$, $12 = 1^3 \cdot 2^2 \cdot 3$, etc.

Divisibility satisfies a number of obvious properties.

LEMMA 1.6. *For $a, b, c \in \mathbb{Z}$, $a \neq 0$:*
(1) $1 \mid b$.
(2) *If $a \mid b$ and $b \mid c$ (here $b \neq 0$), then $a \mid c$.*
(3) *If $a \mid b$ and $a \mid c$, then $a \mid (bx + cy)$ for all $x, y \in \mathbb{Z}$.*
(4) *If $a \mid b$ and $b \neq 0$, then $|a| \leq |b|$. In particular, the set of divisors of b is finite.*
(5) *If $a \mid b$ and $b \mid a$ (here $b \neq 0$), then $a = \pm b$.*

PROOF. As the proofs of parts (1) through (5) are all very similar and make good exercises, we prove part (3) here and leave the rest to Exercise 1.16. Suppose $a \mid b$ and $a \mid c$. Then $b = na$ and $c = ma$ for some $n, m \in \mathbb{Z}$. Thus $bx + cy = (nx + my)\,a$ so that $a \mid (bx + cy)$. □

If $a \in \mathbb{Z}^\times$ is a divisor of both $b, c \in \mathbb{Z}$, then a is called a *common divisor* of b and c. When $b, c \in \mathbb{Z}^\times$, it follows from Lemma 1.6 parts (1) and (4) that there always exists a unique largest common divisor. The study of such gadgets turns out to yield surprisingly powerful results.

DEFINITION 1.7. Let $a, b \in \mathbb{Z}$, not both zero.
(1) The *greatest common divisor* of a and b, written (a, b), is the largest divisor of both a and b.
(2) We say a and b are *relatively prime* if $(a, b) = 1$.

Of course the notion of a greatest common divisor is often introduced in grade school where the alternate notation $\gcd(a, b)$ is often used. As already noted, for $a, b \in \mathbb{Z}^\times$, (a, b) always exists and satisfies

$$1 \leq (a, b) \leq \min\{|a|, |b|\}.$$

If, say, b is allowed to be 0, then $(a, 0)$ still exists and $(a, 0) = |a|$.

As an example, the set of divisors of 12 is $\pm\{1, 2, 3, 4, 6, 12\}$ and the set of divisors of 20 is $\pm\{1, 2, 4, 5, 10, 20\}$ so that the set of common divisors of 12 and 20 is $\pm\{1, 2, 4\}$ and thus $(12, 20) = 4$. On the other hand, the set of divisors of 25 is $\pm\{1, 5, 25\}$ so that the set of common divisors of 12 and 25 is $\pm\{1\}$ and thus $(12, 25) = 1$. In particular 12 and 20 are not relatively prime while 12 and 25 are relatively prime.

Besides simply listing all the common divisors and extracting the largest, there are two other common methods for calculating the greatest common divisor. Although these two techniques are extremely useful for many types of calculations, they are not crucial for the remainder of the text—plus, they make excellent exercises! Therefore the proofs of the two following theorems are outlined in Exercises 1.28 and 1.20.

Note in the following that *product notation* is used where the symbol

$$\prod_{k=1}^{n} a_k$$

is shorthand for $a_1 \cdot a_2 \cdots a_n$ (compare this to the analogous *summation notation*,

$$\sum_{k=1}^{n} a_k,$$

meaning $a_1 + a_2 + \cdots + a_n$). By convention, the empty product equals 1 (just as the empty sum equals 0).

The first method is based upon prime factorizations (cf. §1.5).

THEOREM 1.8. *Let $p_1, \ldots, p_N \in \mathbb{N}$ be distinct primes and let $m_i, n_i \in \mathbb{Z}_{\geq 0}$. If $a = \prod_{i=1}^{N} p_i^{m_i}$ and $b = \prod_{i=1}^{N} p_i^{n_i}$, then $(a, b) = \prod_{i=1}^{N} p_i^{\min\{m_i, n_i\}}$.*

For example, $312 = 2^3 \cdot 3 \cdot 13$ and $180 = 2^2 \cdot 3^2 \cdot 5$ so $(312, 180) = 2^2 \cdot 3^1 \cdot 5^0 \cdot 13^0 = 12$. For easily factored numbers, people usually use Theorem 1.8 to calculate the greatest common divisor.

The second method is called the *Euclidean Algorithm* and is computationally efficient even for fairly large integers. In fact, it is known that the Euclidean Algorithm never needs more steps than five times the number of digits (base 10) of b to calculate (a, b) when $a, b \in \mathbb{N}$ with $a > b$ (Exercise 1.21). This makes the Euclidean Algorithm usable even when a and b have thousands of digits (as is the case in modern cryptography).

THEOREM 1.9 (Euclidean Algorithm). *Let $a, b \in \mathbb{N}$ with $a \geq b$. If $b \mid a$, then $(a, b) = b$. Otherwise, repeatedly use the Division Algorithm to write*

$$\begin{aligned}
a &= bq_0 + r_0 \text{ with } 0 < r_0 < b, \\
b &= r_0 q_1 + r_1 \text{ with } 0 < r_1 < r_0, \\
r_0 &= r_1 q_2 + r_2 \text{ with } 0 < r_2 < r_1, \\
r_1 &= r_2 q_3 + r_3 \text{ with } 0 < r_3 < r_2, \\
r_2 &= r_3 q_4 + r_4 \text{ with } 0 < r_4 < r_3, \\
&\ \vdots \\
r_{N-2} &= r_{N-1} q_N + r_N \text{ with } 0 < r_N < r_{N-1}, \\
r_{N-1} &= r_N q_{N+1} + 0.
\end{aligned}$$

As indicated above, this process arrives at a remainder of 0 *after a finite number of steps. When that happens,*

$$(a,b) = r_N,$$

the last positive remainder.

The Euclidean Algorithm is much more complicated to state than to actually use. Here is an example of repeated use of the Division Algorithm to calculate (703010, 5355):

$$\begin{aligned} 703010 &= 5355 \cdot 131 + 1505, \\ 5355 &= 1505 \cdot 3 + 840, \\ 1505 &= 840 \cdot 1 + 665, \\ 840 &= 665 \cdot 1 + 175, \\ 665 &= 175 \cdot 3 + 140, \\ 175 &= 140 \cdot 1 + 35, \\ 140 &= 35 \cdot 4 + 0 \end{aligned}$$

so that $(703010, 5355) = 35$.

For us, the most important fact about the greatest common divisor (a,b) is that it can be written as a $\mathbb{Z}$-linear combination of a and b. At first glance, this is certainly not an obvious result. More to the point, it may seem like an unimportant triviality. However, we will have many occasions to use this fact and its usefulness can hardly be overestimated.

THEOREM 1.10. *For $a, b \in \mathbb{Z}$, not both 0:*

(1) *There exist $x, y \in \mathbb{Z}$ so that $(a,b) = ax + by$.*

(2) *In fact, $(a,b) = \min\{ax + by \mid x, y \in \mathbb{Z},\ ax + by \geq 1\}$.*

PROOF. The whole trick to this argument is to consider the set $S = \{ax + by \mid x, y \in \mathbb{Z},\ ax + by \geq 1\}$. Using $x = a$ and $y = b$, we see S is nonempty. As S is also bounded below by 1, S has a smallest member. Write $c = \min S$ and choose $x, y \in \mathbb{Z}$ so that $c = ax + by$.

First we show c is a common divisor of a and b. As the argument is the same for a as it is for b, we only show $c \mid a$. To this end, use the Division Algorithm to write $a = cq + r$ for $q, r \in \mathbb{Z}$, $0 \leq r < c$. By way of argument by contradiction, suppose $c \nmid a$. Then $r \neq 0$ so that $1 \leq r < c$. But since

$$\begin{aligned} r &= a - cq \\ &= a - (ax + by)\, q \\ &= a\,(1 - xq) + b\,(-yq), \end{aligned}$$

it follows that $r \in S$. However, the fact that $r < c$ violates the choice of c as the minimum element of S, which is a contradiction. Thus $c \mid a$.

To see c is the largest common divisor, we prove the stronger statement that any common divisor of a and b actually divides c. Thus suppose $d \in \mathbb{Z}^{\times}$ with $d \mid a$ and $d \mid b$. Since $c = ax + by$, it follows from Lemma 1.6 parts (3) and (4) that $d \mid c$ so that $d \leq c$ as desired. □

A remarkable corollary of the proof of Theorem 1.10 is:

COROLLARY 1.11. *Let $a, b \in \mathbb{Z}$, not both 0. Every common divisor of a and b also divides (a,b); i.e., if $c \in \mathbb{Z}^{\times}$ with $c \mid a$ and $c \mid b$, then $c \mid (a,b)$.*

In turn, we get the following equivalent way of looking at greatest common divisors. Because of its use as a model for an eventual generalization (Definition 3.49), some take the following corollary as the definition of a greatest common divisor instead of Definition 1.7.

COROLLARY 1.12. *Let* $a, b \in \mathbb{Z}^{\times}$. *Then* $c \in \mathbb{N}$ *is the greatest common divisor of* a *and* b *if:*

(1) $c \mid a$ *and* $c \mid b$.
(2) *When* $d \in \mathbb{Z}^{\times}$ *with* $d \mid a$ *and* $d \mid b$, *then* $d \mid c$.

1.5. Fundamental Theorem of Arithmetic. It is a complicating fact of life that $a \mid (bc)$ does not always imply $a \mid b$ or $a \mid c$. For instance, $4 \mid (6 \cdot 10)$ but $4 \nmid 6$ and $4 \nmid 10$. However, good things happen when the numbers in question are relatively prime.

THEOREM 1.13. *Let* $a, b, c, p, a_i \in \mathbb{Z}$, $a \neq 0$.

(1) *If* $a \mid (bc)$ *with* a *and* b *relatively prime, then* $a \mid c$.
(2) *If* p *is prime and* $p \mid \prod_{i=1}^{N} a_i$, *then* $p \mid a_i$ *for some* i, $1 \leq i \leq N$.

PROOF. For part (1), we use a technique employed, seemingly, in nearly all arguments dealing with relatively prime integers. Moreover, since it is about the only trick we know so far, it is a good one to learn. Namely, we use the definition of relatively prime and Theorem 1.10 to write

$$1 = ax + by$$

for some $x, y \in \mathbb{Z}$. Multiplying by c shows $c = acx + bcy$. Since $a \mid a$ and $a \mid (bc)$, Lemma 1.6 part (3) shows $a \mid c$ as desired.

For part (2), the case of $N = 1$ is trivial. For the case of $N = 2$, suppose $p \mid (a_1 a_2)$. We must show either $p \mid a_1$ or $p \mid a_2$. To proceed, recall that p is prime so it only has divisors $\pm\{1, p\}$. Therefore (p, a_1) is either 1 or $|p|$, depending on a_1. In the first case, part (1) shows $p \mid a_2$ and we are done. In the second case, it follows that $p \mid a_1$ by the definition of greatest common divisor. In either case we win.

To prove the result in general, use induction on N. For the inductive step, write $\prod_{i=1}^{N+1} a_i = \left(\prod_{i=1}^{N} a_i\right)(a_{N+1})$. Since the theorem is already established in the case of $N = 2$, either $p \mid \prod_{i=1}^{N} a_i$ or $p \mid a_{N+1}$. In the first case, the inductive hypothesis finishes the proof while the in the second case there is nothing left to do. □

With Theorem 1.13 in hand, it is possible to prove the Fundamental Theorem of Arithmetic which roughly says that integers factor uniquely as a product of primes. As the reader will hopefully infer from the august name of the theorem, this is a rather important result that reduces many questions about integers to (sometimes easier) questions about primes. Note that by convention and for the sake of convenience, mathematicians allow themselves to refer to a single prime as a product of primes even though it is all by itself.

Rather than being impressed, students are often underwhelmed with the Fundamental Theorem of Arithmetic. Being familiar with the result since grade school, many feel the theorem is both obvious and barely worth proving. However, the fact that, say, 578,238,902,711,783,464,657 can be written as a unique product of primes is by no means obvious. Try it and see! Moreover, there are plenty of more general algebraic systems for which this result completely fails (cf. §5.1 of Chapter 3).

THEOREM 1.14 (Fundamental Theorem of Arithmetic). *Any $n \in \mathbb{N}$ with $n \neq 1$ can be written as a product of powers of positive primes, i.e.,*

$$n = \prod_{i=1}^{N} p_i^{m_i}$$

for distinct primes $p_1, \ldots, p_N \in \mathbb{N}$ and $m_i \in \mathbb{N}$. Moreover, up to reordering of the factors, the above decomposition is unique; i.e., if also $n = \prod_{i=1}^{M} q_i^{n_i}$ for distinct primes $q_1, \ldots, q_M \in \mathbb{N}$ and $n_i \in \mathbb{N}$, then $M = N$, it is possible to re-index $q_1, \ldots, q_N$ so that $q_i = p_i$ for $1 \leq i \leq N$, and then $m_i = n_i$.

PROOF. We prove existence first by induction on n. In the case of $n = 2$, since it is already prime, there is nothing further to prove. Next, suppose the theorem is true for numbers less than n (and greater than 1). We must show that n also decomposes as a product of primes. If n is prime, again there is nothing to prove. Otherwise, $n = ab$ for some $a, b \in \mathbb{Z}$ with $1 < a, b < n$. The induction hypothesis shows that both a and b are products of primes and thus so is $ab = n$.

We turn now to uniqueness and use induction on $S = \sum_{i=1}^{N} m_i$. If $S = 1$, n is prime and the result is clear. For the inductive step, suppose $\prod_{i=1}^{N} p_i^{m_i} = \prod_{i=1}^{M} q_i^{n_i}$ with $\sum_{i=1}^{N} m_i = S + 1$. Since $p_1 \mid \prod_{i=1}^{M} q_i^{n_i}$, Theorem 1.13 shows that $p_1 \mid q_i$ for some i. As q_i is also prime, this implies $p_1 = q_i$. By reordering the q_i's, we may assume $p_1 = q_1$. Division then gives $p_1^{m_1 - 1} \prod_{i=2}^{N} p_i^{m_i} = p_1^{n_1 - 1} \prod_{i=2}^{M} q_i^{n_i}$. Since now $(m_1 - 1) + \sum_{i=2}^{N} m_i = S$, the inductive hypothesis finishes the argument. □

If we allow $n \in \mathbb{Z}$, $n \neq 0, \pm 1$, and if we allow the primes to be negative, then an analogous theorem holds. The only change of note is a sprinkling in some extra $\pm$'s.

1.6. Exercises 1.1–1.28.

1.6.1. *Basic Properties.*

EXERCISE 1.1. Show the Well-Ordering Axiom is not always true if $\mathbb{Z}$ is replaced by the rational numbers, $\mathbb{Q}$, or by the real numbers, $\mathbb{R}$. *Hint:* Consider sets such as $\{x \in \mathbb{Q} \mid 0 < x < 1\}$.

1.6.2. *Induction.*

EXERCISE 1.2. Use induction to prove:

(a) $1 + 3 + 5 + \cdots + (2n - 1) = n^2$.
(b) $\sum_{k=1}^{n} k^2 = \frac{n(n+1)(2n+1)}{6}$.
(c) $\sum_{k=1}^{n} k^3 = \frac{n^2(n+1)^2}{4}$.
(d) $\sum_{k=0}^{n} x^k = \frac{x^{n+1} - 1}{x - 1}$ for $x \neq 1$.
(e) For $n \geq 4$, $n! > 2^n$.
(f) For $n \in \mathbb{Z}_{\geq 0}$, $3 \mid (n^3 + 2n)$.
(g) $\frac{1}{1 \cdot 3} + \frac{1}{3 \cdot 5} + \frac{1}{5 \cdot 7} + \cdots + \frac{1}{(2n-1)(2n+1)} = \frac{n}{2n+1}$.
(h) For $n, k \in \mathbb{Z}_{\geq 0}$ with $k < n$,

$$\binom{n}{k-1} + \binom{n}{k} = \binom{n+1}{k}$$

where the *binomial coefficient* is defined by $\binom{n}{k} = \frac{n!}{k!(n-k)!}$. Note that the pictorial version of this equation is *Pascal's Triangle*

$$\begin{array}{ccccccccc} & & & & 1 & & & & \\ & & & 1 & & 1 & & & \\ & & 1 & & 2 & & 1 & & \\ & 1 & & 3 & & 3 & & 1 & \\ 1 & & 4 & & 6 & & 4 & & 1 \\ & & & & \vdots & & & & \end{array}$$

where, starting at $n = 0$ and $k = 0$, the nth "row" and k^{th} "column" of the triangle is given by $\binom{n}{k}$. *Truth in advertising*: This is *much* more easily checked by direct algebraic manipulation or combinatorial considerations.

(i) $\binom{n}{k} \in \mathbb{N}$.

(j) (*Binomial Theorem*) $(a+b)^n = \sum_{k=0}^{n} \binom{n}{k} a^k b^{n-k}$.

(k) (*Binet's Formula*) The *Fibonacci Sequence* is defined inductively by $f_1 = 1$, $f_2 = 1$, and $f_{n+2} = f_{n+1} + f_n$ for $n \in \mathbb{N}$. Show

$$f_n = \frac{(1+\sqrt{5})^n - (1-\sqrt{5})^n}{2^n\sqrt{5}}.$$

EXERCISE 1.3. Prove the following variants on induction:

(a) Fix $n_0 \in \mathbb{Z}$. For each $n \in \mathbb{Z}$ with $n \geq n_0$ suppose there is a proposition P_n so that (i) P_{n_0} is true and (ii) whenever P_n is true, then P_{n+1} is also true. Show P_n is true for every $n \in \mathbb{Z}$ with $n \geq n_0$.

(b) (*Complete Induction*) If for each $n \in \mathbb{N}$ there is a proposition P_n so that (i) P_1 is true and (ii) whenever P_k is true for all $1 \leq k \leq n$, then P_{n+1} is also true. Show P_n is true for every $n \in \mathbb{N}$.

EXERCISE 1.4 (*Pólya*). What is wrong with the following "proof" showing all horses are the same color: Induct on the number of horses, n. If $n = 1$, there is clearly only one color. Suppose now any collection of n horses has one color and look at a group of $n+1$ horses, $\{h_1, \ldots, h_{n+1}\}$. Then the two collections $\{h_1, \ldots, h_n\}$ and $\{h_2, \ldots, h_{n+1}\}$ both have n horses each and so the induction hypothesis shows each set of horses has only one color. Since the two sets of horses overlap, there is only one color among all the horses. Thus all horses have the same color.

1.6.3. *Division Algorithm.*

EXERCISE 1.5. Use old-fashioned long division to implement the Division Algorithm and write $a = bq + r$, $0 \leq r < b$, for each a and b listed below.

(a) $a = 20$, $b = 3$.

(b) $a = 54$, $b = 7$.

(c) $a = 163$, $b = 9$.

(d) $a = 211$, $b = 11$.

EXERCISE 1.6. Figure out how to use the basic operations on a calculator to write $a = bq + r$, $0 \leq r < b$, for $a = 5{,}638{,}481$ and $b = 987$. Explain your algorithm and show that it works.

EXERCISE 1.7. Let $a, b, d \in \mathbb{Z}$, $d \neq 0$. Show that division of a by d and division of b by d both leave the same remainder if and only if $d \mid (a - b)$.

EXERCISE 1.8. Show that when the square of an odd integer is divided by 8, the remainder is 1. *Hint:* Remember $2 \mid n(n+1)$.

EXERCISE 1.9. Prove the general statement of the Division Algorithm in Corollary 1.4. *Hint:* First modify the proof of the Division Algorithm to allow $a < 0$. To also work with $b < 0$, begin with $|b|$ and then throw in a $(-1)(-1)$.

EXERCISE 1.10 (*Base m representations*). Let $n, m \in \mathbb{N}$ with $m \neq 1$. Show n can be uniquely written in the form $n = \sum_{k=0}^{N} a_k m^k$ for some $N \in \mathbb{Z}_{\geq 0}$ and $a_k \in \{0, 1, \ldots, (m-1)\}$ with $a_N \neq 0$. *Hint:* Use induction on n and begin by choosing the largest $N \in \mathbb{Z}_{\geq 0}$ so that $m^N \leq n$. Use the Division Algorithm to write $n = a_N m^N + r$ and then apply the inductive hypothesis to r.

EXERCISE 1.11. Let $S \subseteq \mathbb{Z}$. Show there exists an $N \in \mathbb{Z}$ so that $s \leq N$ for all $s \in S$ if and only if there exists a $Q \in \mathbb{Q}$ so that $s \leq Q$ for all $s \in S$.

1.6.4. *Divisors.*

EXERCISE 1.12. List all divisors of the following:

(a) 52,
(b) 110.

EXERCISE 1.13. Let $a, b, d \in \mathbb{Z}$ with $ad \neq 0$, $d \mid a$, and $d \mid b$. Show $a \mid b$ if and only if $\frac{a}{d} \mid \frac{b}{d}$.

EXERCISE 1.14. Evaluate the following:

(a) $(42, 56)$,
(b) $(25, 85)$,
(c) $(18, 108)$,
(d) $(27, 300)$.

EXERCISE 1.15. Find the smallest positive integer of the form $52x + 108y$ for $x, y \in \mathbb{Z}$.

EXERCISE 1.16. Prove the remaining parts of Lemma 1.6.

EXERCISE 1.17. Let $b, q, r \in \mathbb{Z}$ and let $a = bq + r$ with a and b not both 0.

(a) Show a common divisor of a and b is a divisor of r and that a common divisor of b and r is a divisor of a.

(b) Conclude that $(a, b) = (r, b)$.

EXERCISE 1.18. Let $a, b \in \mathbb{Z}^\times$ and $c, d \in \mathbb{Z}$.

(a) If $a \mid c$ and $b \mid d$, then $(ab) \mid (cd)$.

(b) If $a \mid c$ and $b \mid c$, then it may happen that $(ab) \nmid c$.

(c) If $a \mid c$ and $b \mid c$ with $(a, b) = 1$, then $(ab) \mid c$. *Hint:* Write $ax + by = 1$ and multiply by c. Show $(ab) \mid (ac)$ and $(ab) \mid (bc)$.

EXERCISE 1.19. Let $a, b \in \mathbb{Z}$, not both 0.

(a) If $(a, b) = d$, then $(\frac{a}{d}, \frac{b}{d}) = 1$. *Hint:* Write $ax + by = d$ so that $\frac{a}{d}x + \frac{b}{d}y = 1$.

(b) For $c \in \mathbb{Z}^\times$, $(ca, cb) = |c|\,(a, b)$. *Hint:* First show {common divisors of ca and cb} contains $c \cdot$ {common divisors of a and b} to get $(ca, cb) \geq |c|\,(a, b)$. Then show you can solve $cax + cby = |c|\,(a, b)$ for some x and y to get $(ca, cb) \leq |c|\,(a, b)$.

EXERCISE 1.20 (Euclidean Algorithm). This exercise examines the *Euclidean Algorithm* from Theorem 1.9.

(a) Show the Euclidean Algorithm arrives at a remainder of 0 after a finite number of steps.

(b) By repeated use of the equation $(bq + r, b) = (r, b)$ from Exercise 1.17, prove the Euclidean Algorithm.

(c) Use the Euclidean Algorithm to calculate $(336, 128)$ and $(524, 242)$.

(d) Write $d = (a, b)$. By back substitution, the steps of the Euclidean Algorithm can be used to find explicit $x, y \in \mathbb{Z}$ so that $ax + by = d$. For example, applying the Euclidean Algorithm to $(250, 45)$ gives

$$\begin{aligned} 250 &= 45 \cdot 5 + 25, \\ 45 &= 25 \cdot 1 + 20, \\ 25 &= 20 \cdot 1 + 5, \\ 20 &= 5 \cdot 4 + 0 \end{aligned}$$

so that $(250, 45) = 5$. To write $5 = 250x + 45y$, start at the penultimate equation and work your back way up as follows:

$$\begin{aligned} 5 &= 25 - 20 \cdot 1 \\ &= (250 - 45 \cdot 5) - [45 - 25 \cdot 1] \cdot 1 \\ &= (250 - 45 \cdot 5) - [45 - (250 - 45 \cdot 5) \cdot 1] \cdot 1 \\ &= 250\,(1 + 1) + 45\,(-5 - 1 - 5) \\ &= 250\,(2) + 45\,(-11)\,. \end{aligned}$$

Using this technique, find $x, y \in \mathbb{Z}$ so that $336x + 128y = 16$ and find $x', y' \in \mathbb{Z}$ so that $524x' + 242y' = 2$.

EXERCISE 1.21. Suppose the Euclidean Algorithm takes N steps to calculate the greatest common divisor of a pair of numbers $a, b \in \mathbb{N}$, $b < a$.

(a) For a pair of numbers $(x_1, y_1), (x_2, y_2) \in \mathbb{R}^2$, we say that $(x_1, y_1) < (x_2, y_2)$ with respect to the *lexicographic order* if $x_1 < x_2$ or if $x_1 = x_2$ and $y_1 < y_2$. Show that there exists a pair of distinct natural numbers (d, c), $d < c$, with minimal lexicographic order so that the Euclidean Algorithm takes N steps to calculate the greatest common divisor of c and d.

(b) If $N = 1$, show $c = 2$ and $d = 1$. In general, induct on N to show that $c = f_{n+2}$ and $d = f_{n+1}$ where f_n is the nth Fibonacci number (cf. Exercise 1.2).

(c) Show $b \geq f_{N+1} \geq \varphi^N$ where $\varphi = \frac{1+\sqrt{5}}{2} \approx 1.618$ is called the *golden ratio.*

(d) Using the fact that $\log_{10} \varphi > \frac{1}{5}$, show that $N < 5 \log_{10} b$. Conclude that N is bounded by five times the number of digits (base 10) of b.

1.6.5. *Fundamental Theorem of Arithmetic.*

EXERCISE 1.22. Show $p \in \mathbb{Z}$, $p \neq 0, \pm 1$, is prime if and only if it satisfies the property that whenever $p \mid (ab)$ for $a, b \in \mathbb{Z}$, then $p \mid a$ or $p \mid b$.

EXERCISE 1.23. Let $a, b \in \mathbb{Z}$, not both 0, and $n \in \mathbb{Z}_{\geq 0}$.

(a) Show it is possible to have $(a, b) \neq (a, b^n)$.

(b) If $(a, b) = 1$, show $(a, b^n) = 1$. *Hint:* Theorem 1.8 or Theorem 1.13.

EXERCISE 1.24. Let $a, m, n \in \mathbb{Z}$ with $(m, n) = 1$. Show $(a, mn) = (a, m)(a, n)$. *Hint:* Use Theorem 1.8 and recall that m and n have no common divisors.

EXERCISE 1.25. Let p be a positive prime.

(a) If $p \mid a^n$ for $a \in \mathbb{Z}$ and $n \in \mathbb{N}$, then $p^n \mid a^n$. *Hint:* Show $p \mid a$ first.

(b) Show there are no $a, b \in \mathbb{Z}^\times$ satisfying $a^2 = pb^2$. *Hint:* If you could solve $a^2 = pb^2$, show you may also assume $(a, b) = 1$ by dividing. Then show $p \mid a$ and then that $p \mid b$ to get a contradiction.

(c) Show there is no $r \in \mathbb{Q}$ satisfying $r^2 = p$; i.e., show $\sqrt{p} \notin \mathbb{Q}$.

EXERCISE 1.26. Suppose $a = \prod_{i=1}^N p_i^{m_i}$ is a product of powers of distinct primes $p_1, \ldots, p_N \in \mathbb{N}$ with $m_i \in \mathbb{Z}_{\geq 0}$. Show a has $\prod_{i=1}^N (m_i + 1)$ distinct divisors. *Hint:* Use the Fundamental Theorem of Arithmetic to reduce the problem to counting all possible exponents.

EXERCISE 1.27. **(a)** (*Euclid*) Show there are an infinite number of primes. *Hint:* If not, suppose $P = \{p_1, \ldots, p_N\}$ is the set of all primes. Show $p_i \nmid \left(1 + \prod_{j=1}^N p_j\right)$, $1 \leq i \leq N$. Use this to get a contradiction to the assumption that P contains all primes.

(b) Show there exist arbitrarily long gaps in the set of primes by looking at the sequence $n! + 2, n! + 3, \ldots, n! + n$ for $n \geq 2$.

EXERCISE 1.28 (Prime Factorization and Least Common Multiple). For $a, b \in \mathbb{N}$ write $a = \prod_{i=1}^N p_i^{m_i}$ and $b = \prod_{i=1}^N p_i^{n_i}$ as a product of powers of distinct primes $p_1, \ldots, p_N \in \mathbb{N}$ with $m_i, n_i \in \mathbb{Z}_{\geq 0}$.

(a) Show $(a, b) = \prod_{i=1}^N p_i^{\min\{m_i, n_i\}}$.

(b) The *least common multiple* of a and b, written $[a, b]$ or $\operatorname{lcm}(a, b)$, is the smallest positive multiple of a and b; i.e., $[a, b]$ is the smallest positive $m \in \mathbb{Z}$ so that $a \mid m$ and $b \mid m$. Show $[a, b] = \prod_{i=1}^N p_i^{\max\{m_i, n_i\}}$.

(c) Show that whenever $a \mid k$ and $b \mid k$ for some $k \in \mathbb{Z}$, then $[a, b] \mid k$.

(d) Show $(a, b)[a, b] = ab$.

2. Modular Arithmetic

Those of us who use a 12-hour clock never say that the hour after 12 o'clock is 13 o'clock. Instead, we start counting over and say that 1 o'clock comes after 12 o'clock. Similarly, 14 o'clock is really 2 o'clock and 15 o'clock is really 3 o'clock. In other words, we count two different hours as the same if they differ by 12.

When a mathematician counts two integers as basically the same when they differ by a multiple of 12, the mathematician is said to be working with *congruence classes modulo 12*, or, for short, mod 12. These sort of ideas generalize quite nicely and lead to some very powerful mathematics.

2.1. Congruence.

DEFINITION 1.15. Let $a, b, \in \mathbb{Z}$ and $n \in \mathbb{N}$. We say a is *congruent* to b *modulo* n, written $a \equiv b \pmod{n}$, if $n \mid (a - b)$.

Note from the definition that $a \equiv b \pmod{n}$ if and only if

$$a = b + nk$$

for some $k \in \mathbb{Z}$. Thus $a \equiv b \pmod{n}$ if and only if a and b differ by a multiple of n. For example, $13 \equiv 1 \pmod{12}$, $14 \equiv 2 \pmod{12}$, and $26 \equiv 2 \pmod{12}$. When it is clear from the context, the $\pmod{n}$ is often omitted.

LEMMA 1.16. *Let $a, b, c, a_i, b_i \in \mathbb{Z}$ and $n \in \mathbb{N}$. Modulo n:*

(1) $a \equiv a$.
(2) *If $a \equiv b$, then $b \equiv a$.*
(3) *If $a \equiv b$ and $b \equiv c$, then $a \equiv c$.*
(4) *If $a_1 \equiv a_2$ and $b_1 \equiv b_2$, then $a_1 + b_1 \equiv a_2 + b_2$.*
(5) *If $a_1 \equiv a_2$ and $b_1 \equiv b_2$, then $a_1 b_1 \equiv a_2 b_2$.*

PROOF. Since the results are straightforward and good exercises, we only prove parts (3) and (4) and leave the rest to Exercise 1.30. For part (3), use the definition to observe $n \mid (a-b)$ and $n \mid (b-c)$. Thus $n \mid [(a-b)+(b-c)]$ so that $n \mid (a-c)$ as desired. For part (4), use the definition to write $a_1 = a_2 + kn$ and $b_1 = b_2 + k'n$ for some $k, k' \in \mathbb{Z}$. Thus $(a_1 + b_1) = (a_2 + b_2) + (k + k')\, n$ so that $a_1 + b_1 \equiv a_2 + b_2$. □

The first three properties of Lemma 1.16 make the notion of congruence into what is known as an *equivalence relation.* This idea will be pursued further in §2.6.

2.2. Congruence Classes. As is the case of telling time—where congruence modulo 12 is used—mathematicians often want to view two different integers as the "same" when they are really only congruent modulo n. For instance, working with hours modulo 12, people view 3 as the same as 15, which in turn is the same as 27, and so on. In other words, each element of the set $\{\ldots 3, 15, 27, \ldots\}$ is counted as the same hour.

It turns out that lumping together all the "same" things into one great big set is a landmark idea in mathematics. It not only provides the needed rigor to study modular arithmetic, but it is also a new way of thinking, woven through almost every area of higher mathematics.

DEFINITION 1.17. Let $a \in \mathbb{Z}$ and $n \in \mathbb{N}$. The *congruence class of a modulo n*, written as $[a]$, is

$$[a] = \{b \in \mathbb{Z} \mid b \equiv a \pmod{n}\}$$

and a is called a *representative* of the congruence class.

Some authors refer to a congruence class as a *residue class* and write $\overline{a}$ instead of $[a]$. Also, the above equivalence class $[a]$ is occasionally written as $[a]_n$ to emphasize the dependence on n.

It should be emphasized that $[a]$ is a *set.* For instance, mod 12,

$$\begin{aligned} [3] &= \{b \in \mathbb{Z} \mid b \equiv 3 \pmod{12}\} \\ &= \{3 + 12k \mid k \in \mathbb{Z}\} \\ &= \{\ldots -21, -9, 3, 15, 27, 39, 51, \ldots\}. \end{aligned}$$

Notice, for example, that

$$\begin{aligned} [15] &= \{b \in \mathbb{Z} \mid b \equiv 15 \pmod{12}\} \\ &= \{15 + 12k \mid k \in \mathbb{Z}\} \\ &= \{\ldots -21, -9, 3, 15, 27, 39, 51, \ldots\} \\ &= [3]. \end{aligned}$$

Thus, mod 12, $[3] = [15]$. Cumbersome as it seems at first, this is the official way a mathematician makes precise sense out of the statement 3 o'clock is the same as 15 o'clock.

This example also serves as an important warning that two possibly different looking equivalence classes, e.g., $[3]$ and $[15]$ mod 12, may actually be the same. In particular, 3 and 15 are different representatives for the same congruence class. A priori, 3 is just as good as 15, which is just as good as 27, and so on. When dealing with congruence classes, it is very important to remember that the choice of a representative for the class is not unique.

As a final note, it is a very common practice for people to omit the [ ]'s around a representative of a congruence class and to let the reader figure out what is going on from the context. As the reader will soon see, having to write all those [ ]'s is enough to drive a person crazy. Though mathematicians hate to say it too loudly, a little ambiguity in life can be a good thing. Nevertheless, for the sake of clarity, we will attempt to mostly leave them in for the duration of this text.

LEMMA 1.18. *Let $a, b, q, r \in \mathbb{Z}$ and $n \in \mathbb{N}$. Consider congruence classes modulo n.*

(1) $[a] = [b]$ *if and only if* $a \equiv b \pmod{n}$.

(2) *Either $[a] = [b]$ or $[a] \cap [b] = \varnothing$; i.e., congruence classes are either identical or disjoint.*

(3) *$\mathbb{Z}$ is the disjoint union of its congruence classes modulo n.*

(4) *If $a = nq + r$, then $[a] = [r]$.*

(5) *There are exactly n distinct equivalence classes modulo n, namely $[0]$, $[1]$, $[2], \ldots, [n-1]$.*

PROOF. Look at part (1) first. By definition, $b \in [a]$ if and only if $a \equiv b$. Thus $[a] = [b]$ certainly implies $a \equiv b$. Conversely, suppose $a \equiv b$. If $c \in [a]$, then $c \equiv a \equiv b$ so that $c \in [b]$ and so $[a] \subseteq [b]$. By a similar argument, $[b] \subseteq [a]$. Hence $[a] = [b]$ as desired.

For part (2), suppose $[a] \cap [b] \neq \varnothing$ and let $c \in [a] \cap [b]$. Then $a \equiv c \equiv b$ so that $[a] = [b]$ by part (1). For part (3), use part (2) and note that every $k \in \mathbb{Z}$ is a member of some congruence class (e.g., $[k]$).

Part (4) follows from part (1). For part (5), the Division Algorithm coupled with part (4) shows that $[0], [1], [2], \ldots, [n-1]$ is a complete list of congruence classes. They are clearly distinct by part (1). □

DEFINITION 1.19. The set of all equivalence classes modulo n is denoted by $\mathbb{Z}_n$, read "$\mathbb{Z}$ mod n".

In particular, Lemma 1.18 shows

$$\mathbb{Z}_n = \{\, [0], [1], [2], \ldots, [n-1] \,\},$$

though it is important to remember that there is a great deal of flexibility in the choice of a representative for each congruence class.

2.3. Arithmetic. The set $\mathbb{Z}$ comes equipped with two basic operations: addition and multiplication. Since $\mathbb{Z}_n$ is constructed out of pieces of $\mathbb{Z}$, it makes sense to try to carry over some version of these operations. The most obvious definition turns out to work splendidly.

DEFINITION 1.20. For $[a], [b] \in \mathbb{Z}_n$, define addition and multiplication by

$$[a] + [b] = [a+b],$$
$$[a] \cdot [b] = [ab].$$

As is common with multiplication, we often write $[a][b]$ for $[a] \cdot [b]$.

While this definition looks very reasonable, there is a possible flaw that could mess everything up. Namely, our definition depends on the choice of a representative for the equivalence class. If we choose a different representative, does the above formula give a different answer? For instance, modulo 12, $[3] = [15]$. Thus, for example, it ought to be true that

$$[3] + [6] \stackrel{?}{=} [15] + [6].$$

However, the definition says

$$\begin{aligned} [3] + [6] &= [9] \\ \text{while } [15] + [6] &= [21]. \end{aligned}$$

With a cursory glance, these two answers do not appear to be identical. Of course superficial thinking rarely leads to very accurate answers. In fact, a moment's thought actually reveals that $[9] = [21]$ since $9 \equiv 21 \pmod{12}$ and so life is good.

The point is that we need to check that Definition 1.20 is well defined all the time, i.e., that the definition is independent of the choice of representatives for the congruence classes. To see this, suppose $[a_1] = [a_2]$ and $[b_1] = [b_2]$ for $a_i, b_i \in \mathbb{Z}$. We need to verify that $[a_1 + b_1] = [a_2 + b_2]$ and $[a_1 b_1] = [a_2 b_2]$. However, these statements are equivalent to checking $a_1 + b_1 \equiv a_2 + b_2$ and $a_1 b_1 \equiv a_2 b_2$, which are already known to be true by Lemma 1.16.

As an example of modular arithmetic, we can write out the addition and multiplication tables for $\mathbb{Z}_6$. For instance, $[3]+[4] = [7] = [1]$ and $[3][3] = [9] = [3]$. The rest of the calculations are just as easy and the entire tables are given below. The reader is encouraged to verify the results since, unfortunately, watching someone do mathematics is quite different from actually doing mathematics.

(1.21)

+	[0]	[1]	[2]	[3]	[4]	[5]
[0]	[0]	[1]	[2]	[3]	[4]	[5]
[1]	[1]	[2]	[3]	[4]	[5]	[0]
[2]	[2]	[3]	[4]	[5]	[0]	[1]
[3]	[3]	[4]	[5]	[0]	[1]	[2]
[4]	[4]	[5]	[0]	[1]	[2]	[3]
[5]	[5]	[0]	[1]	[2]	[3]	[4]

×	[0]	[1]	[2]	[3]	[4]	[5]
[0]	[0]	[0]	[0]	[0]	[0]	[0]
[1]	[0]	[1]	[2]	[3]	[4]	[5]
[2]	[0]	[2]	[4]	[0]	[2]	[4]
[3]	[0]	[3]	[0]	[3]	[0]	[3]
[4]	[0]	[4]	[2]	[0]	[4]	[2]
[5]	[0]	[5]	[4]	[3]	[2]	[1]

2.4. Structure. By and large, arithmetic on $\mathbb{Z}_n$ is very similar to arithmetic on $\mathbb{Z}$. Though there are some key differences, we start with the similarities. The reader should compare the following theorem to Axiom 1.

THEOREM 1.22. *Let* $[a], [b], [c] \in \mathbb{Z}_n$ *for* $n \in \mathbb{N}$.
The properties of addition are:

(1) Associativity: $([a] + [b]) + [c] = [a] + ([b] + [c])$.
(2) Commutativity: $[a] + [b] = [b] + [a]$.
(3) Identity: $[0] + [a] = [a]$.
(4) Inverses: $[a] + [-a] = [0]$.

The properties of multiplication are:

(1) Associativity: $([a]\,[b])\,[c] = [a]\,([b]\,[c])$.
(2) Commutativity: $[a]\,[b] = [b]\,[a]$.
(3) Identity: $[1]\,[a] = [a]$.

There is also a law that relates addition and multiplication:

(1) Distributive: $[a]([b]+[c]) = [a][b]+[a][c]$.

PROOF. The proof of this theorem is quite elementary, though a good exercise. To see how it works, we prove the Distributive Law and leave the rest for Exercise 1.47. Using the definitions for $\mathbb{Z}_n$ and the axioms for $\mathbb{Z}$,

$$\begin{aligned}[a]([b]+[c]) &= [a][b+c] = [a(b+c)] = [ab+ac] = [ab]+[ac] \\ &= [a][b]+[a][c].\end{aligned}$$

□

These laws allow us to work with $\mathbb{Z}_n$ in nearly the same way that we work with $\mathbb{Z}$. As mentioned though, there are some important differences. First of all, recall that in $\mathbb{Z}$ the only time $ab = 0$ is when either a or b is already 0. However in $\mathbb{Z}_6$, we have already seen that $[2][3] = [0]$ even though $[2] \neq [0]$ and $[3] \neq [0]$. Thus in $\mathbb{Z}_n$ it is sometimes possible to have $[a][b] = [0]$ with $[a] \neq [0]$ and $[b] \neq [0]$!

This is quite a change and an important fact to remember. For instance solving $2x = 0$ in $\mathbb{Z} \subseteq \mathbb{Q}$ is simple since we can divide out by 2 to get $x = 0$ as the only possibility. However, solving the equation $[2]x = [0]$ in $\mathbb{Z}_6$ is harder. If we try to "divide out" by $[2]$, we would only get $x = [0]$ as a solution and miss the solution $x = [3]$. The conclusion is that "dividing" does not always work in $\mathbb{Z}_n$.

Another change depends on the observation that the only multiplicatively invertible integers in $\mathbb{Z}$ are ± 1; i.e., if $a \in \mathbb{Z}$, the only time the equation $ax = 1$ has a solution in $\mathbb{Z}$ is if $a \in \{\pm 1\}$. The good news here is that $\mathbb{Z}_n$ can sometimes be much more flexible than $\mathbb{Z}$. For instance in $\mathbb{Z}_5$, it is easy to see

$$[1][1] = [1],\ [2][3] = [1],\ \text{and}\ [4][4] = [1].$$

Thus in $\mathbb{Z}_5$, the equation $[a]x = [1]$ has a solution for every $[a] \neq [0]$ in $\mathbb{Z}_5$. As an application, solving the equation $[2]x = [3]$ in $\mathbb{Z}_5$ is pretty easy. Simply multiply both sides by $[3]$ (noting that $[2][3] = [1]$ and $[1]x = x$) to get $x = [9] = [4]$ as the (unique) solution.

DEFINITION 1.23. Let $n \in \mathbb{N}$ and $[a] \in \mathbb{Z}_n$ with $[a] \neq [0]$.

(1) $[a]$ is called a *zero divisor* if there exists $[b] \in \mathbb{Z}_n$, $[b] \neq [0]$, so $[a][b] = [0]$.

(2) $[a]$ is called a *unit* or *invertible* if there exists $[b] \in \mathbb{Z}_n$ so that $[a][b] = [1]$. In that case, $[b]$ is called the *(multiplicative) inverse* of $[a]$ and is written $[a]^{-1} = [b]$.

(3) Let $U_n = \{[a] \in \mathbb{Z}_n \mid [a] \text{ is a unit}\}$.

For example, the zero divisors in $\mathbb{Z}_6$ are $\{[2],[3],[4]\}$ since $[2][3] = [0]$ and $[4][3] = [0]$. The set of units in $\mathbb{Z}_6$ is $U_6 = \{[1],[5]\}$ since $[1][1] = [1]$ and $[5][5] = [1]$. In $\mathbb{Z}_5$ it is easy to see there are no zero divisors. There are, however, lots of units in $\mathbb{Z}_5$. In fact, $U_5 = \{[1],[2],[3],[4]\}$ since $[1][1] = [1]$, $[2][3] = [1]$, and $[4][4] = [1]$. In other words, $[1]^{-1} = [1]$, $[2]^{-1} = [3]$, and $[4]^{-1} = [4]$.

As the reader may have already noticed, a possible problem with Definition 1.23, part (2), needs to be clarified. Namely, if $[a] \in \mathbb{Z}_n$ is invertible, we need to know the equation $[a]x = [1]$ has only one solution in $\mathbb{Z}_n$. This is necessary in order for us to write $[a]^{-1}$ without ambiguity. To see this, suppose $[a][b] = [1]$ and $[a][b'] = [1]$. Then

$$[b] = [b][1] = [b][a][b'] = [1][b'] = [b']$$

as desired.

It turns out that deciding when something is a zero divisor or a unit in $\mathbb{Z}_n$ is pretty easy as long as you know about the greatest common divisor. It also shows that $\mathbb{Z}_n$ breaks up into three disjoint subsets: $\{[0]\}$, U_n, and $\{\text{zero divisors}\}$.

THEOREM 1.24. *Let $n \in \mathbb{N}$ and $[a] \in \mathbb{Z}_n$.*
(1) *$[a] = [0]$ if and only if $(a, n) = n$.*
(2) *$[a]$ is a zero divisor if and only if $1 < (a, n) < n$.*
(3) *$[a]$ is a unit if and only if $(a, n) = 1$.*

PROOF. Part (1) follows from the observation that $[a] = [0]$ if and only if a is a multiple of n. Next write $d = (a, n)$ with $a = a_1 d$ and $n = n_1 d$ for some $a_1, n_1 \in \mathbb{Z}$. As part (1) is complete, for the remainder of the proof assume that $1 \leq d < n$ so $[a] \neq [0]$. Equivalently, we have $n \geq n_1 > 1$.

Begin by observing that $1 < d$ if and only if $n > n_1$ if and only if $[n_1] \neq [0]$. In this case, the calculation

$$[a][n_1] = [an_1] = [a_1 d n_1] = [a_1 n] = [0]$$

shows $[a]$ is a zero divisor. On the other hand, when $d = 1$ there is $x, y \in \mathbb{Z}$ so that $ax + ny = 1$. In this case, it follows that

$$[a][x] = [a][x] + [0] = [ax] + [ny] = [ax + ny] = [1]$$

so that $[a]$ is a unit.

Conversely for part (2), if $[a]$ is a zero divisor, we must show $1 < d$. If not, then we would have $d = 1$ so that $[a]$ would also be a unit. Similarly for part (3), if $[a]$ is a unit, we must show $d = 1$. If not, then we would have $1 < d$ so that $[a]$ would also be a zero divisor. In order to finish the proof, it therefore suffices to show that the definitions of unit and zero divisor are mutually exclusive. To see this, suppose $[a]$ is a unit with inverse $[a]^{-1}$. Multiplying both sides of the equation $[a]x = [0]$ by $[a]^{-1}$ shows $x = [0]$. In particular, the only solution to $[a]x = [0]$ is $x = [0]$ and so $[a]$ is not a zero divisor. □

The set of units, U_n, will turn out to be an important subset of $\mathbb{Z}_n$. A few of its properties are listed below.

THEOREM 1.25. *Let $n \in \mathbb{N}$ and $[a] \in \mathbb{Z}_n$.*
(1) *$[a]$ is a unit if and only if for each $[b] \in \mathbb{Z}_n$ the equation $[a]x = [b]$ has a unique solution in $\mathbb{Z}_n$.*
(2) *U_n is closed under multiplication; i.e., if $[a], [b] \in U_n$, then $[a][b] \in U_n$.*
(3) *When p is a positive prime, $U_p = \mathbb{Z}_p \backslash \{[0]\}$.*

PROOF. For part (1), first suppose $[a]$ is a unit. Then $[a]^{-1}$ exists and it follows by multiplication on both sides that the unique solution to $[a]x = [b]$ is $x = [a]^{-1}[b]$. Conversely, if $[a]x = [b]$ can be solved for any $[b]$, then in particular it can be solved for $[b] = [1]$ so that $[a]$ is a unit.

For part (2), simply observe that $([a][b])^{-1} = [b]^{-1}[a]^{-1}$ since clearly

$$([a][b])\left([b]^{-1}[a]^{-1}\right) = [1].$$

Part (3) follows from Theorem 1.24 since $(a, p) = 1$ for $[a] \neq [0]$. □

Note that though U_n is closed under multiplication, it is not closed under addition; see Exercise 1.51.

The above theorem shows that solving the equation $[a]x = [b]$ is extremely easy when $[a]$ is a unit. Namely, $x = [a]^{-1}[b]$. Solving $[a]x = [b]$ for nonunits, however, is more involved. It may happen that solutions do not exist and, even if they exist, they are not unique. For example, in $\mathbb{Z}_4$, the equation $[2]x = [1]$ has no solution since $[2]$ is not a unit while the equation $[2]x = [0]$ has two solutions, namely $x = [0], [2]$. The next theorem tells precisely when $[a]x = [b]$ can be solved in $\mathbb{Z}_n$ and counts the number of solutions.

THEOREM 1.26. *Let $n \in \mathbb{N}$, $[a] \in \mathbb{Z}_n$ with $[a] \neq [0]$, and let $d = (a, n)$.*
(1) *The equation $[a]x = [b]$ has a solution in $\mathbb{Z}_n$ if and only if $d \mid b$.*
(2) *If $d \mid b$, there are exactly d distinct solutions to $[a]x = [b]$ in $\mathbb{Z}_n$.*

PROOF. For part (1), first suppose $[a][s] = [b]$ for some $[s] \in \mathbb{Z}_n$. Then $as = b + nk$ for some $k \in \mathbb{Z}$. Since $d \mid a$ and $d \mid n$, it follows that $d \mid b$. Conversely, suppose $d \mid b$ and write $b = b_1 d$ for $b_1 \in \mathbb{Z}$. Find $u, v \in \mathbb{Z}$ so that $au + nv = d$. Multiply both sides by b_1 to get $aub_1 + nvb_1 = b$, so that $[a][ub_1] = [b]$ as desired.

For part (2), suppose $d \mid b$ and fix a solution $[s_0] \in \mathbb{Z}_n$ to the equation $[a]x = [b]$ (e.g., take $s_0 = ub_1$ from above). An element $[s] \in \mathbb{Z}_n$ is a (another) solution to $[a]x = [b]$ in $\mathbb{Z}_n$ if and only if there is a $k \in \mathbb{Z}$ so that $as = b + nk$. Write $a = a_1 d$ and $n = n_1 d$ for $a_1, n_1 \in \mathbb{Z}$ so that $as = b + nk$ becomes $a_1 ds = b_1 d + n_1 dk$. Division by d shows $[s]$ is a solution to $[a]x = [b]$ if and only if $a_1 s = b_1 + n_1 k$ for some $k \in \mathbb{Z}$. This last condition is equivalent to the statement $[a_1]_{n_1}[s]_{n_1} = [b_1]_{n_1}$ in $\mathbb{Z}_{n_1}$.

Since $(a_1, n_1) = 1$ (Exercise 1.19), Theorem 1.25 shows there is a unique solution to $[a_1]_{n_1} x = [b_1]_{n_1}$ in $\mathbb{Z}_{n_1}$. By assumption, it must be the same as $[s_0]_{n_1}$. As a result, $[s] \in \mathbb{Z}_n$ solves $[a]x = [b]$ in $\mathbb{Z}_n$ if and only if $[s]_{n_1} = [s_0]_{n_1}$. Thus $[s] \in \mathbb{Z}_n$ solves $[a]x = [b]$ in $\mathbb{Z}_n$ if and only if $s = s_0 + mn_1$ for $m \in \mathbb{Z}$.

It remains only to count the number of distinct solutions in $\mathbb{Z}_n$. Since we know every solution is of the form $x = [s_0 + mn_1]$ for some $m \in \mathbb{Z}$, we need to figure out when $[s_0 + mn_1] = [s_0 + m'n_1]$ as elements of $\mathbb{Z}_n$ where $m, m' \in \mathbb{Z}$. But this equality is true if and only if $n \mid n_1(m - m')$ if and only if (dividing by n_1) $d \mid (m - m')$ if and only if $[m]_d = [m']_d$ as elements of $\mathbb{Z}_d$. Since there are exactly d distinct elements of $\mathbb{Z}_d$, part (2) follows. In gory detail, the solutions to $[a]x = [b]$ in $\mathbb{Z}_n$ can thus be written as $\{[s_0 + mn_1] \mid m \in \mathbb{Z},\ 0 \leq m \leq d - 1\}$. □

Finally, we define exponents in the expected fashion.

DEFINITION 1.27. For $k, n \in \mathbb{N}$ and $[a] \in \mathbb{Z}_n$, let

$$[a]^k = \overbrace{[a][a] \cdots [a]}^{k \text{ copies}} = [a^k].$$

If $[a]$ is a unit, let

$$[a]^{-k} = \overbrace{[a]^{-1}[a]^{-1} \cdots [a]^{-1}}^{k \text{ copies}} = [a^k]^{-1}$$

and write $[a]^0 = [1]$.

With these definitions, the usual exponent laws hold. The proof is straightforward and is left to Exercise 1.49.

THEOREM 1.28. *Let $n, k_1, k_2 \in \mathbb{N}$ and $[a], [b] \in \mathbb{Z}_n$. Then*

$$[a]^{k_1}[a]^{k_2} = [a]^{k_1+k_2},$$
$$\left([a]^{k_1}\right)^{k_2} = [a]^{k_1 k_2},$$
$$([a][b])^{k_1} = [a]^{k_1}[b]^{k_1}.$$

If $[a], [b]$ are units, the above equations also hold for $k_1, k_2 \in \mathbb{Z}$.

The only warning needed here is to remember that even though we are working in $\mathbb{Z}_n$, the exponents themselves are honest integers in $\mathbb{Z}$. For instance in $\mathbb{Z}_5$, $1 \equiv 6$ modulo 5, but $[2]^1 = [2]$ while $[2]^6 = [64] = [4]$.

2.5. Applications.

2.5.1. *Chinese Remainder Theorem.*

THEOREM 1.29 (Chinese Remainder Theorem). *Let $n_1, \ldots, n_k \in \mathbb{N}$ so that $(n_i, n_j) = 1$ when $i \neq j$, $1 \leq i, j \leq k$. Then for any choice of $a_1, \ldots, a_k \in \mathbb{Z}$, there exists $x \in \mathbb{Z}$ so that*

$$x \equiv a_1 \pmod{n_1},$$
$$x \equiv a_2 \pmod{n_2},$$
$$\vdots$$
$$x \equiv a_k \pmod{n_k}.$$

Moreover, the solution is unique up to congruence modulo $n_1 n_2 \cdots n_k$.

PROOF. Consider first the case of existence for $k = 2$. Choose $u, v \in \mathbb{Z}$ so that $n_1 u + n_2 v = 1$. Multiplication on both sides by $a_2 - a_1$ and rearrangement show

$$a_1 + n_1 u(a_2 - a_1) = a_2 - n_2 v(a_2 - a_1).$$

Thus $x = a_1 + n_1 u(a_2 - a_1) = a_2 - n_2 v(a_2 - a_1)$ is obviously a solution.

Existence in general is established via induction on k. Assume x_0 is a solution to the first $k-1$ equations of the theorem. Observe $(n_1 \cdots n_{k-1}, n_k) = 1$ since if there were a prime p dividing n_k and $n_1 \cdots n_{k-1}$, then p would have to also divide some n_i, $1 \leq i \leq k-1$, which would violate the condition $(n_i, n_k) = 1$. Therefore, using the $k = 2$ case, there is an $x \in \mathbb{Z}$ so that

$$x \equiv x_0 \pmod{n_1 \cdots n_{k-1}},$$
$$x \equiv a_k \pmod{n_k}.$$

Since a multiple of $n_1 \cdots n_{k-1}$ is certainly also a multiple of n_i, the equivalence $x \equiv x_0 \pmod{n_1 \cdots n_{k-1}}$ implies $x \equiv x_0 \pmod{n_i}$. As the induction hypothesis already guarantees $x_0 \equiv a_i \pmod{n_i}$, the argument is complete. For an alternate proof, see Exercise 1.58.

As far as uniqueness goes, the difference of any two solutions is congruent to 0 modulo n_j, $1 \leq j \leq k$. In turn this means the difference is divisible by each n_j. As the n_j are relatively prime, the difference is therefore divisible by $n_1 n_2 \cdots n_k$ (Exercise 1.18), i.e., two solutions are congruent modulo $n_1 n_2 \cdots n_k$. Conversely any integer congruent modulo $n_1 n_2 \cdots n_k$ to a solution is clearly still a solution. □

Direct applications of the Chinese Remainder Theorem date back to at least the third century. A typical problem of this type might be the following. Suppose I have a certain number of candy bars. If my wife and I divide them among ourselves, one is left over. If I divide them among my three children, two are left over. If I divide them among the whole family, four are left over. How many candy bars do I have?

To solve the problem, write x for the number of candy bars and rephrase it as a system of modular equations

$$x \equiv 1 \pmod 2,$$
$$x \equiv 2 \pmod 3,$$
$$x \equiv 4 \pmod 5.$$

The proof of the Chinese Remainder Theorem (and Exercise 1.20) gives an algorithm for finding x. First consider the equations

$$x_0 \equiv 1 \pmod 2,$$
$$x_0 \equiv 2 \pmod 3.$$

Notice $2\,(-1)+3\,(1) = 1$ so that, from the proof of the Chinese Remainder Theorem, $x_0 = 1 + 2\,(-1)\,(2-1) = -1$ up to multiples of 6. Next consider the equations

$$x \equiv -1 \pmod 6,$$
$$x \equiv 4 \pmod 5.$$

Notice $6\,(1) + 5\,(-1) = 1$ so that $x = (-1) + 6\,(1)\,(4 - (-1)) = 29$ up to multiples of 30. In particular, the smallest number of candy bars I can have is 29, though the most general solution is $29 + 30k$, $k \in \mathbb{Z}_{\geq 0}$.

In terms of theoretical mathematics, the Chinese Remainder Theorem is usually not used in such a silly, albeit fun, way. Usually it is simply used as an existence theorem guaranteeing that a solution exists to a certain family of modular equations. For instance, we will use it below in §2.5.2 to help calculate $|U_n|$.

2.5.2. *Euler's Phi Function.*

DEFINITION 1.30. Let $n \in \mathbb{N}$ with $n > 1$. The *Euler phi function*, or *totient function*, is $\varphi(n) = |U_n|$.

For instance, the set of units in $\mathbb{Z}_2$ is $\{[1]\}$, the set of units in $\mathbb{Z}_3$ is $\{[1],[2]\}$, the set of units in $\mathbb{Z}_4$ is $\{[1],[3]\}$, and the set of units in $\mathbb{Z}_5$ is $\{[1],[2],[3],[4]\}$ so $\varphi(2) = 1$, $\varphi(3) = 2$, $\varphi(4) = 2$, and $\varphi(5) = 4$. Although $\varphi(n)$ may look like a mysterious function at first, it turns out there is a simple formula.

For the following, recall that the *Cartesian product* of two sets S_1 and S_2 is the set of ordered pairs or 2-tuples

$$S_1 \times S_2 = \{(s_1, s_2) \mid s_1 \in S_1,\ s_2 \in S_2\}.$$

Note for finite sets that $|S_1 \times S_2| = |S_1|\,|S_2|$.

THEOREM 1.31. *Let $m, n \in \mathbb{N}$ with $m, n > 1$.*

(1) *If $(m,n) = 1$, there is a bijection, i.e., a one-to-one onto correspondence, between $U_m \times U_n$ and U_{mn}.*

(2) *If $(m,n) = 1$, $\varphi(mn) = \varphi(m)\varphi(n)$.*

(3) *If* $n = \prod_{i=1}^{N} p_i^{m_i}$ *for distinct primes* $p_1, \ldots, p_N \in \mathbb{N}$ *and* $m_i \in \mathbb{N}$, *then*

$$\varphi(n) = \prod_{i=1}^{N} p_i^{m_i - 1} (p_i - 1).$$

PROOF. Consider part (1) first. Given $[a]_m \in U_m$ and $[b]_n \in U_n$, the Chinese Remainder Theorem shows there exists a unique $[c]_{mn} \in \mathbb{Z}_{mn}$ so that

$$c \equiv a \pmod{m},$$
$$c \equiv b \pmod{n}.$$

In fact, we claim that $[c]_{mn} \in U_{mn}$. To see this, first observe that $(c, mn) = (c, m)(c, n)$ since $(m, n) = 1$ (Exercise 1.24). Next note that $(c, m) = (a, m) = 1$ and $(c, n) = (b, n) = 1$ (e.g., Exercise 1.17). Therefore $(c, mn) = 1$ and $[c]_{mn} \in U_{mn}$ as desired. Thus there is a well defined map $f_1 : U_m \times U_n \to U_{mn}$ given by $f_1([a]_m, [b]_n) = [c]_{mn}$.

On the other hand, there is a map that takes any $[c]_{mn} \in \mathbb{Z}_{mn}$ and views it as an element of $\mathbb{Z}_m$ (or $\mathbb{Z}_n$) by mapping $[c]_{mn} \in \mathbb{Z}_{mn}$ to $[c]_m \in \mathbb{Z}_m$ (or $[c]_n \in \mathbb{Z}_n$). It is well defined since multiples of mn are certainly multiples of m (or n). The equality $(c, mn) = (c, m)(c, n)$ again shows that actually $[c]_m \in U_m$ (or $[c]_n \in U_n$) when $[c]_{mn} \in U_{mn}$. Therefore there is a well defined map $f_2 : U_{mn} \to U_m \times U_n$ given by $f_2([c]_{mn}) = ([c]_m, [c]_n)$.

To finish part (1), it suffices to show that f_1 and f_2 invert each other. This follows from the definitions

$$f_1\left(f_2\left([c]_{mn}\right)\right) = f_1\left(([c]_m, [c]_n)\right) = [c]_{mn} \text{ and}$$
$$f_2\left(f_1\left(([a]_m, [b]_n)\right)\right) = f_2\left([c]_{mn}\right) = ([c]_m, [c]_n) = ([a]_m, [b]_n).$$

Part (1) also finishes part (2) since

$$\varphi(mn) = |U_{mn}| = |U_m \times U_n| = |U_m|\,|U_n| = \varphi(m)\varphi(n).$$

For part (3), observe that the only $k \in \mathbb{Z}$, $1 \le k \le p_i^{m_i}$, not relatively prime to $p_i^{m_i}$ are the multiples of p_i given by $k = p_i, 2p_i, 3p_i, \ldots, p_i^{m_i - 1} p_i$. As there are clearly $p_i^{m_i - 1}$ such multiples, it follows that $\left|U_{p_i^{m_i}}\right| = \left|\mathbb{Z}_{p_i^{m_i}}\right| - p_i^{m_i - 1} = p_i^{m_i} - p_i^{m_i - 1}$. Part (3) is now finished by using part (2) and the Fundamental Theorem of Arithmetic. □

Besides its evident aesthetic appeal, the Euler phi function, $\varphi(n)$, makes a big appearance in aspects of number theory and cryptology. Many of those applications stem from what is called Euler's Theorem, which is given below. We should note that though the following proof is both short and elegant, we will soon develop machinery to make the proof completely trivial (Lagrange's Theorem, Theorem 2.28).

THEOREM 1.32 (Euler's Theorem). *Let* $n \in \mathbb{N}$ *with* $n > 1$. *If* $a \in \mathbb{Z}$ *with* $(a, n) = 1$, *then*

$$a^{\varphi(n)} \equiv 1 \pmod{n}.$$

PROOF. Since U_n is multiplicatively closed, the multiplication map $m(x) = [a]x$ can be viewed as a function $m : U_n \to U_n$. The map is one-to-one since $[a]$ is invertible. As the domain and codomain of m both have the same number of elements, $\varphi(n)$, it follows that m is also onto. In particular, the range of m,

$\{[a][u_i] \mid [u_i] \in U_n\}$, is exactly the same as the codomain, $\{[u_i] \mid [u_i] \in U_n\}$, with only the order possibly scrambled. Hence

$$\prod\nolimits_{[u_i]\in U_n} [a][u_i] = \prod\nolimits_{[u_i]\in U_n} [u_i].$$

Grouping together the $\varphi(n)$ copies of $[a]$ from the left-hand side and then dividing both sides by $\prod_{[u_i]\in U_n} [u_i]$ shows $[a^{\varphi(n)}] = [1]$ as desired. □

When n is prime, we obtain the following famous and very useful corollary.

COROLLARY 1.33 (Fermat's Little Theorem). *Let p be a positive prime and let $a \in \mathbb{Z}$. Then $a^p \equiv a \pmod{p}$. If also $p \nmid a$, then*

$$a^{p-1} \equiv 1 \pmod{p}.$$

2.5.3. *Public-Key Cryptography.* Many types of transactions require sensitive information to be passed between two parties. In order to keep that information secure, some type of code is needed. One of the most famous types was developed by R. Rivest, A. Shamir, and L. Adleman and is called the RSA system. It is based on the fact that finding very large primes (with over 300 digits) is computationally extremely difficult—so difficult that today's computers might take many hundreds of years to break the code.

ALGORITHM 1 (RSA Encryption). To code a message:

(1) Two large distinct primes $p \in \mathbb{Z}$ and $q \in \mathbb{Z}$ are chosen and kept *secret.*

(2) Let

$$k = (p-1)(q-1)$$

and pick $d \in \mathbb{Z}$ so that

$$(d, k) = 1.$$

Both k and d are also kept *secret.*

(3) Let

$$n = pq$$

and pick $e \in \mathbb{Z}$ so that

$$ed \equiv 1 \pmod{k}$$

(which exists since $[d]$ is invertible in $\mathbb{Z}_k$). Both n and e are *publicly announced.*

(4) Change your *secret message* into a number by breaking it into blocks and replacing each letter or space by a number. The method is *publicly announced.* A simple version might be given by

$$\text{space} = 00,\ \text{A} = 01,\ \text{B} = 02,\ \ldots,\ \text{Y} = 25,\ \text{Z} = 26.$$

The resulting number is called M for message.

(5) The *publicly announced* encrypted (coded) version of your message is the number

$$C = M^e \pmod{n}.$$

For the sake of illustration, we pick the very small primes $p = 53$ and $q = 59$ so that $n = 53 \cdot 59 = 3127$ and $k = 52 \cdot 58 = 3016$. Suppose our ultra secret message is "HEY DUDE". It is necessary to have $M < n$ so we break the message up into blocks of two as follows:

H	E	Y	<SPACE>	D	U	D	E
08	05	25	00	04	21	04	05

Thus the numerical version of our message is 805, 2500, 421, 405. In order to encode it, first take $d = 5$. This is a legal choice since $5 \nmid k$. Using a computer (though it is easy to do by hand—see the proof of Theorem 1.24 and Exercise 1.20), $[5]^{-1} = [2413]$ in $\mathbb{Z}_k$ so that $e = 2413$. Now we can encode our message. Again using a computer, it is easy to check that modulo n

$$805^e \equiv 2356, \ 2500^e \equiv 1787, \ 421^e \equiv 701, \ 405^e \equiv 1305$$

so that our encrypted message is 2356, 1787, 701, 1305. All the world can see this message now—they can even know what values of n and e we used to calculate it! What we keep secret is the values of p, q, d, and k.

Of course, none of this is worthwhile unless we can invert the encryption. The next theorem tells us how to do that. It turns out that all the recipient needs to know is n and d.

THEOREM 1.34. *Choose p, q, d, k, n, e as in the RSA Encryption Algorithm. Then for each $a \in \mathbb{Z}$,*

$$a^{ed} \equiv a \pmod{n}.$$

PROOF. Pick $j \in \mathbb{Z}$ so $ed = 1 + jk$. Hence

$$a^{ed} = a^{jk+1} = a^{(p-1)(q-1)j}a = \left(a^{p-1}\right)^{(q-1)j} a.$$

When $p \nmid a$, Fermat's Little Theorem thus shows $a^{p-1} \equiv 1 \pmod{p}$ so $a^{ed} \equiv a \pmod{p}$. When $p \mid a$, then $p \mid a^{ed}$ so that $a^{ed} \equiv a \pmod{p}$ in this case as well. Similarly, $a^{ed} \equiv a \pmod{q}$. Thus $p \mid \left(a^{ed} - a\right)$ and $q \mid \left(a^{ed} - a\right)$. Hence, $(pq) \mid \left(a^{ed} - a\right)$ (Exercise 1.18) as desired. □

As a corollary we get our decryption algorithm.

ALGORITHM 2 (RSA Decryption). Choose p, q, d, k, n, e as in the RSA Encryption Algorithm. The original message M is obtained from the coded message C by the formula $M = C^d \pmod{n}$.

Continuing our example, the recipient of our coded message 2356, 1787, 701, 1305 decodes the message by calculating, modulo n, that (recall $d = 5$ and $n = 3127$)

$$2356^d \equiv 0805, \ 1785^d \equiv 2500, \ 701^d \equiv 0421, \ 1305^d \equiv 0405.$$

Substituting the letters for the numbers, the original message of "HEY DUDE" emerges safe and sound.

If a spy could factor n, then he would know p, q, and therefore k. Since he also knows e, he could easily find d and thus decode the message. However as mentioned before, p and q are chosen large enough so that, practically speaking, it is computationally impossible to find p and q from n. On the other hand, all the encryption and decryption computations can be implemented in extremely fast ways. Interested readers should consult texts devoted entirely to the subject of cryptology.

2.6. Equivalence Relations. It turns out that the notion of congruence modulo n generalizes to many other areas of mathematics. It is often the case that studying certain types of problems is made easier by viewing two technically different objects as equivalent. For instance, the statement that 13 o'clock is essentially 1 o'clock can be made precise by writing $13 \equiv 1 \pmod{12}$.

In general, mathematics often needs a way of saying that two things are equivalent even though they are not the same. If we have a set S with $a, b \in S$ and, for some reason, we want to say a and b are equivalent, we will soon write $a \sim b$. In order to do this efficiently and rigorously, we will first make a giant compilation of all the things that are equivalent. This compilation, call it R for *relation set*, will actually be a subset of $S \times S$ and is constructed so that $(a, b) \in R$ if and only if $a \sim b$. As is often the case in mathematics, the official definition below runs the story backwards and starts with R in order to define $\sim$.

DEFINITION 1.35. A *relation*, R, on a set S is a subset $R \subseteq S \times S$. If $(a, b) \in R$, we write

$$a \sim b.$$

Let $a, b, c \in S$. The relation is called an *equivalence relation* if it satisfies the following three properties:

(1) *(Reflexive)* $a \sim a$ for all $a \in S$.
(2) *(Symmetric)* If $a \sim b$, then $b \sim a$.
(3) *(Transitive)* If $a \sim b$ and $b \sim c$, then $a \sim c$.

Now that the official definition is given, in practice people simply specify when two elements are equivalent, $a \sim b$, and never explicitly mention R. For example, let $n \in \mathbb{N}$ and define a relation on $\mathbb{Z}$ by saying that $a \sim b$ if and only if $n \mid (b - a)$. Since this relation is exactly the definition of congruence modulo n, we already know (Lemma 1.16) that it generates an equivalence relation.

As another example, suppose we are working in the plane $\mathbb{R}^2$ and only care about the length of a vector, $\|v\|$ for $v \in \mathbb{R}^2$. We might then define a relation by saying $v \sim w$ for $v, w \in \mathbb{R}^2$ whenever $\|v\| = \|w\|$. Since the relation is clearly reflexive, symmetric, and transitive, the result is an equivalence relation.

As a final example, suppose we want to construct $\mathbb{Q}$ from $\mathbb{Z}$. One way to start the process is to notice $\frac{a}{b} = \frac{c}{d}$ in $\mathbb{Q}$ if and only if $ad - bc = 0$ for $a, b, c, d \in \mathbb{Z}$ with $b \neq 0$ and $d \neq 0$. Thus consider the set $\mathbb{Z} \times \mathbb{Z}^{\times}$ equipped with the relation $(a, b) \sim (c, d)$ if and only if $ad - bc = 0$. It is straightforward to check that this is in fact an equivalence relation (Exercise 1.68). We will return to such types of constructions later in the text (§5.2 of Chapter 3).

Recall that soon after defining congruence modulo n, we lumped all congruent integers together to make congruence classes. We do the same thing with equivalence relations.

DEFINITION 1.36. Let S be a set with an equivalence relation $\sim$.

(1) For $a \in S$, the *equivalence class* of a, denoted $[a]$, is

$$[a] = \{b \in S \mid b \sim a\}.$$

(2) If C is an equivalence class and $[a] = C$, then a is called a *representative* for C.

(3) The set of all equivalence classes is denoted by $S/\tilde{}$.

For instance, consider the set $\mathbb{Z}$ with the equivalence relation $\sim$ given by congruence modulo n. Then $\mathbb{Z}/\tilde{} = \{[0], [1], \ldots [n-1]\} = \mathbb{Z}_n$ and has already been extensively studied. In particular, some of the same warnings given for $\mathbb{Z}_n$ apply to the more general notion of $S/\tilde{}$. For instance, representatives of equivalence classes are usually not unique.

As another example, look at $\mathbb{R}^2$ with the equivalence relation given by length. In this case, each equivalence class is a circle centered about the origin. Thus each congruence class can be uniquely specified by a radius $r \in \mathbb{R}$, $r \geq 0$. Alternately, notice that every vector in $\mathbb{R}^2$ can be uniquely rotated around to the nonnegative x-axis (without changing its length). It follows that every equivalence class can be written as $[(r, 0)]$ for some unique $r \in \mathbb{R}$ with $r \geq 0$. Thus $\mathbb{R}^2/^\sim = \{[(r,0)] \mid r \in \mathbb{R}, r \geq 0\}$ and we (again) see that $\mathbb{R}^2/^\sim$ can be indexed by the closed half line.

Finally, look at the set $\mathbb{Z} \times \mathbb{Z}^\times$ equipped with the relation $(a, b) \sim (c, d)$ if and only if $ad - bc = 0$. As we have already seen, two elements are equivalent if and only if they correspond to the same fraction. Thus $(\mathbb{Z} \times \mathbb{Z}^\times)/^\sim$ can be indexed by $\mathbb{Q}$. For instance, associate to $\frac{1}{2}$ the equivalence class $[(1, 2)]$ or, equivalently, any other representative such as $[(3, 6)]$ or $[(50, 100)]$.

Just as in Lemma 1.18, equivalence classes partition S into disjoint subsets. For instance in the case of $\mathbb{R}^2$ with the equivalence relation given by length, this is geometrically obvious. The equivalence classes are the set of circles centered around the origin and clearly exhaustively partition the plane into disjoint subsets. For a general equivalence relation, the proof that $S/^\sim$ partitions S is nearly identical to Lemma 1.18 and left to Exercise 1.69.

THEOREM 1.37. *Let S be a set with an equivalence relation $\sim$ and let $a, b \in S$.*

(1) $[a] = [b]$ *if and only if* $a \sim b$.

(2) *Either* $[a] = [b]$ *or* $[a] \cap [b] = \varnothing$; *i.e., equivalence classes are identical or disjoint.*

(3) *S is the disjoint union of its equivalence classes; i.e.,* $S = \coprod_{[a] \in S/^\sim} [a]$.

It is worth noting that every partition of S gives rise to a unique equivalence relation on S so that the equivalence classes are the partitions (Exercise 1.72). Thus the notions of equivalence relation and partition are essentially equivalent ideas.

2.7. Exercises 1.29–1.72.

2.7.1. *Congruence.*

EXERCISE 1.29. Which of the following statements are true?

(a) $6 \stackrel{?}{\equiv} 42 \pmod 2$.

(b) $6 \stackrel{?}{\equiv} 43 \pmod 2$.

(c) $7 \stackrel{?}{\equiv} 108 \pmod{10}$.

(d) $7 \stackrel{?}{\equiv} 117 \pmod{10}$.

(e) $2 \stackrel{?}{\equiv} 54 \pmod 3$.

(f) $2 \stackrel{?}{\equiv} 56 \pmod 3$.

EXERCISE 1.30. Prove the remaining parts of Lemma 1.16. *Hint:* For (5), write $a_1 = a_2 + kn$ and $b_1 = b_2 + k'n$ so that $a_1 b_1 = (a_2 + kn)(b_2 + k'n) = a_2 b_2 + n(kb_2 + k'a_2 + nkk')$.

EXERCISE 1.31. Let $m, n \in \mathbb{N}$ and $a, b \in \mathbb{Z}$. If $a \equiv b \pmod m$ and $n \mid m$, show $a \equiv b \pmod n$.

EXERCISE 1.32. **(a)** If $x \equiv 2 \pmod 5$, what is $3x^4 + x^3 + 2x - 6$ congruent to modulo 5?

(b) If $x \equiv 3 \pmod 6$, what is $2x^{47,891} + 5x^3 + 2x + 1$ congruent to modulo 6?

EXERCISE 1.33. Use modular arithmetic to determine the remainder of $5^{100,000}$ after division by 6. *Hint:* Note $5 \equiv -1 \pmod 6$.

EXERCISE 1.34. Find an example of the following (cf. Exercise 1.50).
(a) $ab \equiv 0 \pmod n$ but $a, b \not\equiv 0$.
(b) $ab \equiv ac \pmod n$ with $a \not\equiv 0$ and $b \not\equiv c$.
(c) $a^2 \equiv b^2 \pmod n$ but $a \not\equiv \pm b$. *Hint:* Look in a nonprime modulus.

EXERCISE 1.35. Let $a, b \in \mathbb{Z}$. Show $a = b$ if and only if $a \equiv b \pmod p$ for all positive primes p. *Hint:* Show $a - b \neq 0$ if and only if there is a prime p so that $p \nmid (a - b)$.

EXERCISE 1.36 (Divisibility Tricks). Recall that our decimal system is a base 10 system. For example, this means that 5672 is the decimal representation of $5 \cdot 10^3 + 6 \cdot 10^2 + 7 \cdot 10^1 + 2 \cdot 10^0$. The numbers $5, 6, 7, 2$ are called the *digits* of the number. In general, let $n \in \mathbb{Z}_{\geq 0}$ and write its decimal representation as $a_N \ldots a_2 a_1 a_0$ so that $n = \sum_{k=0}^{N} a_k 10^k$. Use modular arithmetic to verify the following rules.
(a) For $k \in \mathbb{N}$, show $k \mid n$ if and only if $n \equiv 0 \pmod k$.
(b) Show $2 \mid n$ if and only if 2 divides the last digit of n, i.e., if and only a_0 is even.
(c) Show $3 \mid n$ if and only if 3 divides the sum of the digits of n, i.e., if and only if $3 \mid \sum_{k=0}^{N} a_k$. *Hint:* Notice $10 \equiv 1 \pmod 3$.
(d) Show $4 \mid n$ if and only if $4 \mid (2a_1 + a_0)$.
(e) Show $5 \mid n$ if and only if 5 divides the last digit of n, i.e., if and only $a_0 \in \{0, 5\}$.
(f) Show $6 \mid n$ if and only if $6 \mid (2\sum_{k=1}^{N} a_k - a_0)$.
(g) If n has a decimal representation of $a_4 a_3 a_2 a_1 a_0$, show $7 \mid n$ if and only if $7 \mid (-3a_4 - a_3 + 2a_2 + 3a_1 + a_0)$. Describe the general pattern.
(h) Show $8 \mid n$ if and only if $8 \mid (a_0 + 2a_1 + 4a_2)$.
(i) Show $9 \mid n$ if and only if 9 divides the sum of the digits of n, i.e., if and only if $9 \mid \sum_{k=0}^{N} a_k$.
(j) Show $11 \mid n$ if and only if 11 divides the alternating sum of the digits of n, i.e., if and only if $11 \mid \sum_{k=0}^{N} (-1)^k a_k$.

EXERCISE 1.37. A number of the form a^2 for $a \in \mathbb{Z}$ is called a *perfect square*. Show that the last digit of a perfect square lies in $\{0, 1, 4, 5, 6, 9\}$. *Hint:* Use modulo 10.

EXERCISE 1.38. Let $m, n \in \mathbb{N}$ with $(m, n) = 1$ and $a, b \in \mathbb{Z}$.
(a) Show $a \equiv b \pmod m$ and $a \equiv b \pmod n$ if and only if $a \equiv b \pmod{mn}$. *Hint:* To go left to right, first show $m, n \mid (a - b)$. Then use $(m, n) = 1$ and Exercise 1.18 to get $(mn) \mid (a - b)$.
(b) Show part (a) can be false if m and n are not relatively prime.

2.7.2. *Congruence Classes.*

EXERCISE 1.39. **(a)** In $\mathbb{Z}_5$, which sets are the same as $[2]$:

(i) $\{12 + 5k \mid k \in \mathbb{Z}\}$, (ii) $\{14 + 5k \mid k \in \mathbb{Z}\}$, or (iii) $\{27 - 5k \mid k \in \mathbb{Z}\}$?

(b) In $\mathbb{Z}_3$, which sets are the same as $[2]$:

(i) $\{\ldots 28, 31, 34, 37, \ldots\}$, (ii) $\{\ldots 41, 44, 47, 50, \ldots\}$, or (iii) $\{\ldots -7, -4, -1, \ldots\}$?

(c) Find four different representatives for $[2] \in \mathbb{Z}_7$.

EXERCISE 1.40. Show $\mathbb{Z}_3 = \{[0], [2], [2^2]\}$ and $\mathbb{Z}_5 = \{[0], [2], [2^2], [2^3], [2^4]\}$ but $\mathbb{Z}_4 \neq \{[0], [2^1], [2^2], [2^3]\}$.

EXERCISE 1.41. Let $m, n \in \mathbb{N}$. Attempt to define a map $f : \mathbb{Z}_m \to \mathbb{Z}_n$ by setting $f([a]_m) = [a]_n$ for $[a]_m \in \mathbb{Z}_m$.

(a) Show this supposed function may not be well defined by finding an example in which $[a]_n \neq [a + m]_n$.

(b) Show f is a well defined function if and only if $n \mid m$. *Hint:* Show f is well defined if and only if, for each $k \in \mathbb{Z}$ (especially $k = 1$), $a + km = a + jn$ for some $j \in \mathbb{Z}$.

2.7.3. *Arithmetic.*

EXERCISE 1.42. Write out the entire addition and multiplication tables for:
(a) $\mathbb{Z}_4$,
(b) $\mathbb{Z}_5$,
(c) $\mathbb{Z}_7$,
(d) $\mathbb{Z}_8$.

EXERCISE 1.43. **(a)** If $x = [2] \in \mathbb{Z}_5$, simplify $[3]x^4 + x^3 + [2]x - [6]$.
(b) If $x = [3] \in \mathbb{Z}_6$, simplify $[2]x^{47,891} + [5]x^3 + [2]x + [1]$.

EXERCISE 1.44. Since $\mathbb{Z}_n$ has only n elements, it is possible to solve an explicit equation simply by substituting in all possible values of $\mathbb{Z}_n$ and checking for success. Use this method to find all solutions to the following equations.
(a) $x^3 + x^2 + x = [0]$ in $\mathbb{Z}_4$.
(b) $x^3 + x^2 + x = [0]$ in $\mathbb{Z}_5$.

EXERCISE 1.45. **(a)** Try using a calculator to evaluate $2^{10,000}$ in hopes of calculating $[2^{10,000}]$ in $\mathbb{Z}_{43}$. Any luck?

(b) Evaluate $[2^{10,000}]$ in $\mathbb{Z}_{43}$ by hand. *Hint:* Write $[x]^n$ as $[x^2]^{\frac{n}{2}}$ or as $[x][x^2]^{\frac{n-1}{2}}$ and repeat a whole bunch.

EXERCISE 1.46. Let $n \in \mathbb{N}$ and consider solutions to the equation $x^2 = [1]$ in $\mathbb{Z}_n$.

(a) Let $n = p^m$ for p an odd prime. Show there are only two solutions to the equation $x^2 = [1]$, namely $x = [\pm 1]$. *Hint:* If $[y]^2 = [1]$, $[y] \neq [\pm 1]$, show $[y \pm 1]$ are both zero divisors so that $y = 1 + pk$ and $y = -1 + pk'$. Work modulo p to get a contradiction.

(b) Let $n = \prod_{i=1}^N p_i^{m_i}$ be odd with distinct prime factors $p_1, \ldots, p_N$. Show there are 2^N solutions to the equation $x^2 = [1]$ (see Exercise 1.38). *Hint:* Show $[y^2]_n = [1]_n$ is equivalent to $[y^2]_{p_i^{m_i}} = [1]_{p_i^{m_i}}$ for each i.

(c) Let $n = 2^m$ for $m \in \mathbb{N}$. Show solutions to the equation $x^2 = [1]$ have the form $x = [\pm 1], [\pm 1 + 2^{m-1}]$. Conclude that there is one solution to $x^2 = [1]$ when $m = 1$, two solutions when $m = 2$, and four solutions when $m \geq 3$. *Hint:* Similarly to (a), check that $y = 1 + 2k$, that $4k(1+k) \equiv 0$, and then that $k = 2^{m-2}j$ or $k = 2^{m-2}j - 1$. Reduce to $j = 0, 1$ by replacing j with $j - 2$.

(d) Let $n = 2^m \prod_{i=1}^N p_i^{m_i}$ be even with distinct prime factors $2, p_1, \ldots, p_N$. Show there are 2^N solution when $m = 1$, 2^{N+1} solutions when $m = 2$, and 2^{N+2} solutions when $m \geq 3$.

2.7.4. *Structure.*

EXERCISE 1.47. Prove the remaining parts of Theorem 1.22.

EXERCISE 1.48. List all units and (separately) all zero divisors of:
(a) $\mathbb{Z}_3$,
(b) $\mathbb{Z}_4$,
(c) $\mathbb{Z}_7$,
(d) $\mathbb{Z}_8$,
(e) $\mathbb{Z}_{12}$,
(f) $\mathbb{Z}_{15}$.

EXERCISE 1.49. Prove Theorem 1.28.

EXERCISE 1.50. Let $a, b \in \mathbb{Z}$ and let p be a positive prime (cf. Exercise 1.34).
(a) Show that $ab \equiv 0 \pmod{p}$ implies $a \equiv 0$ or $b \equiv 0$. *Hint:* If $[a] \neq 0$, use $[a]^{-1}$.
(b) Show that $ab \equiv ac \pmod{p}$ with $a \not\equiv 0$ implies $b \equiv c$. *Hint:* Rewrite as $[a][b - c] = [0]$.
(c) Show that $a^2 \equiv b^2 \pmod{p}$ implies $a \equiv \pm b$. *Hint:* Rewrite as $([a] - [b])([a] + [b]) \equiv 0$.

EXERCISE 1.51. Show by example that U_n is not closed under addition, i.e., there exists $[a], [b] \in U_n$ so that $[a] + [b] \notin U_n$. *Hint:* Can you get $[0]$?

EXERCISE 1.52. Determine how many solutions there are to the following equations.
(a) $[2]x = [9]$ in $\mathbb{Z}_7$.
(b) $[15]x - [1] = [5]$ in $\mathbb{Z}_{24}$.
(c) $[2]x + [2] = [3]$ in $\mathbb{Z}_6$.
(d) $[35]x - [20] = [20]$ in $\mathbb{Z}_{100}$.

EXERCISE 1.53. Using the proof of Theorem 1.26, write down all solutions to the following equations.
(a) $[3]x - [1] = [2]$ in $\mathbb{Z}_{12}$.
(b) $[2]x + [1] = [5]$ in $\mathbb{Z}_{12}$.

EXERCISE 1.54. Show the set $\mathbb{Z}_n \backslash U_n = \{[0]\} \cup \{\text{zero divisors}\}$ is closed under multiplication. *Hint:* Show $[a] \in \mathbb{Z}_n \backslash U_n$ if and only if $(a, n) > 1$ and that $(ak, n) \geq (a, n)$.

EXERCISE 1.55 (Wilson's Theorem). Let $n \in \mathbb{N}$, $n \geq 2$. This exercise proves *Wilson's Theorem*, which says $(n - 1)! \equiv -1 \pmod{n}$ if and only if n is prime.
(a) If n is not prime, show that a factor of n appears in $(n - 1)!$. Conclude that $[(n - 1)!]$ is a zero divisor and that $(n - 1)! \not\equiv -1$.
(b) If $n = p$ is prime, show the only solutions to $x^2 = [1]$ in $\mathbb{Z}_p$ are $[1]$ and $[p - 1] = [-1]$ (cf. Exercise 1.50 or Exercise 1.46).
(c) Show that every element on the list $[2], [3], \ldots, [(p - 2)]$ uniquely pairs with its multiplicative inverse somewhere else on the same list. Conclude that $(n - 1)! \equiv -1$.
(d) If $p = 4k + 1$ is a prime for $k \in \mathbb{N}$, show $x = [(2k)!]$ solves $x^2 = [-1]$ in $\mathbb{Z}_p$. *Hint:* Notice $-a \equiv p - a$.

2.7.5. *Applications.*

EXERCISE 1.56. Find the smallest solution in $\mathbb{Z}_{\geq 0}$ to the following systems of equations.

(a) $x \equiv 1 \pmod 5$, $x \equiv 4 \pmod 7$.
(b) $x \equiv 3 \pmod{20}$, $x \equiv 10 \pmod{39}$.
(c) $x \equiv 2 \pmod 3$, $x \equiv 6 \pmod{10}$, $x \equiv 5 \pmod{11}$.
(d) $x \equiv 2 \pmod 4$, $x \equiv 3 \pmod 9$, $x \equiv 4 \pmod 5$, $x \equiv 5 \pmod 7$.

EXERCISE 1.57. Skyler is writing a breathtaking new treatise on *Hamlet* and decides each chapter must have an equal number of pages. Skyler first tries to make each chapter ten pages long but ends up with five pages left over. Next Skyler removes two pages from the text and tries to make each chapter nine pages long but ends up with four pages left over. Finally Skyler meets with success by removing one more page and making each chapter seven pages long. How long was Skyler's original paper?

EXERCISE 1.58 (*Alternate Proof of the Chinese Remainder Theorem*). Let $n_1, \ldots, n_k \in \mathbb{N}$ so that $(n_i, n_j) = 1$ when $i \neq j$, $1 \leq i, j \leq k$, and fix $a_1, \ldots, a_k \in \mathbb{Z}$.

(a) Let $N = \prod_{i=1}^k n_i$ and $N_i = \frac{N}{n_i}$. Show there exists $[d_i] \in \mathbb{Z}_{n_i}$ so $[N_i][d_i] = [1]$.

(b) Show $x = \sum_{i=1}^k a_i d_i N_i$ is a solution to

$$\begin{aligned} x &\equiv a_1 \pmod{n_1}, \\ x &\equiv a_2 \pmod{n_2}, \\ &\vdots \\ x &\equiv a_k \pmod{n_k}. \end{aligned}$$

EXERCISE 1.59 (General Chinese Remainder Theorem). Suppose $m, n \in \mathbb{N}$ with $d = (m, n)$.

(a) For $a, b \in \mathbb{Z}$, show there is a solution to

$$\begin{aligned} x &\equiv a \pmod m, \\ x &\equiv b \pmod n \end{aligned}$$

if and only if $a \equiv b \pmod d$. *Hint:* If x is a solution, write $x = a + mk$ and $x = b + nk'$. Subtract and show $d \mid (a - b)$. For the converse, first show you can solve $x' \equiv 0 \pmod{\frac{m}{d}}$ and $x' \equiv \frac{b-a}{d} \pmod{\frac{n}{d}}$ (cf. Exercise 1.19).

(b) If $a \equiv b \pmod d$, show the solution is unique up to congruence modulo $[m, n]$ (cf. Exercise 1.28). *Hint:* Look at the difference of two solutions. Show it is divisible by n and m and therefore divisible by $[m, n]$.

EXERCISE 1.60. Let $m, N \in \mathbb{N}$ and pick distinct primes $p_1, \ldots, p_N$. Show there exists $x \in \mathbb{Z}$ so that $p_k^m \mid (x + k)$ for $1 \leq k \leq N$. *Hint:* Chinese Remainder Theorem.

EXERCISE 1.61. Calculate:

(a) $|U_{26}|$,
(b) $|U_{72}|$,
(c) $|U_{3240}|$.

EXERCISE 1.62. Show there exist $m, n \in \mathbb{N}$, $m, n > 1$, so that $\varphi(mn) \neq \varphi(m)\varphi(n)$.

EXERCISE 1.63. Let $m, n \in \mathbb{N}$.

(a) Show the map $f : \mathbb{Z}_m \times \mathbb{Z}_n \to \mathbb{Z}_{mn}$ given by $f([a]_m, [b]_n) = [na + mb]_{mn}$ is well defined.

(b) If $(m, n) = 1$, find an inverse to f. *Hint:* Write $[n_0] = [n]^{-1} \in \mathbb{Z}_m$ and $[m_0] = [m]^{-1} \in \mathbb{Z}_n$.

EXERCISE 1.64 (First Supplementary Law of Quadratic Reciprocity). Let p be a positive odd prime. This exercise shows $x^2 = [-1]$ has a solution in $\mathbb{Z}_p$ if and only if $p \equiv 1 \pmod 4$ (cf. Exercise 3.199).

(a) If $x^2 = [-1]$, show $[1] = [-1]^{\frac{p-1}{2}}$. Conclude $p \equiv 1 \pmod 4$. *Hint:* Square $x^{\frac{p-1}{2}}$ and use Fermat's Little Theorem.

(b) If $p \equiv 1 \pmod 4$, show $x^2 = [-1]$ has a solution. *Hint:* Exercise 1.55.

EXERCISE 1.65. **(a)** Using $p = 53$, $q = 59$, $d = 5$, and a block size of 2, use the RSA Encryption Algorithm to encode the message: "AM I COOL".

(b) Using the same values of p, q, and d, decode the message: "0117".

EXERCISE 1.66. The UPC (Universal Product Code) consists of a string of 12 digits $a_1a_2\ldots a_{12}$, often encoded as a bar code. The first six digits identify the manufacturer, the second five identify the product, and the last digit is a check digit chosen so that

$$3a_1 + a_2 + 3a_3 + a_4 + \cdots + 3a_{11} + a_{12} \equiv 0 \pmod{10}.$$

(a) Is 123456789128 a valid UPC?

(b) Is 098765432111 a valid UPC?

(c) Show that the check digit will detect an error if you mistakenly enter the wrong digit in exactly one spot.

(d) What transposition errors will it detect?

EXERCISE 1.67. Before 2007, the ISBN (International Standard Book Number) consisted of a string of 10 digits $a_1a_2\ldots a_{10}$. The last digit is again a check digit and is chosen so that

$$10a_1 + 9a_2 + 8a_3 + \cdots + 2a_9 + a_{10} \equiv 0 \pmod{11},$$

where an X is used if the last digit needs to be 10.

(a) Is 123456789X a valid ISBN?

(b) Is 9876543219 a valid ISBN?

(c) Show that the check digit will detect an error if you mistakenly enter the wrong digit in exactly one spot.

(d) What transposition errors will it detect?

2.7.6. *Equivalence Relations.*

EXERCISE 1.68. For each set S and relation $\sim$ below, (i) determine if it is an equivalence relation and (ii) if it is an equivalence relation, describe the congruence classes.

(a) $S = \mathbb{R}^2$ with $(x, y) \sim (x', y')$ when $x = x'$.

(b) $S = \{\text{all lines in } \mathbb{R}^2\}$ with $l_1 \sim l_2$ when l_1 and l_2 are parallel.

(c) $S = \mathbb{Z}$ with $a \sim b$ when $b = 2^k a$ for some $k \in \mathbb{Z}_{\geq 0}$.

(d) $S = \mathbb{Z}$ with $a \sim b$ when $b = 2^k a$ for some $k \in \mathbb{Z}$.
(e) $S = \{\text{functions } f : \mathbb{R} \to \mathbb{R}\}$ with $f \sim g$ when $f(0) = g(0)$.
(f) $S = \mathbb{R}$ with $x \sim y$ when $|x - y| \leq 1$.
(g) $S = \mathbb{R}$ with $x \sim y$ when $x - y \in \mathbb{Z}$.
(h) $S = \mathbb{Z} \times \mathbb{Z}^\times$ with $(a, b) \sim (c, d)$ when $ad - bc = 0$.

EXERCISE 1.69. Prove Theorem 1.37.

EXERCISE 1.70. Consider the equivalence relation on $\mathbb{Z} \times \mathbb{Z}^\times$ given by $(a, b) \sim (c, d)$ when $ad - bc = 0$. Examine the following attempts at defining a function $f : (\mathbb{Z} \times \mathbb{Z}^\times)/\tilde{} \to \mathbb{Q}$ and determine which are actually well defined.
(a) $f([(a, b)]) = \frac{a}{b}$.
(b) $f([(a, b)]) = a$.
(c) $f([(a, b)]) = \frac{a^2}{a^2+b^2}$.

EXERCISE 1.71. Let A and B be sets and suppose $f : A \to B$ is a function. For $a, a' \in A$, write $a \sim a'$ when $f(a) = f(a')$.
(a) Show $\sim$ is an equivalence relation.
(b) What is $[a]$?
(c) Show $\widetilde{f} : (A/\sim) \to B$ given by $\widetilde{f}([a]) = f(a)$ is well defined.
(d) Show $\widetilde{f}$ is injective.
(e) If f is surjective, show $\widetilde{f}$ is surjective.

EXERCISE 1.72. Let S be a set. A *partition* of S is a collection of disjoint subsets $S_\alpha \subseteq S$, $\alpha \in \mathcal{A}$ an index set, so that $S = \bigcup_{\alpha \in \mathcal{A}} S_\alpha$. Sometimes this is written as the disjoint union $S = \coprod_{\alpha \in \mathcal{A}} S_\alpha$.
(a) Show the partition $S = \coprod_{\alpha \in \mathcal{A}} S_\alpha$ gives rise to an equivalence relation on S by setting $a \sim b$ if and only if there is an $\alpha \in \mathcal{A}$ so that $a, b \in S_\alpha$.
(b) Show the equivalence classes for this relation are the sets $\{S_\alpha \mid \alpha \in \mathcal{A}\}$.

CHAPTER 2

Groups

The integers have two basic operations: addition and multiplication. It turns out that many mathematical objects also come equipped with generalizations of addition or multiplication. For instance, $\mathbb{Z}_n$ has both an addition and multiplication structure while U_n only has a multiplication operation. Further examples abound in mathematics and this chapter begins their study.

Roughly speaking, gadgets with one operation are called *groups* and gadgets with two operations are called *rings*. We start by studying groups since there is less to worry about with only one operation running amuck.

1. Definitions and Examples

1.1. Binary Operations. First off, we need a definition capable of handling various generalizations of addition or multiplication. If we think about the meaning of, say, addition on $\mathbb{Z}$, what we really have is a map taking a pair of numbers and sputtering out their sum. In fancy language, addition is nothing more than a very special function from $\mathbb{Z} \times \mathbb{Z}$ to $\mathbb{Z}$.

Given an arbitrary set S, the natural way to generalize this is to take a function $B : S \times S \to S$ (B for binary) and to use it to define a souped-up version of addition or multiplication on S. For instance, if we want to use B to define a type of addition on S, we can define $s_1 + s_2 = B(s_1, s_2)$ for $s_1, s_2 \in S$. If we want to think of B as defining a type of multiplication, we can instead write $s_1 \cdot s_2 = B(s_1, s_2)$ or even $s_1 s_2 = B(s_1, s_2)$ if we are lazy. If we prefer to show no bias towards addition or multiplication, we use the generic symbol $*$ and write $s_1 * s_2 = B(s_1, s_2)$.

DEFINITION 2.1. Given a set S, a *binary operation* $*$ on S is a function $B_* : S \times S \to S$. For $s_1, s_2 \in S$, define $s_1 * s_2$ by

$$s_1 * s_2 = B_*(s_1, s_2).$$

For example, in $\mathbb{Z}_n$, consider the map $B_+ : \mathbb{Z}_n \times \mathbb{Z}_n \to \mathbb{Z}_n$ given by $B_+([a],[b]) = [a+b]$. This is the binary operation we used to define addition in $\mathbb{Z}_n$ since, by definition, $[a]+[b] = B_+([a],[b])$. Similarly, the map $B_. : \mathbb{Z}_n \times \mathbb{Z}_n \to \mathbb{Z}_n$ given by $B_.([a],[b]) = [ab]$ is the binary operation we used to define multiplication on $\mathbb{Z}_n$.

As a less useful example, consider the binary operation $B_* : \mathbb{Z} \times \mathbb{Z} \to \mathbb{Z}$ given by $B_*(a,b) = a^2 b$ for $a, b \in \mathbb{Z}$. Using this operation, we calculate that $5 * 3 = B_*(5,3) = 5^2 3 = 75$.

About the only way to mess up a binary operation is to forget that the codomain of a binary operation on S is S. In other words, the result of our generalized addition or multiplication must still be in the original set S. This property of binary operations is sometimes referred to as *closure*.

For instance, the map $B_{*_1} : \mathbb{Z} \times \mathbb{Z} \to \mathbb{Z}$ given by $B_{*_1}(a,b) = 2ab$ is a binary operation since $2ab \in \mathbb{Z}$. However, the "map" $B_{*_2} : \mathbb{Z} \times \mathbb{Z} \to \mathbb{Z}$ given by $B_{*_2}(a,b) =$

$\frac{ab}{2}$ is not a binary operation since $\frac{ab}{2}$ is not always in $\mathbb{Z}$. On the other hand, $B_{*3} : \mathbb{Q} \times \mathbb{Q} \to \mathbb{Q}$ given by $B_{*3}(a, b) = \frac{ab}{2}$ is a perfectly fine binary operation since $\frac{ab}{2} \in \mathbb{Q}$.

1.2. Definition of a Group. A group is simply an object with a sufficiently nice generalized version of addition or multiplication.

DEFINITION 2.2. A *group* is a set G with a binary operation $*$ on G satisfying the properties:

(1) (*Associativity*) For $g_1, g_2, g_3 \in G$,

$$g_1 * (g_2 * g_3) = (g_1 * g_2) * g_3.$$

(2) (*Identity*) There exists an element $e \in G$ so that

$$e * g = g * e = g$$

for all $g \in G$. The element e is called the *identity element.*

(3) (*Inverses*) For every element $g \in G$, there exists an $h \in G$ so that

$$g * h = h * g = e.$$

The element h is called the *inverse* of g.

Some notes are in order before moving on to examples. The first is that when people think of the binary operation $*$ as an additive sort of operation, the symbol $+$ is usually used instead of $*$. In this case, the associative property looks like

$$g_1 + (g_2 + g_3) = (g_1 + g_2) + g_3,$$

the identity property looks like

$$e + g = g + e = g,$$

and the inverse property looks like

$$g + h = h + g = e.$$

Because this setting looks so familiar to normal addition on $\mathbb{Z}$, people sometimes write 0_G, or even just 0, instead of e for the identity element. In that case, the identity property looks like

$$0 + g = g + 0 = g$$

and the inverse property looks like

$$g + h = h + g = 0.$$

For the same reason, people in this setting usually write $-g$ for the inverse of g. Thus the inverse property says that for each $g \in G$, there is an element $-g \in G$ so that

$$g + (-g) = (-g) + g = 0.$$

Of course people also abbreviate statements such as $g_1 + (-g_2)$ with the notation $g_1 - g_2$. When people adopt parts of this sort of notation for a group, they say they are writing a group *additively.*

Similarly, when people think of $*$ as a multiplicative sort of operation, they often use $\cdot$ instead of $*$. In this situation, the associative property looks like

$$g_1 \cdot (g_2 \cdot g_3) = (g_1 \cdot g_2) \cdot g_3,$$

the identity property looks like

$$e \cdot g = g \cdot e = g,$$

and the inverse property looks like

$$g \cdot h = h \cdot g = e.$$

Even more commonly, just as we do with ordinary multiplication, the $\cdot$ is omitted altogether. In that case, it is simply understood that two adjacent elements are "multiplied" with the binary operation. In that setting, the associative property looks like

$$g_1(g_2g_3) = (g_1g_2)g_3,$$

the identity property looks like

$$eg = ge = g,$$

and the inverse property looks like

$$gh = hg = e.$$

Because this setting looks so familiar to normal multiplication, people sometimes write 1_G, or even just 1, instead of e for the identity element. In that case, the identity property looks like

$$1 \cdot g = g \cdot 1 = g$$

and the inverse property looks like

$$gh = hg = 1.$$

For the same reason, people in this setting usually write g^{-1} for the inverse of g. Thus the inverse property says that for each $g \in G$, there is an element $g^{-1} \in G$ so that

$$gg^{-1} = g^{-1}g = 1.$$

When people adopt parts of this sort of notation for a group, they say they are writing a group *multiplicatively*. Multiplicative notation is the default notation for most generic groups.

Our final remark addresses the uniqueness of an inverse in a group. We will shortly see (Theorem 2.5) that inverses turn out to be unique and so it is fair to say "the inverse" instead of "an inverse". A similar remark holds for the identity element.

1.3. Examples.

1.3.1. *Additive Arithmetic:* $(\mathbb{Z},+)$, $(\mathbb{Q},+)$, $(\mathbb{R},+)$, $(\mathbb{R}^n,+)$, $(\mathbb{C},+)$, *and* $(\mathbb{Z}_n,+)$. To be honest, these are rather boring examples even if they are extremely important. Start with $\mathbb{Z}$. Groups have one operation while $\mathbb{Z}$ comes equipped with two. Therefore, if you want to get a group here, it is important to specify which binary operation is going to be used. In order to emphasize that the additive operation is being used, people often write $(\mathbb{Z},+)$. The reader has probably known since first grade that the set $\mathbb{Z}$ along with the binary operation $+$ satisfies the three properties of a group. Using additive notation (of course), the associative law follows from Axiom 1. The (additive) identity element is $e = 0$ since $n + 0 = 0 + n = 0$ for all $n \in \mathbb{Z}$. Finally, (additive) inverses exist since $n - n = -n + n = 0$. Thus $(\mathbb{Z},+)$ is a group.

Along with $(\mathbb{Z},+)$, there are a number of analogous examples of groups: $(\mathbb{Q},+)$, $(\mathbb{R},+)$, $(\mathbb{R}^n,+)$, $(\mathbb{C},+)$, and $(\mathbb{Z}_n,+)$. Here $\mathbb{C}$ is the set of complex numbers and

$\mathbb{R}^n$ is the set of vectors in n-space equipped with vector addition. The verification that each of these objects is a group is almost identical to the argument for $(\mathbb{Z}, +)$ The hardest is $(\mathbb{Z}_n, +)$, but Theorem 1.22 takes care of it all.

Notice, however, that objects such as $(\mathbb{Z}_{\geq 0}, +)$ are not groups. While $\mathbb{Z}_{\geq 0}$ is associative and has an identity, $e = 0$, it fails to have enough inverses. For example, there is no inverse to 2 in $\mathbb{Z}_{\geq 0}$ since $-2 \notin \mathbb{Z}_{\geq 0}$.

1.3.2. *Multiplicative Arithmetic:* $(\mathbb{Q}^\times, \cdot)$, $(\mathbb{Q}^+, \cdot)$, $(\mathbb{R}^\times, \cdot)$, $(\mathbb{R}^+, \cdot)$, $(\mathbb{C}^\times, \cdot)$, *and* $(U_n, \cdot)$. A reasonable question is to ask whether $(\mathbb{Z}, \cdot)$ also forms a group (obvious notation here). The answer is no. With respect to the binary operation of multiplication, $\mathbb{Z}$ is clearly associative (Axiom 1) and there exists a (multiplicative) identity, $e = 1$. However, $(\mathbb{Z}, \cdot)$ fails to satisfy the inverse property. For instance, 2 has no (multiplicative) inverse in $\mathbb{Z}$. Of course if we worked with $\mathbb{Q}$ the whole time instead of $\mathbb{Z}$, then 2 would have an inverse, namely $\frac{1}{2}$. However, even scrapping $(\mathbb{Z}, \cdot)$ and passing to $(\mathbb{Q}, \cdot)$ still does not quite fix the problem since 0 has no (multiplicative) inverse.

Since the only thing preventing $(\mathbb{Q}, \cdot)$ from being a group is the number 0, let us get rid of it. Write $\mathbb{Q}^\times$ for the set of invertible elements in $\mathbb{Q}$; i.e., $\mathbb{Q}^\times = \mathbb{Q} \backslash \{0\}$. Then $\mathbb{Q}^\times$ is closed under the binary operation of multiplication. Moreover, $\mathbb{Q}^\times$ contains all its inverses. Thus $(\mathbb{Q}^\times, \cdot)$ is a group. Similar examples (with the obvious notation) include $(\mathbb{R}^\times, \cdot)$, $(\mathbb{C}^\times, \cdot)$, and $(U_n, \cdot)$. Again, the verification that each of these objects is a group is almost identical to the argument for $(\mathbb{Q}^\times, \cdot)$. The only one that requires work is $(U_n, \cdot)$. However, Theorem 1.25 shows closure of the binary operation, Theorem 1.22 gives associativity and the identity, and the existence of inverses follows from the definition.

Finally, notice that the inverse of a positive number is still positive. Thus if $\mathbb{Q}^+ = \{q \in \mathbb{Q} \mid q > 0\}$ and similarly for $\mathbb{R}^+$, it easily follows that $(\mathbb{Q}^+, \cdot)$ and $(\mathbb{R}^+, \cdot)$ are groups as well. The negative numbers, though, do not form a group. For instance, multiplication on $\mathbb{Q}^-$ is not a binary operation since the product of two numbers lies in $\mathbb{Q}^+$ instead of $\mathbb{Q}^-$. Another problem with $\mathbb{Q}^-$ is that it has no identity element since $1 \notin \mathbb{Q}^-$.

1.3.3. *The Circle:* $(S^1, \cdot)$. Write S^1 for the circle in the plane (this notation comes from calculus where S^n is the n-dimensional sphere in $\mathbb{R}^{n+1}$). It turns out there are many ways to look at S^1 (Exercise 2.8). One of the easiest ways is to identify $\mathbb{R}^2$ with $\mathbb{C}$ by the correspondence $(x, y) \leftrightarrow x + iy$. In that setting, define

$$S^1 = \{z \in \mathbb{C} \mid |z| = 1\}$$

where $|z| = \sqrt{a^2 + b^2}$ is the *length* (also called the *modulus* or *norm*) of $z = a + ib$ with $a, b \in \mathbb{R}$. Recall that complex multiplication satisfies $|z_1 z_2| = |z_1| |z_2|$ for $z_1, z_2 \in \mathbb{C}$. Thus, if $z_1, z_2 \in S^1$,

$$|z_1 z_2| = |z_1| |z_2| = 1 \cdot 1 = 1$$

so that the product of two elements on S^1 still lies in S^1. Therefore complex multiplication is a binary operation on S^1. To check that $(S^1, \cdot)$ is a group, first notice that multiplication is associative since it is already associative on all of $\mathbb{C}$. As far as the identity goes, it is clear that $e = 1$ works. Finally, to find inverses, write $\overline{z} = a - ib$ for the *complex conjugate* of z and recall that $|\overline{z}| = |z|$ and $z\overline{z} = |z|^2$.

In particular, if $z \in S^1$, then $\overline{z} \in S^1$ since $|\overline{z}| = |z| = 1$. Moreover, $z^{-1} = \overline{z}$ since

$$\overline{z}z = z\overline{z} = |z|^2 = 1^2 = 1.$$

1.3.4. *Linear Algebra: $GL(n,\mathbb{F})$, $SL(n,\mathbb{F})$, $SO(n,\mathbb{F})$, and $O(n,\mathbb{F})$.* Write $M_{n,m}(\mathbb{F})$ for the set of $n \times m$ matrices with entries in $\mathbb{F}$ where $\mathbb{F}$ (for now) is $\mathbb{Q}$, $\mathbb{R}$, $\mathbb{C}$, or $\mathbb{Z}_p$ for $p \in \mathbb{N}$ prime. The reader is surely familiar with the basic properties of matrix algebra when $\mathbb{F} = \mathbb{R}$. Most likely, the reader also saw that most results still hold when $\mathbb{F} = \mathbb{C}$. In fact, almost all basic properties work even when $\mathbb{F} = \mathbb{Q}$ or $\mathbb{F} = \mathbb{Z}_p$. In any case, operations such as matrix addition, multiplication, inverses, and determinants all go through as expected for any choice of $\mathbb{F}$. For instance, under matrix addition, $M_{n,m}(\mathbb{F})$ forms a group that is of course written additively.

However, the more interesting operation is matrix multiplication. Since groups need inverses, define the *General Linear Group* as

$$GL(n,\mathbb{F}) = \{g \in M_{n,n}(\mathbb{F}) \mid g \text{ is invertible}\}.$$

Recall from linear algebra that there are a number of equivalent criteria for determining if $g \in M_{n,n}(\mathbb{F})$ is invertible. The first definition is that g is *invertible* if there is a matrix $h \in M_{n,n}(\mathbb{F})$ so that $gh = hg = I_n$ where I_n is the *identity matrix*

$$I_n = \operatorname{diag}(1,\ldots,1) = \begin{pmatrix} 1 & 0 & & 0 \\ 0 & 1 & & 0 \\ & & \ddots & \\ 0 & 0 & & 1 \end{pmatrix}$$

(here 1 means $[1]$ and 0 means $[0]$ in the case of $\mathbb{F} = \mathbb{Z}_p$).

However, as the reader will no doubt remember, one of the theoretically (if not always practically) most useful ways to test for invertibility is to check that $\det g \neq 0$. Hence if $g_1, g_2 \in GL(n,\mathbb{F})$, then $\det(g_1g_2) = \det g_1 \det g_2 \neq 0$ so that $g_1g_2 \in GL(n,\mathbb{F})$. In particular, matrix multiplication is a binary operation on $GL(n,\mathbb{F})$. Associativity is known from linear algebra. The identity element is clearly the identity matrix. Finally, inverses exist by definition. Thus, $GL(n,\mathbb{F})$ is a group under matrix multiplication.

There are quite a few variants on this theme. To name only a few, consider the *Special Linear Group* defined by

$$SL(n,\mathbb{F}) = \{g \in GL(n,\mathbb{F}) \mid \det g = 1\},$$

the *Orthogonal Group* defined by

$$O(n,\mathbb{F}) = \{g \in GL(n,\mathbb{F}) \mid gg^T = I_n\}$$

where g^T refers to the transpose of a matrix, and the *Special Orthogonal Group* defined by

$$SO(n,\mathbb{F}) = \{g \in O(n,\mathbb{F}) \mid \det g = 1\}.$$

Using properties of the determinant, it is easy to check that each one of these objects really is a group (see Exercise 2.18). As a final remark, it is worth noting that matrix groups are some of the very most important groups around.

1.3.5. *Symmetric Group, S_n.* Roughly speaking, the *symmetric group* or *permutation group* on n-letters is the set of all permutations or rearrangements of n objects. It is written as S_n and is also one of the most important groups. To avoid trivialities, we always tacitly assume $n \geq 2$. By relabeling, we may take the n objects to be the numbers $1, 2, \ldots, n$. For example, one element of S_3 rearranges the ordered set $\{1, 2, 3\}$ to $\{2, 3, 1\}$.

In order to make these notions precise, we think of each permutation as a map $\varphi : \{1, 2, \ldots, n\} \to \{1, 2, \ldots, n\}$ that tells how to reorder things. For instance, the permutation $\{1, 2, 3\} \to \{2, 3, 1\}$ is realized by the map (of unordered sets) $\sigma : \{1, 2, 3\} \to \{1, 2, 3\}$ defined by $\sigma(1) = 2$, $\sigma(2) = 3$, and $\sigma(3) = 1$. Notice that since permutations only rearrange the order, the map σ must be one-to-one and onto—in other words, a bijection.

This leads to our official definition:

$$S_n = \{\sigma : \{1, 2, \ldots, n\} \to \{1, 2, \ldots, n\} \mid \sigma \text{ is bijective}\}.$$

If we desire to define an element $\sigma \in S_n$, we have n possible choices for the value of $\sigma(1)$. Since σ is bijective, that leaves $(n-1)$ remaining choices for $\sigma(2)$, then $(n-2)$ choices for $\sigma(3)$, and so on. In particular,

$$|S_n| = n\,(n-1)\,(n-2)\cdots 1 = n!.$$

There is a visual way of writing elements of S_n that is commonly used. Associate to $\sigma \in S_n$ the array

$$\begin{pmatrix} 1 & 2 & \cdots & n \\ \sigma(1) & \sigma(2) & \cdots & \sigma(n) \end{pmatrix}.$$

For instance, the permutation $\{1, 2, 3\} \to \{2, 3, 1\}$ is represented by $\begin{pmatrix} 1 & 2 & 3 \\ 2 & 3 & 1 \end{pmatrix}$. Conversely, the array $\begin{pmatrix} 1 & 2 & 3 \\ 1 & 3 & 2 \end{pmatrix}$ represents the permutation $\{1, 2, 3\} \to \{1, 3, 2\}$.

In order to make S_n into a group, a bilinear operation is still needed. This is provided by composition of functions. In particular for $\sigma_1, \sigma_2 \in S_n$, define

$$\sigma_1\sigma_2 = \sigma_1 \circ \sigma_2.$$

Since the composition of bijective functions is still bijective, this is a closed operation on S_n. As an example, suppose $\sigma_1 = \begin{pmatrix} 1 & 2 & 3 \\ 2 & 3 & 1 \end{pmatrix}$ and $\sigma_2 = \begin{pmatrix} 1 & 2 & 3 \\ 1 & 3 & 2 \end{pmatrix}$. Then

$$\begin{aligned}
(\sigma_1\sigma_2)\,(1) &= (\sigma_1 \circ \sigma_2)\,(1) = \sigma_1(\sigma_2(1)) = \sigma_1(1) = 2, \\
(\sigma_1\sigma_2)\,(2) &= (\sigma_1 \circ \sigma_2)\,(2) = \sigma_1(\sigma_2(2)) = \sigma_1(3) = 1, \\
(\sigma_1\sigma_2)\,(3) &= (\sigma_1 \circ \sigma_2)\,(3) = \sigma_1(\sigma_2(3)) = \sigma_1(2) = 3
\end{aligned}$$

so that

$$\begin{pmatrix} 1 & 2 & 3 \\ 2 & 3 & 1 \end{pmatrix}\begin{pmatrix} 1 & 2 & 3 \\ 1 & 3 & 2 \end{pmatrix} = \begin{pmatrix} 1 & 2 & 3 \\ 2 & 1 & 3 \end{pmatrix}.$$

To see that the axioms of a group are satisfied, first recall that composition of functions is always associative. Next observe that the identity function $\mathrm{Id} \in S_n$

defined by $\mathrm{Id}(k) = k$, $1 \le k \le n$, clearly serves as the identity element e. In other words,

$$\mathrm{Id} = \begin{pmatrix} 1 & 2 & \cdots & n \\ 1 & 2 & \cdots & n \end{pmatrix}.$$

Finally, note that any $\sigma \in S_n$ has an inverse function, σ^{-1}, since it is bijective. By definition, this means $\sigma \circ \sigma^{-1} = \sigma^{-1} \circ \sigma = \mathrm{Id}$. In particular, σ^{-1} is the required group inverse. For example, the inverse to the permutation $\begin{pmatrix} 1 & 2 & 3 \\ 2 & 3 & 1 \end{pmatrix}$ sends $\{2,3,1\} \to \{1,2,3\}$. Reordering, we see

$$\begin{pmatrix} 1 & 2 & 3 \\ 2 & 3 & 1 \end{pmatrix}^{-1} = \begin{pmatrix} 1 & 2 & 3 \\ 3 & 1 & 2 \end{pmatrix}.$$

1.3.6. *Dihedral Groups, D_n.* A *rigid motion* or *isometry* of the plane is a map $\varphi : \mathbb{R}^2 \to \mathbb{R}^2$ that preserves the distance between any two points; i.e., if $x, y \in \mathbb{R}^2$, then the distance from x to y is the same as the distance from $\varphi(x)$ to $\varphi(y)$. Using composition of functions as the bilinear operation, it is easy to see the set of rigid motions of the plane forms a group (cf. Exercise 2.29). Perhaps surprisingly, it turns out there are only four kinds of rigid motions: parallel translations, rotations about a point, reflections across a line, and glide reflections (a reflection about a line l followed by a translation parallel to l).

Given $S \subseteq \mathbb{R}^2$, the *symmetry group of* S is the set of rigid motions of the plane mapping S to S. It is straightforward to see that composition of functions makes the symmetry group of S into a group (Exercise 2.22). An important example occurs by taking S to be a regular n-gon, $n \ge 3$. In that case, the symmetry group is written D_n and is called the *dihedral group.*

For instance, D_3 is the set of rigid motions that preserve the triangle (the dotted lines are added to help visualize certain reflections):

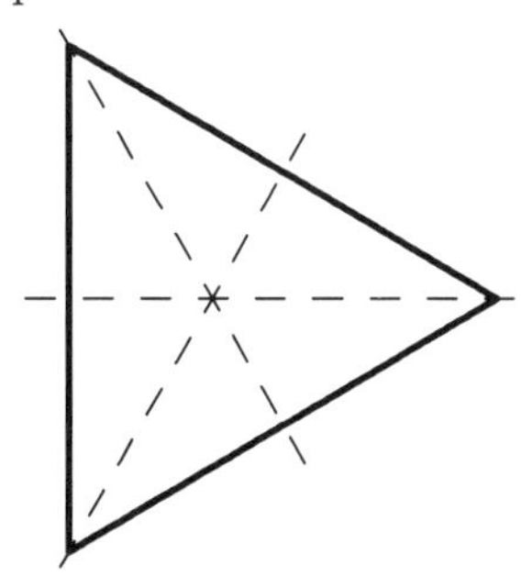

There are six such motions: the identity map, rotations by 60° and 120°, and the three reflections across the dotted lines. While this is a wonderfully explicit geometric description, the notation can become a bit cumbersome when we turn to D_n for larger values of n.

To motivate notation suitable for D_n, observe that the six linear maps comprising D_3 are completely determined by their action on the three vertices of the triangle. In other words, instead of specifying an element of D_3 by giving an explicit rigid motion of the entire plane, it suffices to simply say where each of the three vertices are mapped. In order to make use of this observation, label the vertices with elements of $\mathbb{Z}_3$. Specifically, refer to the right-most vertex as [1], the top-most vertex as [2], and the bottom-most vertex as [3].

For instance, rotation by $60°$ maps vertex [1] to [2], [2] to [3], and [3] to $[1] = [4]$. In particular, rotation by $60°$ is realized by the map $[k] \to [k+1]$ on $\mathbb{Z}_3$. Similarly, rotation by $120°$ is realized by the map $[k] \to [k+2]$. The identity map is of course realized by the map $[k] \to [k]$. The reflections also have simple formulas. For instance, reflection across the horizontal dotted line maps $[1] \to [1]$, $[2] \to [3]$, and $[3] \to [2]$. As the reader can readily check, this is realized by the map $[k] \to [2-k]$. Similarly, the other two reflections are given by the maps $[k] \to [1-k]$ and $[k] \to [3-k]$.

Therefore, let $R_j : \mathbb{Z}_3 \to \mathbb{Z}_3$ be defined by $R_j([k]) = [k+j]$ and let $W_j : \mathbb{Z}_3 \to \mathbb{Z}_3$ be defined by $R_j([k]) = [j-k]$. Given our above identifications, we see that

$$D_3 = \{R_0, R_1, R_2, W_0, W_1, W_2\}$$

with the binary operation given by composition of functions. For instance since

$$(R_1 \circ W_1)([k]) = R_1([1-k]) = [2-k] = W_2([k]),$$

it follows that $R_1 W_1 = W_2$. Similarly since $R_1(R_1([k])) = [2+k] = R_2([k])$, it follows that $R_1 R_1 = R_2$.

In a similar spirit, D_4 is the set of rigid motions that preserve the square (again, the dotted lines are included to help visualize certain reflections):

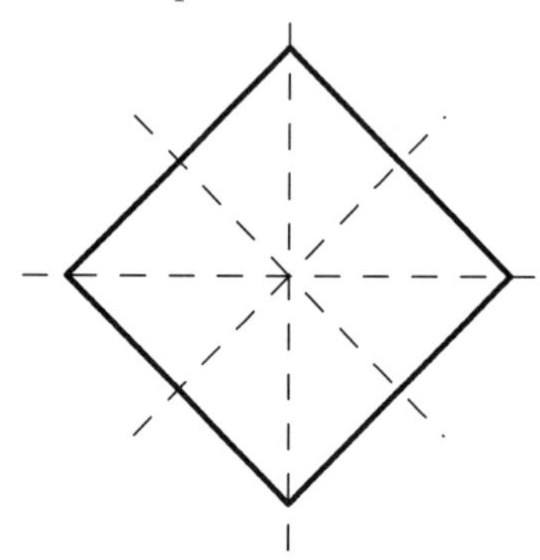

There are eight such motions: rotations by $0°$, $90°$, $180°$, and $270°$ degrees and the four reflections across the dotted lines. Restricting our attention again to the vertices and labeling them by elements of $\mathbb{Z}_4$, it is easy to see that the rotations are realized by the maps $[k] \to [k+j]$ and that the reflections are realized by the maps $[k] \to [j-k]$ on $\mathbb{Z}_4$ for $j \in \{0, 1, 2, 3\}$. Thus defining $R_j : \mathbb{Z}_4 \to \mathbb{Z}_4$ by $R_j([k]) = [k+j]$ and $W_j : \mathbb{Z}_4 \to \mathbb{Z}_4$ by $W_j([k]) = [j-k]$, it follows that

$$D_4 = \{R_0, R_1, R_2, R_3, W_0, W_1, W_2, W_3\}.$$

The binary operation is composition of functions. For instance, here $(R_1 \circ W_1)([k]) = R_1([1-k]) = [2-k] = W_2([k])$ so that $R_1 W_1 = W_2$.

The general case can be handled by similar geometric considerations. Instead of belaboring the point, we jump to the punch line and give a more computationally friendly realization of the dihedral groups.

DEFINITION 2.3. Fix $n \in \mathbb{Z}$, $n \geq 3$.

(1) For $j \in \mathbb{Z}$, let $R_j : \mathbb{Z}_n \to \mathbb{Z}_n$ be defined by $R_j([k]) = [j+k]$ and let $W_j : \mathbb{Z}_n \to \mathbb{Z}_n$ be defined by $W_j([k]) = [j-k]$.

(2) The *dihedral group* is

$$D_n = \{R_j, W_j \mid j \in \mathbb{Z}\}.$$

The binary operation is composition of functions.

THEOREM 2.4. *Fix $n \in \mathbb{Z}$, $n \geq 3$. The dihedral group, D_n, is a group with*

$$|D_n| = 2n$$

satisfying the following relations:

$$R_i R_j = R_{i+j} \text{ and } R_j^n = \mathrm{Id},$$
$$W_i W_j = R_{i-j} \text{ and } W_j^2 = \mathrm{Id},$$
$$R_i W_j = W_{i+j} \text{ and } W_i R_j = W_{i-j}.$$

In particular, the identity element is R_0 and we can also write

$$D_n = \{R_1^j, R_1^j W_0 \mid 0 \leq j \leq n-1\}.$$

PROOF. To see that D_n is a group, it suffices to prove the above relations since (i) associativity always holds for composition of functions, (ii) R_0 clearly serves as an identity, and (iii) $R_j^{-1} = R_{-j}$ and $W_j^{-1} = W_j$. The verification of the identities is straightforward. For instance, to show $R_i W_j = W_{i+j}$, simply calculate that

$$R_i(W_j([k])) = R_i([j-k]) = [i+j-k] = W_{i+j}([k]).$$

The rest of the relations are handled similarly and are left to the reader (Exercise 2.26). □

1.4. Exercises 2.1–2.29.

1.4.1. *Arithmetic.*

EXERCISE 2.1. **(a)** With respect to addition, show that the set of even integers, $2\mathbb{Z}$, is a group but that the set of odd integers, $2\mathbb{Z}+1$, is not a group.

(b) With respect to complex multiplication, show $\pm\{1, i\}$ is a group.

(c) With respect to multiplication, show $\{5^a \mid a \in \mathbb{Q}\}$ is a group.

(d) With respect to addition, show $\{a \in \mathbb{Z} \mid a \equiv 0 \bmod(3)\}$ is a group but that $\{a \in \mathbb{Z} \mid a \equiv 1 \bmod(3)\}$ is not.

(e) With respect to multiplication in $\mathbb{R}$, show $(0,1]$ is not a group.

(f) Show that $U_9 = \mathbb{Z}_9 \backslash \{[0],[3],[6]\}$ and that $\{[1],[4],[7]\} \subseteq U_9$ is a group with respect to multiplication.

EXERCISE 2.2. Define a binary operation on $\mathbb{R}$ by $x * y = x + y + 1$.

(a) Find $e \in \mathbb{R}$ so that $x * e = e * x = x$ for all $x \in \mathbb{R}$.

(b) Given $x \in \mathbb{R}$, find $y \in \mathbb{R}$ so that $x * y = y * x = e$.

(c) Show $*$ makes $\mathbb{R}$ into a group.

EXERCISE 2.3. Define a binary operation on $\mathbb{R}$ by $x * y = x + xy$. Show $*$ is not associative and that there is no identity element. In particular, $*$ does not make $\mathbb{R}$ into a group.

EXERCISE 2.4. Let $n \in \mathbb{N}$ and let $\zeta(n) = \{z \in \mathbb{C} \mid z^n = 1\}$.

(a) Show $\zeta(n)$ is a group under complex multiplication.

(b) Show $\zeta(n) = \{e^{\frac{2k\pi i}{n}} \mid k = 0, 1, \ldots, (n-1)\}$.

(c) Show there is a bijection $\varphi : \mathbb{Z}_n \to \zeta(n)$ so that $\varphi([a]+[b]) = \varphi([a])\varphi([b])$.

EXERCISE 2.5 ($\mathbb{Z}_n[x]$). Let $\mathbb{Z}_n[x]$, $n \in \mathbb{N}$, be the set of polynomials with coefficients in $\mathbb{Z}_n$; i.e., $\mathbb{Z}_n[x] = \{\sum_{k=0}^{\infty} a_k x^k \mid a_k \in \mathbb{Z}_n,$ all but finitely many $a_k = [0]\}$. Define addition and multiplication of polynomials in the usual fashion. Namely, if $f = \sum_k a_k x^k$ and $g = \sum_k a'_k x^k$,

$$f + g = \sum_k (a_k + a'_k)\, x^k \quad \text{and} \quad fg = \sum_k \left(\sum_{j=0}^{k} a_j a'_{k-j} \right) x^k.$$

(a) Show the above definitions of addition and multiplication are well defined binary operations on $\mathbb{Z}_n[x]$.

(b) Show $\mathbb{Z}_n[x]$ is a group under addition.

(c) Show $\mathbb{Z}_n[x]$ is not a group under multiplication.

EXERCISE 2.6 (Baby Division Alogrithm for Polynomials). If $f = \sum_k a_k x^k \in \mathbb{Z}_n[x]$ with $f \neq [0]$, $a_m \neq [0]$, and $a_k = [0]$ for $k > m$, we say the *degree* of f is $\deg(f) = m$ (cf. Exercise 2.5). If $a \in \mathbb{Z}_n$, write $f(a) = \sum_{k=0}^{m} a_k a^k \in \mathbb{Z}_n$.

(a) Fix $a \in \mathbb{Z}_n$ and $f \in \mathbb{Z}_n[x]$ with $\deg(f) = m$. Show there is $h_{m-1} \in \mathbb{Z}_p[x]$ with $\deg(h_{m-1}) \leq (m-1)$ so $f = a_m (x-a)^m + h_{m-1}$.

(b) Show $f = \sum_{k=0}^{m} b_k (x-a)^k$ for some $b_k \in \mathbb{Z}_n$ with $b_m = a_m$ and $b_0 = f(a)$.

(c) Show f can be written as $f = (x-a)g + f(a)$ for some $g \in \mathbb{Z}_n[x]$ with $\deg(g) = m - 1$.

EXERCISE 2.7 (Zeros). If $f \in \mathbb{Z}_n[x]$ with $f \neq [0]$ and $a \in \mathbb{Z}_n$, we say a is a *zero* of f if $f(a) = [0]$ (cf. Exercise 2.6).

(a) If $\deg(f) = m$, show $a \in \mathbb{Z}_p$ is a zero of f if and only if $f = (x-a)g$ for some $g \in \mathbb{Z}_n[x]$ with $\deg(g) = m - 1$.

(b) If p is a prime and $f \in \mathbb{Z}_p[x]$ has $\deg(f) = m$, show f has at most m (distinct) zeros in $\mathbb{Z}_p$.

(c) Show by example that if p is replaced by a nonprime, then part (b) can be false. *Hint:* In $\mathbb{Z}_8$, look at $x^2 - [1]$.

1.4.2. *Circle.*

EXERCISE 2.8. Recall that we defined $S^1 = \{z \in \mathbb{C} \mid |z| = 1\}$. Let $S' = \{e^{i\theta} \mid \theta \in \mathbb{R}\}$, $S = \{(x,y) \in \mathbb{R}^2 \mid x^2 + y^2 = 1\}$, and $SO(2,\mathbb{R}) = \{ \begin{pmatrix} \cos\theta & \sin\theta \\ -\sin\theta & \cos\theta \end{pmatrix} \mid \theta \in \mathbb{R}\}$.

(a) Show $S^1 = S'$ and establish a bijection between S^1 and S.

(b) Show there is a bijection $\varphi : S^1 \to SO(2,\mathbb{R})$ so that $\varphi(z_1 z_2) = \varphi(z_1)\varphi(z_2)$.

1.4.3. *Linear Algebra.*

EXERCISE 2.9. Let $G = GL(2, \mathbb{Z}_2)$.

(a) Write down all elements in G.

(b) Multiply $\begin{pmatrix} [1] & [1] \\ [0] & [1] \end{pmatrix} \begin{pmatrix} [0] & [1] \\ [1] & [1] \end{pmatrix}$.

(c) Find $\begin{pmatrix} [0] & [1] \\ [1] & [1] \end{pmatrix}^{-1}$.

(d) Find all $g \in G$ so that $g^2 = I_2$.

EXERCISE 2.10. Let $G = GL(2, \mathbb{Z}_3)$.

(a) Write down all elements in G.

(b) Multiply $\begin{pmatrix} [1] & [2] \\ [0] & [1] \end{pmatrix} \begin{pmatrix} [0] & [1] \\ [1] & [2] \end{pmatrix}$.

(c) Find $\begin{pmatrix} [0] & [1] \\ [1] & [2] \end{pmatrix}^{-1}$.

(d) Find all $g \in G$ so that $g^2 = I_2$.

EXERCISE 2.11. Show the following subsets of $GL(n, \mathbb{F})$ are groups under matrix multiplication.

(a) $D(n, \mathbb{F}) = \{\text{diag}(c_1, \ldots, c_n) \mid c_i \neq 0,\ 1 \leq i \leq n\}$.

(b) $U(n, \mathbb{F}) = \{$upper triangular matrices with 1's on the main diagonal$\}$. *Hint:* Use the cofactor method of computing the inverse to see that $U(n, \mathbb{F})$ contains its inverses.

(c) $N(n, \mathbb{F}) = \{$upper triangular matrices with nonzero entries on the main diagonal$\}$.

EXERCISE 2.12. Let $G = \{\begin{pmatrix} a & b \\ -b & a \end{pmatrix} \mid a^2 + b^2 \neq 0,\ a, b \in \mathbb{R}\}$.

(a) Show each $g \in G$ can be uniquely written as $\lambda \begin{pmatrix} \cos\theta & \sin\theta \\ -\sin\theta & \cos\theta \end{pmatrix}$ for $\lambda \in \mathbb{R}^+$ and $\theta \in [0, 2\pi)$.

(b) With respect to matrix multiplication, show G is a group.

EXERCISE 2.13 (Quaternion Group, Q). Define the symbols $\mathbf{1}, \mathbf{i}, \mathbf{j}, \mathbf{k}$ as the following elements of $GL(2, \mathbb{C})$:

$$\mathbf{1} = \begin{pmatrix} 1 & 0 \\ 0 & 1 \end{pmatrix},\ \mathbf{i} = \begin{pmatrix} i & 0 \\ 0 & -i \end{pmatrix},\ \mathbf{j} = \begin{pmatrix} 0 & 1 \\ -1 & 0 \end{pmatrix},\ \mathbf{k} = \begin{pmatrix} 0 & i \\ i & 0 \end{pmatrix}.$$

Using matrix multiplication, show $Q = \pm\{\mathbf{1}, \mathbf{i}, \mathbf{j}, \mathbf{k}\}$ is a group satisfying $\mathbf{i}^2 = \mathbf{j}^2 = \mathbf{k}^2 = -\mathbf{1}$, $\mathbf{ij} = -\mathbf{ji} = \mathbf{k}$, $\mathbf{jk} = -\mathbf{kj} = \mathbf{i}$, and $\mathbf{ki} = -\mathbf{ik} = \mathbf{j}$. The group Q is called the *Quaternion Group*.

EXERCISE 2.14 (Heisenberg Group). **(a)** Let $H_3(\mathbb{R}) = \{ \begin{pmatrix} 1 & x & z \\ 0 & 1 & y \\ 0 & 0 & 1 \end{pmatrix} \mid x, y, z \in \mathbb{R}\}$. Show $H_3(\mathbb{R})$ is a group under matrix multiplication. It is called the 3-dimensional *Heisenberg Group*.

(b) Writing matrices in $1 \times n \times 1$ block form, let $H_{2n+1}(\mathbb{R}) = \{ \begin{pmatrix} 1 & x & z \\ 0 & I_n & y^T \\ 0 & 0 & 1 \end{pmatrix} \mid x, y \in \mathbb{R}^n \text{ and } z \in \mathbb{R}\}$ (here $\cdot^T$ denotes *transpose*). Show $H_{2n+1}(\mathbb{R})$ is a group under matrix multiplication. It is called the $(2n+1)$-dimensional *Heisenberg group*.

(c) Find all $Z \in H_{2n+1}(\mathbb{R})$ so that $gZ = Zg$ for all $g \in H_{2n+1}(\mathbb{R})$.

EXERCISE 2.15. Let $(x, y), (x', y') \in \mathbb{R}^\times \times \mathbb{R}$ and define $(x, y) * (x', y') = (xx', xy' + yx')$. Let $M(2, \mathbb{R}) = \{ \begin{pmatrix} x & y \\ 0 & x \end{pmatrix} \mid x, y \in \mathbb{R},\ x \neq 0\}$.

(a) Show $*$ makes $\mathbb{R}^\times \times \mathbb{R}$ into a group.

(b) Show $M(2, \mathbb{R})$ is a group with respect to matrix multiplication.

(c) Show there is a bijection $\varphi : \mathbb{R}^\times \times \mathbb{R} \to M(2, \mathbb{R})$ so that $\varphi(g_1 g_2) = \varphi(g_1)\varphi(g_2)$.

EXERCISE 2.16. Let $A = \left\{ \begin{pmatrix} 1 & x \\ 0 & 0 \end{pmatrix} \mid x \in \mathbb{R} \right\}$.

(a) Show that matrix multiplication gives an associative binary operation on A.

(b) Show A has a *left identity* and *right inverse*; i.e., show there exits $e_L \in A$ so that (i) $e_L a = a$ for all $a \in A$ and (ii) for each $a \in A$, there is a $b \in A$ so $ab = e_L$.

(c) Nevertheless, show A is not a group.

EXERCISE 2.17 ($SL(2,\mathbb{Z})$). Let $SL(2,\mathbb{Z}) = \left\{ \begin{pmatrix} a & b \\ c & d \end{pmatrix} \mid ad - bc = 1 \text{ and } a,b,c,d \in \mathbb{Z} \right\}$. Show $SL(2,\mathbb{Z})$ is a group.

EXERCISE 2.18. Show $SL(n,\mathbb{F})$, $O(n,\mathbb{F})$, and $SO(n,\mathbb{F})$ are groups.

1.4.4. *Symmetric Group.*

EXERCISE 2.19. Multiply the following elements of S_3 (parts (a)–(c)) or S_5 (parts (d) and (e)):

(a) $\begin{pmatrix} 1 & 2 & 3 \\ 3 & 2 & 1 \end{pmatrix} \begin{pmatrix} 1 & 2 & 3 \\ 2 & 3 & 1 \end{pmatrix}$,

(b) $\begin{pmatrix} 1 & 2 & 3 \\ 2 & 3 & 1 \end{pmatrix} \begin{pmatrix} 1 & 2 & 3 \\ 3 & 2 & 1 \end{pmatrix}$,

(c) $\begin{pmatrix} 1 & 2 & 3 \\ 3 & 2 & 1 \end{pmatrix} \begin{pmatrix} 1 & 2 & 3 \\ 2 & 1 & 3 \end{pmatrix} \begin{pmatrix} 1 & 2 & 3 \\ 1 & 3 & 2 \end{pmatrix}$,

(d) $\begin{pmatrix} 1 & 2 & 3 & 4 & 5 \\ 4 & 3 & 1 & 5 & 2 \end{pmatrix} \begin{pmatrix} 1 & 2 & 3 & 4 & 5 \\ 1 & 3 & 2 & 5 & 4 \end{pmatrix}$,

(e) $\begin{pmatrix} 1 & 2 & 3 & 4 & 5 \\ 4 & 5 & 1 & 2 & 3 \end{pmatrix} \begin{pmatrix} 1 & 2 & 3 & 4 & 5 \\ 5 & 2 & 4 & 3 & 1 \end{pmatrix}$.

EXERCISE 2.20. Find the inverse to the following elements of S_3 (parts (a) and (b)) or S_5 (parts (c) and (d)):

(a) $\begin{pmatrix} 1 & 2 & 3 \\ 3 & 2 & 1 \end{pmatrix}$,

(b) $\begin{pmatrix} 1 & 2 & 3 \\ 2 & 1 & 3 \end{pmatrix}$,

(c) $\begin{pmatrix} 1 & 2 & 3 & 4 & 5 \\ 1 & 3 & 2 & 5 & 4 \end{pmatrix}$,

(d) $\begin{pmatrix} 1 & 2 & 3 & 4 & 5 \\ 4 & 5 & 1 & 2 & 3 \end{pmatrix}$.

EXERCISE 2.21. Find all $\sigma \in S_3$ so that:

(a) $\sigma^2 = \mathrm{Id}$.

(b) $\sigma^3 = \mathrm{Id}$.

EXERCISE 2.22. Given nonempty $S \subseteq \mathbb{R}^2$, show that the set of rigid motions of the plane mapping S to S forms a group under composition of functions.

1.4.5. *Dihedral groups.*

EXERCISE 2.23. Multiply the following elements of D_3 and write them in the form R_1^j or $R_1^j W_1$ for $0 \le j \le 2$:

(a) $R_1 W_2$,

(b) $W_2 R_1$,

(c) $W_1R_2W_2$,
(d) R_1^{20}.

EXERCISE 2.24. Multiply the following elements of D_4 and write them in the form R_1^j or $R_1^jW_1$ for $0 \leq j \leq 3$:
(a) R_3W_2,
(b) W_3R_2,
(c) $R_2W_2R_1$.

EXERCISE 2.25. Find the inverse of the following elements of D_5 and write them in the form R_1^j or $R_1^jW_1$ for $0 \leq j \leq 4$:
(a) W_1R_4,
(b) $R_2W_3R_1$.

EXERCISE 2.26. Verify the remaining portions of Theorem 2.4.

1.4.6. *Miscellaneous.*

EXERCISE 2.27. For a set S, define a binary operation on the power set $\mathcal{P}(S)$ by $X * Y = (X \cup Y) \backslash (X \cap Y)$. Show that $*$ makes $\mathcal{P}(S)$ into a group.

EXERCISE 2.28. Write $\mathcal{F}(\mathbb{R})$ for the set of all functions $f : \mathbb{R} \to \mathbb{R}$. For $f, g \in \mathcal{F}(\mathbb{R})$, define binary operations $(f+g)$ and $(f \cdot g)$ by $(f+g)(x) = f(x)+g(x)$ and $(f \cdot g)(x) = f(x)g(x)$ for $x \in \mathbb{R}$.

(a) Show $\mathcal{F}(\mathbb{R})$ is a group under addition but not multiplication.

(b) Let $\mathcal{C}^0(\mathbb{R}) = \{f \in \mathcal{F}(\mathbb{R}) \mid f \text{ is continuous}\}$. Show $\mathcal{C}^0(\mathbb{R})$ is a group under addition. *Calculus fact:* the sum of continuous functions is continuous.

(c) Let $\mathcal{C}^\infty(\mathbb{R}) = \{f \in \mathcal{F}(\mathbb{R}) \mid f \text{ is infinitely differentiable}\}$. Show $\mathcal{C}^\infty(\mathbb{R})$ is a group under addition. *Calculus fact:* the sum of infinitely differentiable functions is infinitely differentiable.

(d) Let $\mathcal{F}^{\text{ev}}(\mathbb{R}) = \{f \in \mathcal{F}(\mathbb{R}) \mid f(-x) = f(x),\ x \in \mathbb{R}\}$. Show $\mathcal{F}^{\text{ev}}(\mathbb{R})$ is a group under addition.

(e) Let $\mathcal{F}^+(\mathbb{R}) = \{f \in \mathcal{F}(\mathbb{R}) \mid f(x) > 0,\ x \in \mathbb{R}\}$. Show $\mathcal{F}^+(\mathbb{R})$ is a group under multiplication but not addition.

(f) Show $\{f \in \mathcal{F}(\mathbb{R}) \mid f(0) = 0\}$ is a group under addition but that $\{f \in \mathcal{F}(\mathbb{R}) \mid f(0) = 1\}$ is not.

EXERCISE 2.29 (Motions of $\mathbb{R}^n$). A *motion* of $\mathbb{R}^n$ is a bijection of $\mathbb{R}^n$ onto $\mathbb{R}^n$. The set of motions is a group under composition of functions. Arbitrary motions of $\mathbb{R}^n$ are usually not very useful objects; however, certain subsets of motions arise frequently in applications.

(a) A *translation* of $\mathbb{R}^n$ is a motion of the form $T_v : \mathbb{R}^n \to \mathbb{R}^n$ where $T_v(x) = x + v$ for $x, v \in \mathbb{R}^n$. Show the set of translations is a group.

(b) A *linear motion* of $\mathbb{R}^n$ is a motion of the form $T_g : \mathbb{R}^n \to \mathbb{R}^n$ where $T_g(x) = gx$ for $x \in \mathbb{R}^n$ and $g \in GL(n, \mathbb{R})$ where gx is given by matrix multiplication. Show the set of linear motions is a group.

(c) An *affine motion* of $\mathbb{R}^n$ is a motion of the form $T_{g,v} : \mathbb{R}^n \to \mathbb{R}^n$ where $T_{g,v}(x) = gx + v$ for $x, v \in \mathbb{R}^n$ and $g \in GL(n, \mathbb{R})$. Show the set of affine motions is a group.

(d) Show the set of rigid motions of the plane forms a group.

2. Basic Properties and Order

In this section we study some basic properties of groups. The first result justifies use of the phrases "the identity" and "the inverse" instead of "an identity" and "an inverse."

THEOREM 2.5. *Let G be a group.*
(1) *The identity element e is unique.*
(2) *Inverses are unique.*

PROOF. For part (1), suppose $e_1, e_2 \in G$ both satisfy the identity property that $e_i g = g e_i = g$ for all $g \in G$. Then using this property first for e_2 and then for e_1, we see

$$e_1 = e_1 e_2 = e_2$$

as desired. For part (2), suppose $g \in G$ and that $h_1, h_2 \in G$ both satisfy the inverse property for g, namely $gh_i = h_i g = e$. Then

$$h_1 = h_1 e = h_1(gh_2) = (h_1 g)h_2 = eh_2 = h_2$$

as desired. □

One of the most basic properties of a group is a measurement of its size.

DEFINITION 2.6. A group G is *finite* if $|G| < \infty$. In that case, $|G|$ is called the *order* of G.

For instance, the groups $(\mathbb{Z}_n, +)$, $(U_n, \cdot)$, $GL(n, \mathbb{Z}_p)$, S_n, and D_n are all finite. However, groups such as $(\mathbb{Z}, +)$, $(\mathbb{R}^+, \cdot)$, S^1, and $GL(n, \mathbb{R})$ are infinite.

Another fundamental property of a group is whether elements commute.

DEFINITION 2.7. A group G is *abelian* or *commutative* if $gh = hg$ for all $g, h \in G$.

For instance, the groups $(\mathbb{Z}, +)$, $(\mathbb{Z}_n, +)$, $(U_n, \cdot)$, $(\mathbb{R}^+, \cdot)$, and S^1 are all abelian. Most groups, however, are not abelian. For instance, $GL(n, \mathbb{R})$, S_n, and D_n are not abelian. If working with a group that is not abelian, it is very important to preserve the order of elements. In such cases writing ab instead of ba quickly leads to nonsense. After all, order usually matters in life: "put the lid on the blender, turn the power on" or "open the window, stick your head out".

Next we define exponentials just as we did with U_n.

DEFINITION 2.8. Let G be a group written multiplicatively and let $g \in G$. For $n \in \mathbb{N}$, let

$$\begin{aligned} g^n &= \overbrace{gg \cdots g}^{n \text{ copies}}, \\ g^{-n} &= g^{-1} g^{-1} \cdots g^{-1}, \\ g^0 &= e. \end{aligned}$$

THEOREM 2.9. *Let G be a group, $g, h \in G$, and $n, m \in \mathbb{Z}$.*
(1) *If G is written multiplicatively, then*

$$\begin{aligned} g^n g^m &= g^{n+m}, \\ (g^n)^m &= g^{nm}. \end{aligned}$$

In particular, $(g^{-1})^{-1} = g$.

(2) *The inverse of a product is given by*

$$(gh)^{-1} = h^{-1}g^{-1}.$$

(3) *If G is abelian, then $(gh)^n = g^n h^n$.*

PROOF. We leave parts (1) and (3) to Exercise 2.33. For part (2), notice

$$(gh)(h^{-1}g^{-1}) = g(hh^{-1})g^{-1} = geg^{-1} = gg^{-1} = e.$$

Similarly, $(h^{-1}g^{-1})(gh) = e$ so that $h^{-1}g^{-1}$ is the inverse of gh as desired. $\square$

As was mentioned, the above group is written *multiplicatively*—the default notation. In cases where G is written *additively*, slightly different notation is adopted for Definition 2.8. In such cases, define

$$\begin{aligned} ng &= \overbrace{g + g + \cdots + g}^{n \text{ copies}}, \\ (-n)\,g &= -g - g - \cdots - g, \\ 0g &= 0_G \end{aligned}$$

for $n \in \mathbb{N}$. Then Theorem 2.9, part (1), looks similar to the usual rules of multiplication instead of exponentiation:

$$\begin{aligned} ng + mg &= (n+m)g, \\ m\,(ng) &= (nm)\,g. \end{aligned}$$

Returning to the standard multiplicative notation, we look at the analog of roots of unity.

DEFINITION 2.10. Let G be a group and let $g \in G$. If there is no $n \in \mathbb{N}$ so that $g^n = e$, then g is said to have *infinite order* and we write $|g| = \infty$. Otherwise, we say g has *finite order* and we write $|g|$ for the smallest $n \in \mathbb{N}$ so that $g^n = e$. In that case, $|g|$ is called the *order* of g.

For instance, in $(\mathbb{C}^\times, \cdot)$ the order of i is 4 since $i^4 = 1$ with $i^k \neq 1$ for $k = 1, 2, 3$. However, the order of $2 \in \mathbb{R}^\times$ is infinite since $2^n \neq 1$ for all $n \in \mathbb{N}$. As another example, consider $\sigma = \begin{pmatrix} 1 & 2 & 3 \\ 3 & 1 & 2 \end{pmatrix} \in S_3$. Then it is easy to check that $\sigma^2 = \begin{pmatrix} 1 & 2 & 3 \\ 2 & 3 & 1 \end{pmatrix}$ and $\sigma^3 = \begin{pmatrix} 1 & 2 & 3 \\ 1 & 2 & 3 \end{pmatrix} = \mathrm{Id}$ so that $|\sigma| = 3$.

THEOREM 2.11. *Let G be a group with $g \in G$.*

(1) *g has infinite order if and only if all elements in $\{g^n \mid n \in \mathbb{Z}\}$ are distinct.*

(2) *If g has finite order, then $g^{n_1} = g^{n_2}$ for $n_1, n_2 \in \mathbb{Z}$ if and only if $n_1 \equiv n_2 \pmod{|g|}$. In particular, $g^n = e$ if and only if $|g| \mid n$.*

(3) *If g has finite order, then $|g^n| = \frac{|g|}{(n,|g|)}$.*

(4) *If G is finite, then $|g| \leq |G|$.*

PROOF. For part (1), first suppose all g^n are distinct. In particular, $g^n \neq g^0 = e$ when $n \neq 0$ so that g has infinite order. On the other hand, suppose $g^{n_1} = g^{n_2}$ for some $n_1 \neq n_2$. Relabeling if necessary, we may assume $n_2 > n_1$ so that $n_2 - n_1 > 0$. Then multiplying by g^{-n_1} on both sides shows $e = g^0 = g^{n_2 - n_1}$ so that the order of g is bounded by $n_2 - n_1$ and is therefore finite.

For part (2), notice again that $g^{n_1} = g^{n_2}$ if and only if $g^{n_2 - n_1} = e$. Writing $m = n_2 - n_1$ and noting that $n_1 \equiv n_2 \pmod{|g|}$ if and only if $m \equiv 0 \pmod{|g|}$, it suffices to show $g^m = e$ if and only if $m \equiv 0 \pmod{|g|}$. For this, use the Division Algorithm to write $m = |g|\, q + r$ with $0 \le r < |g|$ for some $r, q \in \mathbb{Z}$. Then

$$g^m = g^{|g|q+r} = (g^{|g|})^q g^r = eg^r = g^r.$$

Thus $g^m = e$ if and only if $g^r = e$. But since $|g|$ is the minimal positive exponent taking g to e and since $0 \le r < |g|$, we see $g^r = e$ if and only if $r = 0$. By construction, $r = 0$ if and only if $m \equiv 0 \pmod{|g|}$ and we are done.

For part (3), let $k \in \mathbb{Z}$ and observe that part (2) shows $(g^n)^k = g^{nk} = e$ if and only if $|g|$ divides (nk). Write $n = (n, |g|)n_1$ and $|g| = (n, |g|)m$ (so $m = \frac{|g|}{(n,|g|)}$). By division we see that $|g|$ divides (nk) if and only if $m \mid (n_1 k)$ (Exercise 1.13). Since m and n_1 are relatively prime (Exercise 1.19), $m \mid (n_1 k)$ if and only if $m \mid k$. Thus the only exponents taking g^n to e are multiples of m. Clearly the minimal positive multiple is m as desired.

For part (4), consider the set $\{g, g^2, \ldots, g^{|G|+1}\}$. Since G cannot have $|G|+1$ distinct elements, at least two elements must agree. Thus there is $1 \le n_1 < n_2 \le |G|+1$ so that $g^{n_1} = g^{n_2}$. In other words $g^{n_2 - n_1} = e$ with $1 \le n_2 - n_1 \le |G|$ as desired. □

Part (4) in Theorem 2.11 will be greatly strengthened in Corollary 2.29 below to show that, in fact, $|g| \mid |G|$.

2.1. Exercises 2.30–2.52.

EXERCISE 2.30. Let be a group. For $g, h, k \in G$, show the following are equivalent:

(a) $g = h$.
(b) $kg = kh$.
(c) $gk = hk$.

EXERCISE 2.31. Let G be a group and let $g, h \in G$. If $gh = e$, show $hg = e$. *Hint:* Take the inverse of both sides of the equation and then multiply both sides by h and g on the left and right, respectively.

EXERCISE 2.32. Let G be a group and let $g, h \in G$. Consider the equations $gx = h$ and $yg = h$.

(a) Show the equation $gx = h$ has a unique solution for x in G.
(b) Show the equation $yg = h$ has a unique solution for y in G.
(c) Find an example of a (nonabelian) group for which $x \ne y$. *Hint:* Try a matrix group.

EXERCISE 2.33. **(a)** Prove the remaining parts of Theorem 2.9.

(b) Give an example of a (nonabelian) group G with $g, h \in G$ so that $(gh)^2 \ne g^2h^2$.

(c) Let G be a group. Show G is abelian if and only if $(gh)^2 = g^2h^2$ for all $g, h \in G$.

EXERCISE 2.34. Let G be a group and let $g, h \in G$. Show $gh = hg$ if and only if $ghg^{-1} = h$.

EXERCISE 2.35. Let G be a group with $g, h \in G$. Show the following maps are bijections:

(a) $l_g : G \to G$ by $l_g(h) = gh$.

(b) $r_g : G \to G$ by $r_g(h) = hg$.
(c) $c_g : G \to G$ by $c_g(h) = ghg^{-1}$.

EXERCISE 2.36. Let G be a group with $a, b, c, d \in G$. Use the associative property to show the following elements are equal. As a result, we simply write $abcd$ for any of the following equivalent expressions. It is a fact that such relations hold in general.
(a) $a(b(cd))$,
(b) $a((bc)d)$,
(c) $(ab)(cd)$,
(d) $(a(bc))d$,
(e) $((ab)c)d$.

EXERCISE 2.37. For each finite group G listed below, determine the order of G, $|G|$, and whether or not G is abelian. If G is not abelian, find two elements that fail to commute.
(a) $(\mathbb{Z}_n, +)$,
(b) $(U_n, \cdot)$,
(c) S_n,
(d) D_n,
(e) $GL(2, \mathbb{Z}_2)$ (cf. Exercise 2.9),
(f) $GL(2, \mathbb{Z}_3)$ (cf. Exercise 2.10).

EXERCISE 2.38. For each element g of the listed groups below, find the order of g, $|g|$.
(a) $[3] \in (\mathbb{Z}_{15}, +)$.
(b) $[3] \in (U_{10}, \cdot)$.
(c) $[2] \in (U_7, \cdot)$.
(d) $\begin{pmatrix} 1 & 2 & 3 \\ 3 & 1 & 2 \end{pmatrix} \in S_3$.
(e) $\begin{pmatrix} 1 & 2 & 3 & 4 & 5 \\ 2 & 1 & 3 & 5 & 4 \end{pmatrix} \in S_5$.
(f) $R_2 \in D_3$.
(g) $\begin{pmatrix} [1] & [1] \\ [0] & [1] \end{pmatrix} \in GL(2, \mathbb{Z}_2)$.
(h) $\begin{pmatrix} [1] & [1] \\ [0] & [1] \end{pmatrix} \in GL(2, \mathbb{Z}_3)$.

EXERCISE 2.39. In each infinite group below, find all elements of finite order:
(a) $(\mathbb{R}, +)$,
(b) $(\mathbb{R}^\times, \cdot)$,
(c) $(\mathbb{C}^\times, \cdot)$,
(d) $D(n, \mathbb{R}) = \{\mathrm{diag}(c_1, \ldots, c_n) \mid c_i \in \mathbb{R}^\times,\ 1 \le i \le n\}$.

EXERCISE 2.40. Let G be a group and let $g \in G$.
(a) Show $|g^{-1}| = |g|$.
(b) For $h \in G$, show $|hgh^{-1}| = |g|$.
(c) If $|g| < \infty$, show $g^{-1} = g^{|g|-1}$.

EXERCISE 2.41. Let G be a group and let $g, h \in G$. Show $|gh| = |hg|$.

EXERCISE 2.42. Let G be a group and let $g \in G$.

(a) If $g^{100} = e$, what are the possible orders of g?

(b) If $|g| = 100$, what are the orders of g^2, g^3, g^4, and g^5?

(c) If $g^{p^n} = e$ for $p \in \mathbb{N}$ a prime and $n \in \mathbb{Z}_{\geq 0}$, show $|g| = p^m$, some $0 \leq m \leq n$.

EXERCISE 2.43. Let G be a group and let $g, h \in G$.

(a) If g and h commute, show $|gh|$ divides $[g, h]$ (cf. Exercise 1.28).

(b) Show this may be false if g and h fail to commute. *Hint:* One good place to look for such g and h is in $GL(2, \mathbb{R})$.

EXERCISE 2.44. Let G be a group with $g_i \in G$, $1 \leq i \leq N$. Suppose $g_i g_j = g_j g_i$ and that the $\{|g_i|\}$ are mutually relatively prime. This exercise shows

$$\left| \prod_{i=1}^{N} g_i \right| = \prod_{i=1}^{N} |g_i| \, .$$

(a) Show $\{g_i^k \mid k \in \mathbb{Z}\} \cap \{g_j^k \mid k \in \mathbb{Z}\} = \{e\}$ when $i \neq j$.

(b) Look at $(g_i g_j)^{|g_i g_j|}$ to show $g_i^{|g_i g_j|} = g_j^{-|g_i g_j|} = e$ so that $|g_i|$ and $|g_j|$ divide $|g_i g_j|$ (cf. Exercise 2.43).

(c) Prove the result for $N = 2$.

(d) Use induction to prove the result in general.

EXERCISE 2.45 ($g^2 = e \Longrightarrow$ Abelian). Suppose G is a group so that $g^2 = e$ for every $g \in G$. Show G is abelian. *Hint:* Show $g \in G$ implies $g^{-1} = g$ and then apply this fact to the product of two elements.

EXERCISE 2.46. In Corollary 2.29 below we will see $|g|$ divides $|G|$ for G a finite group and any $g \in G$.

(a) Using this corollary, show that if $g \in U_{53}$, then $|g| \in \{1, 2, 4, 13, 26, 52\}$.

(b) It is straightforward, though not especially enjoyable, to verify that $[2]^{26} = [-1]$ in U_{53}. Assuming this fact, show $|[2]| = 52$.

(c) Prove that $U_{53} = \{[2]^k \mid 0 \leq k < 52\}$.

(d) Use part (c) to describe all elements of U_{53} with order 1, 2, 4, 13, 26, and 52, respectively.

EXERCISE 2.47 ($n = \sum_{d|n} \varphi(d)$). Let $n \in \mathbb{N}$.

(a) For $[k] \in \mathbb{Z}_n$, show $|[k]| = \frac{n}{(k,n)}$. Conclude $|[k]|$ divides n. *Hint:* Transfer Theorem 2.11 from multiplicative to additive notation.

(b) If $d \in \mathbb{N}$ with $d \mid n$, show $|[k]| = d$ if and only if $k = \frac{n}{d} m$ with $m \in \mathbb{Z}$ and $(m, d) = 1$.

(c) For $m, m' \in \mathbb{Z}$, show $\left[\frac{n}{d} m\right] = \left[\frac{n}{d} m'\right]$ if and only if $m \equiv m' \pmod{d}$.

(d) Conclude that $|\{[k] \in \mathbb{Z}_n \mid |[k]| = d\}| = \varphi(d)$ and that $n = \sum_{d|n} \varphi(d)$.

EXERCISE 2.48. Let $G = \{e, g_1, \ldots, g_n\}$ be a finite group of order $n + 1$.

(a) If G is abelian, show $(g_1 \cdots g_n)^2 = e$. *Hint:* Write g_i^{-1} as $g_{\sigma(i)}$ for some $\sigma \in S_n$. Then rearrange the terms of $(g_1 \cdots g_n)^2$ to get g_i next to $g_{\sigma(i)}$.

(b) Show part (a) can be false if G is not abelian. *Hint:* Try D_6.

(c) If G is possibly nonabelian but n is odd, show there is an i, $1 \leq i \leq n$, so that $g_i^2 = e$ (cf. Theorem 2.69). *Hint:* Pair up each g_i with its inverse—can each $g_i \neq g_i^{-1}$?

EXERCISE 2.49. Let $p \in \mathbb{N}$ be an odd prime and let $n \in \mathbb{N}$.
(a) Use the Binomial Theorem to show $(1+p)^{p^{n-1}} \equiv 1 \bmod(p^n)$.
(b) Show $(1+p)^{p^{n-2}} \equiv 1+p^{n-1} \bmod(p^n)$ for $n \geq 2$.
(c) Use Exercise 2.42, part (3), to conclude $|[1+p]| = p^{n-1}$ in U_{p^n}.

EXERCISE 2.50. Let $n \in \mathbb{N}$.
(a) Show $(1+2)^{2^{n-1}} \equiv 1 \bmod(2^n)$.
(b) Show $(1+2)^{2^{n-2}} \equiv \begin{cases} 3 \bmod(2^2), & n=2, \\ 1 \bmod(2^n), & n \geq 3. \end{cases}$
(c) Show $(1+2)^{2^{n-3}} \equiv \begin{cases} 3 \bmod(2^3), & n=3, \\ 1-2^{n-1} \bmod(2^n), & n \geq 4. \end{cases}$
(d) Conclude $|[3]| = \begin{cases} 1, & n=1, \\ 2, & n=2, \\ 2^{n-2}, & n \geq 3, \end{cases}$ in U_{2^n}.
N.B.: Since $|U_{2^n}| = 2^{n-1}$, powers of 3 account for half of U_{2^n} when $n \geq 3$.

EXERCISE 2.51. Let $n \in \mathbb{N}$.
(a) Show $(1+2^2)^{2^{n-1}} \equiv 1 \bmod(2^n)$.
(b) Show $(1+2^2)^{2^{n-2}} \equiv 1$ for $n \geq 2$.
(c) Show $(1+2^2)^{2^{n-3}} \equiv 1+2^{n-1} \bmod(2^n)$ for $n \geq 3$.
(d) Conclude $|[5]| = \begin{cases} 1, & n=1,2, \\ 2^{n-2}, & n \geq 3, \end{cases}$ in U_{2^n}.

EXERCISE 2.52. Let G be a group equipped with a binary operation $*$. Define a new binary operation # on G by $a\#b = b*a$ and write G^{opp} for the set G equipped with #. Show G^{opp} is a group.

3. Subgroups and Direct Products

Given a group G, there are many ways to manufacture new groups. We examine some of these methods in this section.

3.1. Subgroups.

DEFINITION 2.12. Let G be a group and let $H \subseteq G$, $H \neq \varnothing$. We say H is a *subgroup* if H is a group with respect to the binary operation of G restricted to H.

It is common to write $H \leq G$ to indicate that H is a subgroup of G. Of course every group comes equipped with at least two subgroups, neither one terribly exciting in its own right: the first is G itself and the second is $\{e\}$, the *trivial subgroup.* The more interesting subgroups lie strictly between these two possibilities and are called *proper subgroups.*

We have already seen a number of examples of proper subgroups. For instance, $\mathbb{Z} \subseteq \mathbb{Q} \subseteq \mathbb{R} \subseteq \mathbb{C}$ is a chain of subgroups as are $\mathbb{Q}^\times \subseteq \mathbb{R}^\times \subseteq \mathbb{C}^\times$, $S^1 \subseteq \mathbb{C}^\times$, and $SO(n,F) \subseteq SL(n,F) \subseteq GL(n,F)$. Of course most subsets of a group are not subgroups. For instance, the set of odd numbers is not a subgroup of the integers since it is not even closed under the binary operation of addition. The next theorem tells which subsets really are subgroups.

THEOREM 2.13. *Let G be a group and let $H \subseteq G$.*
(1) *H is a subgroup if and only if*
(a) *H is closed under the bilinear operation of G, i.e., if $h_1, h_2 \in H$, then $h_1h_2 \in H$,*

(b) *$e \in H$, and*

(c) *when $h \in H$, then $h^{-1} \in H$.*

(2) *Alternately, H is a subgroup if and only if $H \neq \varnothing$ and whenever $h_1, h_2 \in H$, then $h_1 h_2^{-1} \in H$.*

PROOF. Part (1) follows immediately from the definitions of binary operation and group. For part (2), fix $h \in H$. By the hypothesis, $e = hh^{-1} \in H$. It then follows that $h^{-1} = eh^{-1} \in H$. Finally, if $h_1, h_2 \in H$, then $h_1 h_2 = h_1(h_2^{-1})^{-1} \in H$. Thus part (1) shows H is a subgroup. □

For example, let $\mathcal{R}_n$ be the set of rotations in D_n so that $\mathcal{R}_n = \{R_j \mid j \in \mathbb{Z}\}$. Clearly $\mathcal{R}_n \neq \varnothing$ and since $R_i(R_j)^{-1} = R_{i-j} \in \mathcal{R}_n$, it follows from Theorem 2.13, part (2), that $\mathcal{R}_n$ is a subgroup of D_n. In particular, $\mathcal{R}_n$ is a group in its own right.

As another example, consider $U(2,\mathbb{R}) = \{ \begin{pmatrix} 1 & x \\ 0 & 1 \end{pmatrix} \mid x \in \mathbb{R}\} \subseteq GL(2,\mathbb{R})$. Then $U(2,\mathbb{R}) \neq \varnothing$ and

$$\begin{pmatrix} 1 & x \\ 0 & 1 \end{pmatrix} \begin{pmatrix} 1 & y \\ 0 & 1 \end{pmatrix}^{-1} = \begin{pmatrix} 1 & x-y \\ 0 & 1 \end{pmatrix} \in U(2,\mathbb{R})$$

so that $U(2,\mathbb{R})$ is a subgroup of $GL(2,\mathbb{R})$.

Turning to an example where the group is written additively, let $n \in \mathbb{Z}$ and consider $n\mathbb{Z} = \{nk \mid k \in \mathbb{Z}\} \subseteq \mathbb{Z}$. Clearly $n\mathbb{Z} \neq \varnothing$ and $nk_1 - nk_2 = n(k_1 - k_2) \in n\mathbb{Z}$ so that $n\mathbb{Z}$ is a subgroup of $\mathbb{Z}$. Though $n\mathbb{Z}$ is an elementary example of a subgroup, it will turn out to be very useful.

Turning to a nonexample, the set of reflections in D_n, $\{W_j \mid j \in \mathbb{Z}\}$, is not a subgroup. One way to see this is to observe that $W_1 W_1^{-1} = R_0$ is not a reflection.

In addition to specific examples, there are a number of general constructions of subgroups that arise frequently.

DEFINITION 2.14. Let G be a group with $S \subseteq G$, $S \neq \varnothing$.

(1) The *centralizer* of S in G is

$$Z(S) = Z_G(S) = \{g \in G \mid gs = sg \text{ for all } s \in S\}.$$

In the special case of $S = G$,

$$Z(G) = \{g \in G \mid gh = hg \text{ for all } h \in G\}$$

is called the *center* of G.

(2) Let $\mathcal{S} = \{\text{subgroups } H \subseteq G \mid S \subseteq H\}$. The *subgroup generated by* S is

$$\langle S \rangle = \bigcap_{H \in \mathcal{S}} H.$$

The elements of S are called *generators* of $\langle S \rangle$.

It is clear from the definition that $Z(G) = G$ if and only if G is abelian. In general, the center is smaller. For example, it is a well-know fact from linear algebra that $Z(GL(n,\mathbb{R}))$ is the set of constant diagonal matrices (Exercise 2.85).

As a more explicit example, we calculate $Z_{D_n}(R_1)$ (technically we should write $Z_{D_n}(\{R_1\})$, but that notation is simply too hideous). From Theorem 2.4, $R_i R_1 = R_1 R_i$ so $\mathcal{R}_n \subseteq Z(R_1)$. On the other hand, $R_1 W_i = W_{1+i}$ while $W_i R_1 = W_{i-1}$. Therefore $W_i \in Z(R_1)$ if and only if $1 + i \equiv i - 1 \pmod{n}$ if and only if $2 \equiv 0 \pmod{n}$ which is impossible ($n \geq 3$). Therefore $\mathcal{R}_n = Z(R_1)$.

Examples of a subgroup of G generated by a set S are more subtle. In fact, our definition of $\langle S \rangle$ is nearly impossible to use in most real-life situations. For one thing, finding all subgroups H with the property $S \subseteq H$ can easily be an intractable problem.

Odd as it seems at first, such horribly abstract definitions are rather common in mathematics. For example, here, even though our definition of $\langle S \rangle$ does not provide a practical means of explicitly calculating $\langle S \rangle$, it will allow us to easily prove certain important properties about $\langle S \rangle$ such as existence and the fact that it is a subgroup. With the aid of such tools, it is then possible to develop techniques to facilitate the explicit calculation of $\langle S \rangle$ in many situations.

THEOREM 2.15. *Let G be a group with $S \subseteq G$, $S \neq \varnothing$.*

(1) *$Z(S)$ (in particular, $Z(G)$) and $\langle S \rangle$ are subgroups of G.*

(2) *$\langle S \rangle$ is the smallest subgroup of G containing S and can alternately be written*

$$\langle S \rangle = \{s_1 s_2 \cdots s_k \mid k \in \mathbb{N} \text{ with } s_i \in S \cup S^{-1}\}$$

where $S^{-1} = \{s^{-1} \mid s \in S\}$.

PROOF. To see $Z(S)$ is a subgroup, first notice $Z(S) \neq \varnothing$ since $e \in Z(S)$. Next, if $g \in Z(G)$ and $s \in S$, then $gs = sg$. Multiplying by g^{-1} on both the left and right sides of the equation shows $sg^{-1} = g^{-1}s$ so that $g^{-1} \in Z(G)$. Finally, if $g_1, g_2 \in Z(S)$, then $g_1 g_2^{-1} s = g_1 s g_2^{-1} = s g_1 g_2^{-1}$. Thus $g_1 g_2^{-1} \in Z(S)$ and so $Z(S)$ is a subgroup.

To see $\langle S \rangle$ is a subgroup, first note that $\mathcal{S} \neq \varnothing$ since $G \in \mathcal{S}$. As e is an element of every subgroup, it follows that $e \in \langle S \rangle$. Next, suppose $g_1, g_2 \in \langle S \rangle$. Then for each $H \in \mathcal{S}$, $g_1, g_2 \in H$ so that $g_1 g_2^{-1} \in H$. As a result, $g_1 g_2^{-1} \in \langle S \rangle$ and so $\langle S \rangle$ is a subgroup.

For part (2), observe $\langle S \rangle$ is the smallest subgroup of G containing S since, by definition, it is contained in any subgroup that contains S. Finally, let $H = \{s_1 s_2 \cdots s_k \mid k \in \mathbb{N} \text{ with } s_i \in S \cup S^{-1}\}$. Since the definition of $\langle S \rangle$ forces $S \subseteq \langle S \rangle$, the fact that $\langle S \rangle$ is a group forces $H \subseteq \langle S \rangle$. Thus it remains only to see that $\langle S \rangle \subseteq H$. Noting that $S \subseteq H$ and using the definition of $\langle S \rangle$, it therefore suffices to show H is a subgroup of G. However, this is clear from the definition of H and Theorem 2.13. □

COROLLARY 2.16. *Let G be a group with $g \in G$. Then $\langle g \rangle = \{g^k \mid k \in \mathbb{Z}\}$.*

PROOF. Using Theorem 2.15, part (2), it follows that $\langle g \rangle = \{g^{\varepsilon_1} \ldots g^{\varepsilon_k} \mid k \in \mathbb{N}$ with $\varepsilon_i \in \{\pm 1\}\}$. Since $g^{\varepsilon_1} \ldots g^{\varepsilon_k} = g^{\varepsilon_1 + \cdots + \varepsilon_k}$ and since $\varepsilon_1 + \cdots + \varepsilon_k$ varies over all integers as k varies over $\mathbb{N}$ with $\varepsilon_i \in \{\pm 1\}$, the result follows. □

For example, in the case of D_n, it follows that $\langle R_1 \rangle$ is the rotation subgroup, $\mathcal{R}_n$. Similarly, $\langle \{R_1, W_1\} \rangle = D_n$ since clearly $\mathcal{R}_n \subseteq \langle \{R_1, W_1\} \rangle$ and $R_i W_1 = W_{i+1}$. For an additive example, it follows that $\langle 1 \rangle = \mathbb{Z}$ since any integer can be written as a sum of 1's and -1's. Similarly, $\mathbb{Z}_n = \langle [1] \rangle$. Groups such as $\mathcal{R}_n$, $\mathbb{Z}$, or $\mathbb{Z}_n$ that are generated by a single element get a special name (cf. Theorem 2.21).

DEFINITION 2.17. Let G be a group. If there exists $g \in G$ so that $G = \langle g \rangle$, then G is called *cyclic* and g is called a *generator* of G.

Observe that cyclic subgroups are abelian by Theorem 2.9. An outgrowth of their study in Section 6 will be the classification of all finite abelian groups. Also notice that generators are usually not unique. For instance in the case of a finite cyclic group G, it follows from Theorem 2.11 that if g is a generator, then g^n, $n \in \mathbb{Z}$, is a generator if and only if $(n, |g|) = 1$. In the infinite cyclic case, only g and g^{-1} generate the group (Exercise 2.66).

3.2. Direct Products. The following construction allows us to take groups and glue them together in an infinite number of new ways. It is analogous to the construction of $\mathbb{R}^n$ from n-copies of $\mathbb{R}$ and will also be used in the classification of finite abelian groups. Moreover, it will allow us to recognize when a complicated group is really just pieced together from "simpler" groups.

DEFINITION 2.18. Let G_i, $1 \leq i \leq N$, be groups. Define a binary operation on the *direct product* $\prod_{i=1}^N G_i = G_1 \times G_2 \times \cdots \times G_N$ by

$$(g_1, g_2, \ldots, g_N)(g_1', g_2', \ldots, g_N') = (g_1 g_1', g_2 g_2', \ldots, g_N g_N')$$

for $g_i, g_i' \in G_i$.

THEOREM 2.19. *With the above structure,* $\prod_{i=1}^N G_i$ *is a group.*

PROOF. Associativity follows from the associativity of G_i. The identity element is $(e_{G_1}, e_{G_2}, \ldots, e_{G_N})$ and the inverse to $(g_1, g_2, \ldots, g_N)$ is clearly $(g_1^{-1}, g_2^{-1}, \ldots, g_N^{-1})$. □

For instance, in $U_3 \times U_5$,

$$([1]_3, [3]_5) \cdot ([2]_3, [4]_5) = ([1]_3 \cdot [2]_3, [3]_5 \cdot [4]_5) = ([2]_3, [2]_5).$$

It is worth emphasizing that the default multiplicative notation is used in Definition 2.18. When the groups are written additively, an alternate notation is sometimes employed for the direct product. In that case, people might write $\sum_{i=1}^N G_i$ or $\bigoplus_{i=1}^N G_i$ for $\{(g_1, g_2, \ldots, g_N) \mid g_i \in G_i\}$ and refer to it as the *direct sum*. In this setting, the group law takes on the form

$$(g_1, g_2, \ldots, g_N) + (g_1', g_2', \ldots, g_N') = (g_1 + g_1', g_2 + g_2', \ldots, g_N + g_N').$$

For instance, in $\mathbb{Z}_3 \times \mathbb{Z}_4$, alternately written $\mathbb{Z}_3 \oplus \mathbb{Z}_4$,

$$([2]_3, [3]_4) + ([1]_3, [2]_4) = (([2]_3 + [1]_3, [3]_4 + [2]_4) = ([0]_3, [1]_4).$$

3.3. Exercises 2.53–2.91.

EXERCISE 2.53. Show the following subsets are subgroups:

(a) $\pm\{1, i\} \subseteq \mathbb{C}^\times$.
(b) $\{5^a \mid a \in \mathbb{Q}\} \subseteq \mathbb{R}^\times$.
(c) $k\mathbb{Z}_n = \{k[m] \mid [m] \in \mathbb{Z}_n\} \subseteq \mathbb{Z}_n$ where $k \in \mathbb{Z}$ and $n \in \mathbb{N}$.
(d) $\{(m, n) \mid 2m - 3n = 0\} \subseteq \mathbb{Z} \times \mathbb{Z}$.
(e) $\{[1], [4], [7]\} \subseteq U_9$.
(f) $\{\begin{pmatrix} 1 & 0 \\ 0 & 1 \end{pmatrix}, \begin{pmatrix} -1 & 0 \\ 0 & -1 \end{pmatrix}, \begin{pmatrix} 0 & 1 \\ -1 & 0 \end{pmatrix}, \begin{pmatrix} 0 & -1 \\ 1 & 0 \end{pmatrix}\} \subseteq SL(2, \mathbb{R})$.
(g) $SL(n, \mathbb{R}) \subseteq GL(n, \mathbb{R})$.
(h) $\{T \in GL(n, \mathbb{R}) \mid Tv_0 = \lambda v_0,\ \lambda \in \mathbb{R}^+\} \subseteq GL(n, \mathbb{R})$ for some fixed $v_0 \in \mathbb{R}^n$.
(i) $\{\sigma \in S_n \mid \sigma(n) = n\} \subseteq S_n$.
(j) $\{\begin{pmatrix} 1 & 2 & 3 \\ 1 & 2 & 3 \end{pmatrix}, \begin{pmatrix} 1 & 2 & 3 \\ 2 & 3 & 1 \end{pmatrix}, \begin{pmatrix} 1 & 2 & 3 \\ 3 & 1 & 2 \end{pmatrix}\} \subseteq S_3$.
(k) $\{g \in D_n \mid gW_1 g^{-1} = W_1\} \subseteq D_n$.
(l) $\{z \in \mathbb{C} \mid z^n = 1 \text{ for some } n \in \mathbb{N}\} \subseteq \mathbb{C}^\times$.

EXERCISE 2.54. Show the following are not subgroups:
(a) $\{5^a \mid a \in \mathbb{Q}^+\} \subseteq \mathbb{R}^\times$.
(b) $\{(m,n) \mid 2m - 3n = 1\} \subseteq \mathbb{Z} \times \mathbb{Z}$.
(c) $\{[1],[4],[5]\} \subseteq U_9$.
(d) $\{\begin{pmatrix} 1 & 0 \\ 0 & 1 \end{pmatrix}, \begin{pmatrix} 1 & 0 \\ 0 & 2 \end{pmatrix}, \begin{pmatrix} 2 & 0 \\ 0 & 1 \end{pmatrix}, \begin{pmatrix} 2 & 0 \\ 0 & 2 \end{pmatrix}\} \subseteq SL(2,\mathbb{R})$.
(e) For $v_0 \in \mathbb{R}^n$, $\{T \in GL(n,\mathbb{R}) \mid Tv_0 = 2v_0\} \subseteq GL(n,\mathbb{R})$.
(f) $\{\sigma \in S_6 \mid \sigma(6) = 1\} \subseteq S_6$.
(g) $\{\sigma \in S_n \mid \sigma(1) \neq 1\} \subseteq S_n$.
(h) $\{\sigma \in S_5 \mid \sigma^2 = \mathrm{Id}\} \subseteq S_5$.
(i) $\{R_0, R_1\} \subseteq D_3$.

EXERCISE 2.55. Calculate the centralizers of the following subsets:
(a) $\{\begin{pmatrix} 2 & 0 \\ 0 & 1 \end{pmatrix}\} \subseteq GL(2,\mathbb{R})$.
(b) $\{\begin{pmatrix} 1 & 1 \\ 0 & 1 \end{pmatrix}\} \subseteq GL(2,\mathbb{R})$.
(c) $\{\begin{pmatrix} 1 & 2 & 3 & 4 \\ 4 & 2 & 3 & 1 \end{pmatrix}\} \subseteq S_4$.
(d) $\{R_2\} \subseteq D_n$. *Hint:* The answer depends on n.

EXERCISE 2.56. Show the following:
(a) $\langle [2] \rangle = U_5$.
(b) $\langle 3, 11 \rangle = \mathbb{Z}$.
(c) $\left\langle \begin{pmatrix} 1 & 2 & 3 \\ 2 & 1 & 3 \end{pmatrix}, \begin{pmatrix} 1 & 2 & 3 \\ 1 & 3 & 2 \end{pmatrix} \right\rangle = S_3$.
(d) $\langle R_1, W_2 \rangle = D_n$.
(e) $\langle (1,[0]), (1,[2]) \rangle = \mathbb{Z} \times \mathbb{Z}_5$.
(f) $\langle \mathbf{j}, \mathbf{k} \rangle = Q$ (cf. Exercise 2.13).

EXERCISE 2.57. Evaluate the following:
(a) $([3]_7,[5]_6) \cdot ([2]_7,[5]_6) \in U_7 \times U_6$.
(b) $([2]_4,[3]_5,[6]_7) + ([3]_4,[2]_5,[1]_7) \in \mathbb{Z}_4 \times \mathbb{Z}_5 \times \mathbb{Z}_7$.

EXERCISE 2.58. Show the following subsets of $GL(n,\mathbb{F})$ are subgroups.
(a) $D(n,\mathbb{F}) = \{\mathrm{diag}(c_1,\ldots,c_n) \mid c_i \neq 0,\ 1 \leq i \leq n\}$.
(b) $U(n,\mathbb{F}) = \{$upper triangular matrices with 1's on the main diagonal$\}$.
(c) $N(n,\mathbb{F}) = \{$upper triangular matrices with nonzero entries on the main diagonal$\}$.

EXERCISE 2.59. Show $\{\begin{pmatrix} a & b \\ -b & a \end{pmatrix} \mid a^2 + b^2 = 1,\ a,b \in \mathbb{R}\} \subseteq GL(2,\mathbb{R})$ is a subgroup.

EXERCISE 2.60 (Dicyclic Group, Q_n). Let $n \in 2\mathbb{N}$, $n = 2m$, and $\zeta = e^{\frac{2\pi i}{n}}$. Define the elements of $GL(2,\mathbb{C})$

$$a = \begin{pmatrix} \zeta & 0 \\ 0 & \zeta^{-1} \end{pmatrix} \quad \text{and} \quad b = \begin{pmatrix} 0 & i \\ i & 0 \end{pmatrix}$$

and let $Q_n = \langle a, b \rangle$, called the *Dicyclic Group* or the *Generalized Quaternion Group*.
(a) Show $|a| = n$, $aba = b$, $b^2 = a^m$, $bab^{-1} = a^{-1}$, and $|ba^k| = 4$.
(b) Conclude $Q_n = \{I_2, a, \ldots, a^{n-1}, b, ba, \ldots, ba^{n-1}\}$. Conclude that $|Q_n| = 2n$.

EXERCISE 2.61. Let G be a group and H_α a subgroup, α in some index set $\mathcal{A}$. Show $H = \bigcap_{\alpha\in\mathcal{A}} H_\alpha$ is a subgroup of G.

EXERCISE 2.62. **(a)** Show by example that the union of subgroups need not be a subgroup.

(b) If H, K are subgroups of G, show $H \cup K$ is a subgroup if and only if either $H \subseteq K$ or $K \subseteq H$. *Hint:* By way of contradiction, suppose $H \cap K$ is a proper subset of H and K. Now look at the product of an element of $H\backslash (H \cap K)$ and $K\backslash (H \cap K)$.

(c) Suppose G is a group and H_n is a subgroup with $H_n \subseteq H_{n+1}$, $n \in \mathbb{Z}_{\geq 0}$. Show $\bigcup_{n=0}^{\infty} H_n$ is a subgroup.

EXERCISE 2.63 (Subgroups of $\mathbb{Z}$ and $\mathbb{Z}_n$). **(a)** If $H \neq \{0\}$ is a subgroup of $\mathbb{Z}$, let k be the minimal element of $\mathbb{N} \cap H$. Show k exists and that $H = \langle k \rangle$. *Hint:* If $m \in H$, use the Euclidean Algorithm to write $m = kq + r$ with $0 \leq r < k$ and show $kq \in H$ so that $r \in H$.

(b) Conclude that the set of subgroups of $\mathbb{Z}$ is $\{\langle k \rangle = k\mathbb{Z} \mid k \in \mathbb{Z}_{\geq 0}\}$.

(c) For $n \in \mathbb{N}$, show that every subset of $\mathbb{Z}_n$ is of the form $\langle [k] \rangle$ for $0 \leq k < n$.

(d) Show $\langle [k] \rangle = \langle [(k, n)] \rangle$. *Hint:* For $\subseteq$, use $(k, n) \mid k$. For $\supseteq$, write $(k, n) = kx + ny$.

(e) Conclude that the set of subgroups of $\mathbb{Z}_n$ is $\{\langle [k] \rangle \mid k \in \mathbb{N} \text{ and } k \mid n\}$.

(f) Find all subgroups of $\mathbb{Z}_{15}$.

EXERCISE 2.64. Use Exercise 2.63 to show $n = \sum_{k|n} \varphi(k)$ (cf. Exercise 2.47).

EXERCISE 2.65. **(a)** Show U_5 is cyclic.

(b) Show D_n is not cyclic ($n \geq 3$).

EXERCISE 2.66 (Generators of Cyclic Groups). Let G be a cyclic group with generator g. Let $n \in \mathbb{Z}$.

(a) If $|G| = \infty$, show g^n is a generator if and only if $n \in \{\pm 1\}$.

(b) If $|G| < \infty$, show g^n is a generator if and only if $(n, |g|) = 1$.

EXERCISE 2.67 (Subgroups of Cyclic Groups are Cyclic). Let $G = \langle g \rangle$ be a cyclic group. If H is a subgroup of G, show H is cyclic. *Hint:* For nontrivial H, take $k \in \mathbb{N}$ minimal so $g^k \in H$. If $g^n \in H$, use the Euclidean Algorithm and the subgroup property to show $k \mid n$.

EXERCISE 2.68. Let G be a group with $g, h \in G$. Show $\langle g \rangle \subseteq \langle h \rangle$ if and only if $g = h^k$ for $k \in \mathbb{Z}$.

EXERCISE 2.69. If a nontrivial group G has no proper subgroups (i.e., the only subgroups are $\{e\}$ and G), show G is finite and cyclic of prime order. *Hint:* Pick $g \neq e$ and look at $\langle g \rangle$.

EXERCISE 2.70. If G is a finite group, p is a positive prime, and c_p is the number of cyclic subgroups of order p, show G has $(p-1)c_p$ elements of order p.

EXERCISE 2.71. Let G be a group, let $g \in G$, and let H be a subgroup. Show $gHg^{-1} = \{ghg^{-1} \mid h \in H\}$ is a subgroup, called a *conjugate* subgroup.

EXERCISE 2.72 (Normalizer). Let G be a group and let $S \subseteq G$ be nonempty. The *normalizer* of S in G is $N(S) = N_G(S) = \{g \in G \mid gSg^{-1} = S\}$.

(a) Show $N(S)$ is a subgroup of G.

(b) For $G = GL(2,\mathbb{Q})$, $H = \{\begin{pmatrix} 1 & k \\ 0 & 1 \end{pmatrix} \mid k \in \mathbb{Z}\}$, and $g = \mathrm{diag}(2,1)$, show $gHg^{-1} \subseteq H$ but $g^{-1}Hg \not\subseteq H$.

(c) In general, show $gSg^{-1} = S$ if and only if $gSg^{-1} \subseteq S$ and $g^{-1}Sg \subseteq S$.

(d) If S is finite, show $gSg^{-1} = S$ if and only if $gSg^{-1} \subseteq S$. *Hint*: Show the map $s \to gsg^{-1}$ is injective.

EXERCISE 2.73 (Commutator). Suppose G is a group and let G' be the subgroup generated by $\{g_1g_2g_1^{-1}g_2^{-1} \mid g_i \in G\}$, called the *commutator subgroup* of G.

(a) Show G is abelian if and only if $G' = \{e\}$.

(b) If $s \in G'$ and $g \in G$, show $gsg^{-1} \in G'$.

(c) Calculate S_3'.

(d) Calculate D_4'.

EXERCISE 2.74 (Torsion). Let G be an abelian group and let T be the set of all elements of G with finite order.

(a) Show T is a subgroup, called the *torsion subgroup.*

(b) Show by example that T may not be a subgroup if the abelian hypothesis is dropped. *Hint:* Look in $GL(2,\mathbb{R})$.

EXERCISE 2.75. Let p be prime and let $R_p = \{[k]^2 \mid [k] \in U_p\}$.

(a) Show R_p is a subgroup, called the *quadratic residues modulo* p.

(b) Show $|R_2| = 1$.

(c) Show $|R_p| = \frac{p-1}{2}$ for p odd.

EXERCISE 2.76. Let $\mathrm{Nil}(\mathbb{Z}_n) = \{[k] \in \mathbb{Z}_n \mid [k]^m = [0] \text{ for some } m \in \mathbb{N}\}$.

(a) Show $\mathrm{Nil}(\mathbb{Z}_n)$ is a subgroup of $\mathbb{Z}_n$. *Hint:* Use prime factorization or the Binomial Theorem.

(b) Calculate $\mathrm{Nil}(\mathbb{Z}_{12})$.

(c) Calculate $\mathrm{Nil}(\mathbb{Z}_{15})$.

EXERCISE 2.77. Let H and K be subgroups of G.

(a) If G is abelian, show $HK = \{hk \mid h \in H,\, k \in K\}$ is a subgroup.

(b) If G is not abelian, show HK may not be a subgroup.

EXERCISE 2.78. Show the set of translations, the set of linear motions, and the set of affine motions are each subgroups of the group of motions on $\mathbb{R}^n$ (cf. Exercise 2.29).

EXERCISE 2.79. Let G be a group and let $\Delta = \{(g,g) \mid g \in G\} \subseteq G \times G$. Show Δ is a subgroup, called the *diagonal subgroup.*

EXERCISE 2.80. Suppose H is a subgroup of G and H' is a subgroup of G'. Show $H \times H'$ is a subgroup of $G \times G'$.

EXERCISE 2.81. Suppose G_i is a group, $1 \leq i \leq N$, and $g_i \in G_i$ with $|g_i| = n_i$. Show $|(g_1,\ldots,g_N)|$ in $\prod_{i=1}^N G_i$ is $\mathrm{lcm}(n_1,\ldots,n_N)$. Here the *least common multiple* of $n_1,\ldots,n_N$ is defined to be the smallest positive multiple of $n_1,\ldots,n_N$ (cf. Exercise 1.28).

EXERCISE 2.82. Let G be a group.

(a) Show $Z(G) = \bigcap_{g\in G} Z(g)$.

(b) Show $g \in Z(G)$ if and only if $Z(g) = G$.

EXERCISE 2.83 (Center of D_n). **(a)** For n odd, show $Z(D_n) = \{R_0\}$.
(b) For n even, show $Z(D_n) = \{R_0, R_{\frac{n}{2}}\}$.

EXERCISE 2.84 (Center of S_n). **(a)** Show $Z(S_2) = S_2$.
(b) Let $n \geq 3$ and fix $\sigma \in S_n$, $\sigma \neq \mathrm{Id}$. Show there exist distinct $i, j, k \in \{1, \ldots, n\}$ so that $\sigma(i) = j$.
(c) Let $\tau \in S_n$ interchange i and k and fix the remaining elements of $\{1, \ldots, n\}$. Show $(\tau\sigma)(i) = j$ and $(\sigma\tau)(i) = \sigma(k)$.
(d) Use the fact that σ is injective to show $\sigma(k) \neq j$. Conclude $Z(S_n) = \{\mathrm{Id}\}$.

EXERCISE 2.85 (Center of $GL(n, \mathbb{F})$). Show $Z(GL(n, \mathbb{F})) = \{\lambda I_n \mid \lambda \in \mathbb{F}^\times\}$. *Hint:* First check that the only matrices commuting with all diagonal matrices is the set of diagonal matrices. Finish up by tossing in permutation matrices.

EXERCISE 2.86. Suppose G_i is a group, $1 \leq i \leq N$. Show $Z(\prod_{i=1}^N G_i) = \prod_{i=1}^N Z(G_i)$.

EXERCISE 2.87 (Finite Subgroup Test). Suppose G is a group and $H \subseteq G$ is a finite nonempty set. Let $h \in H$.
(a) If H is closed, i.e., $h_1, h_2 \in H$ implies $h_1 h_2 \in H$, show there exists $n, m \in \mathbb{N}$ so that $h^n = h^{n+m}$. Conclude $h^m = e$ and $h^{-1} \in H$.
(b) Show H is a subgroup if and only if H is closed.

EXERCISE 2.88. Let G be a group with $g, h \in G$ commuting distinct elements of order 2. Show $|\langle g, h\rangle| = 4$.

EXERCISE 2.89. Let $SL(2, \mathbb{Z}) = \left\{ \begin{pmatrix} a & b \\ c & d \end{pmatrix} \mid a, b, c, d \in \mathbb{Z} \text{ and } ad - bcd = 1 \right\}$ and let $T = \begin{pmatrix} 1 & 1 \\ 0 & 1 \end{pmatrix}$ and $S = \begin{pmatrix} 0 & -1 \\ 1 & 0 \end{pmatrix}$.
(a) Show $SL(2, \mathbb{Z})$ is a subgroup of $SL(2, \mathbb{R})$.
(b) Show $\langle T, S\rangle$ contains all upper triangular matrices.
(c) For $\begin{pmatrix} a & b \\ c & d \end{pmatrix} \in SL(2, \mathbb{Z})$ with $c \neq 0$, show there exists $n \in \mathbb{Z}$ so that $\begin{pmatrix} a & b \\ c & d \end{pmatrix} T^n S = \begin{pmatrix} a' & b' \\ c' & d' \end{pmatrix}$ with $|c'| < |c|$.
(d) Iterate part (c) to prove that T and S generate $SL(2, \mathbb{Z})$.

EXERCISE 2.90. The *lattice diagram* of a group G is a list of all subgroups of G with two subgroups $K, H \subseteq G$ connected by a line whenever $K \subseteq H$.
(a) Show the lattice diagram for $\mathbb{Z}_4$ is

$$\begin{array}{c} \mathbb{Z}_4 \\ | \\ \langle [2] \rangle \\ | \\ \{[0]\}. \end{array}$$

(b) Show the lattice diagram for $\mathbb{Z}_2 \times \mathbb{Z}_2$, the *Klein group*, is

$$\begin{array}{ccccc} & & \mathbb{Z}_2 \times \mathbb{Z}_2 & & \\ & / & | & \backslash & \\ \langle([1],[0])\rangle & & \langle([1],[1])\rangle & & \langle([0],[1])\rangle \;. \\ & \backslash & | & / & \\ & & \{([0],[0])\} & & \end{array}$$

(c) Show the lattice diagram for Q (cf. Exercise 2.13) is

$$\begin{array}{ccccc} & & Q & & \\ & / & | & \backslash & \\ \langle \mathbf{i} \rangle & & \langle \mathbf{j} \rangle & & \langle \mathbf{k} \rangle . \\ & \backslash & | & / & \\ & & \langle -\mathbf{1} \rangle & & \\ & & | & & \\ & & \{\mathbf{1}\} & & \end{array}$$

(d) Determine the lattice diagram for S_3.

EXERCISE 2.91 (Infinite Products). If G_a, $\alpha \in \mathcal{A}$, is a possibly infinite collection of groups, there are two different product structures. In this setting,

$$\prod_{\alpha \in \mathcal{A}} G_\alpha = \{(g_\alpha) \mid g_\alpha \in G_\alpha\}$$

is called the *direct product* and

$$\sum_{\alpha \in \mathcal{A}} G_\alpha = \{(g_\alpha) \mid g_\alpha \in G_\alpha \text{ and } g_a = e_{G_a} \text{ for all but finitely many } \alpha\}$$

is called the *direct sum*. For many applications, the infinite direct sum is more useful since it turns out that the infinite direct product is often too big and pathological.

(a) Show $\prod_{\alpha \in \mathcal{A}} G_\alpha$ and $\sum_{\alpha \in \mathcal{A}} G_\alpha$ are groups.

(b) Show $\sum_{\alpha \in \mathcal{A}} G_\alpha$ is a subgroup of $\prod_{\alpha \in \mathcal{A}} G_\alpha$ with equality if and only if $|\mathcal{A}| < \infty$.

(c) (Requires knowledge of the concept of *countability*) Show $\sum_{n \in \mathbb{N}} \mathbb{Z}$ is countable but that $\prod_{n \in \mathbb{N}} \mathbb{Z}$ is uncountable. *Hint:* Using decimals, show $[0, 1)$ injects into $\prod_{n \in \mathbb{N}} \mathbb{Z}$.

4. Morphisms

4.1. Introduction. It is a rather common occurrence in mathematics that two objects that initially look completely different are, in fact, the same object dressed up in different clothes. As an example, consider the two groups D_3 and S_3 which initially appear very different, though the order of both is 6. Recall our discussion of D_3 in §1.3.6. Simplifying notation a bit to make patterns stand out, write r for R_1 (rotation by $60°$), t for the W_2 (reflection across the x-axis), and I for R_0. As in Theorem 2.4, it is easy to check that $D_3 = \{I, r, r^2, t, tr, tr^2\}$ and that the group multiplication table for D_3 is then given by

$\cdot$	I	r	r^2	t	tr	tr^2
I	I	r	r^2	t	tr	tr^2
r	r	r^2	I	tr^2	t	tr
r^2	r^2	I	r	tr	tr^2	t
t	t	tr	tr^2	I	r	r^2
tr	tr	tr^2	t	r^2	I	r
tr^2	tr^2	t	tr	r	r^2	I

Actually, the above calculation can be accomplished fairly quickly if one simply notes that $r^3 = I$, $t^2 = I$, and $rt = tr^2$. In any case, we leave the details to the reader.

Similarly in S_3, write $a = \begin{pmatrix} 1 & 2 & 3 \\ 2 & 3 & 1 \end{pmatrix}$ and $b = \begin{pmatrix} 1 & 2 & 3 \\ 1 & 3 & 2 \end{pmatrix}$. It is straightforward (though not especially enjoyable) to verify that $S_3 = \{I, a, a^2, b, ba, ba^2\}$ and that the group multiplication table is given by

$\cdot$	I	a	a^2	b	ba	ba^2
I	I	a	a^2	b	ba	ba^2
a	a	a^2	I	ba^2	b	ba
a^2	a^2	I	a	ba	ba^2	b
b	b	ba	ba^2	I	a	a^2
ba	ba	ba^2	b	a^2	I	a
ba^2	ba^2	b	ba	a	a^2	I

Again, this calculation is made easier by noting that $a^3 = I$, $b^2 = I$, and $ab = ba^2$.

Notice that relabeling $r \to a$ and $t \to b$ completely transforms the group D_3 into the group S_3. Moreover, the multiplication tables become identical. Therefore, in every way that really matters, D_3 and S_3 are the same group up to cosmetic relabeling.

The word we will use to describe this phenomenon is *isomorphic*. To make it precise, the idea of relabeling will be replaced by a function from one group to another. For instance, in our case we can use $\varphi : D_3 \to S_3$ defined by $\varphi(r^n t^m) = a^n b^m$. The map φ is simply a fancy way to say we are relabeling things. The fact that the multiplication tables correspond is contained in the observation that we get the same answer by multiplying in D_3 and then relabeling as we do by first relabeling and then multiplying in S_3. In other words, $\varphi(gh) = \varphi(g)\varphi(h)$ for $g, h \in D_3$.

4.2. Definitions and Examples.

DEFINITION 2.20. Let G and H be groups and let $\varphi : G \to H$ be a function.

(1) If

$$\varphi(g_1 g_2) = \varphi(g_1)\varphi(g_2)$$

for all $g_1, g_2 \in G$, then φ is called a *homomorphism*.

(2) A bijective (one-to-one and onto) homomorphism is called an *isomorphism*. In that case G and H are said to be *isomorphic* and we write $G \cong H$.

(3) If $G = H$, an isomorphism is also called an *automorphism*.

Note that, as usual, we wrote the groups multiplicatively in the above definition. If both G and H are written additively, then the homomorphism condition takes the form

$$\varphi(g_1 + g_2) = \varphi(g_1) + \varphi(g_2).$$

If G is written multiplicatively and H additively, then the homomorphism condition looks like

$$\varphi(g_1 g_2) = \varphi(g_1) + \varphi(g_2).$$

If G is written additively and H multiplicatively, then the homomorphism condition looks like

$$\varphi(g_1 + g_2) = \varphi(g_1)\varphi(g_2).$$

Of course, our map $\varphi : D_3 \to S_3$ defined above by $\varphi(r^n t^m) = a^n b^m$ is an example of an isomorphism. Thus $D_3 \cong S_3$. This is the only occurrence of an isomorphism between D_n and S_n. To see this, note that $|D_n| = 2n$ while $|S_n| = n!$ and, at the very least, isomorphic groups have the same number of elements.

Another example of an isomorphism occurs with $\mathbb{Z}_2 \times \mathbb{Z}_2$, sometimes called the *Klein group*, and U_8. Recall $U_8 = \{[1], [3], [5], [7]\}$ and define $\varphi : \mathbb{Z}_2 \times \mathbb{Z}_2 \to U_8$ by $\varphi(([a], [b])) = [3^a 5^b]$. This is well defined since $[3^2] = [1] = [5^2]$ so that φ does not depend on the choice of representative for $[a], [b] \in \mathbb{Z}_2$. To check φ is a homomorphism, observe

$$\begin{aligned} \varphi\left(([a_1], [b_1]) + ([a_2], [b_2])\right) &= \varphi\left([a_1 + a_2], [b_1 + b_2]\right) \\ &= [3^{a_1+a_2} 5^{b_1+b_2}] \\ &= [3^{a_1} 5^{b_1}][3^{a_2} 5^{b_2}] \\ &= \varphi([a_1], [b_1])\varphi([a_2], [b_2]). \end{aligned}$$

Finally, notice φ maps $([0], [0]) \to [1]$, $([1], [0]) \to [3]$, $([0], [1]) \to [5]$, and $([1], [1]) \to [7]$ so that φ is a bijection. Thus $\mathbb{Z}_2 \times \mathbb{Z}_2 \cong U_8$.

As a nonexample, consider the map $\varphi : \mathbb{Z} \to \mathbb{Z}$ given by $\varphi(x) = x^2$. It is not an isomorphism since

$$\varphi(x + y) = (x + y)^2 = x^2 + 2xy + y^2 \neq x^2 + y^2 = \varphi(x) + \varphi(y).$$

As an example of the power of isomorphisms to cut through unimportant differences, we are able to classify all cyclic subgroups. It turns out that cyclic groups have little flexibility—essentially the only examples are $\mathbb{Z}$ or $\mathbb{Z}_n$, $n \in \mathbb{N}$.

THEOREM 2.21 (Classification of Cyclic Groups). *Let G be a group with $g \in G$.*
(1) *If $|g| = \infty$, then $\langle g \rangle \cong \mathbb{Z}$.*
(2) *If $|g| < \infty$, then $\langle g \rangle = \{e, g, g^2, \ldots, g^{|g|-1}\}$ and $\langle g \rangle \cong \mathbb{Z}_{|g|}$.*

PROOF. For part (1), Theorem 2.11 shows the map $\varphi : \langle g \rangle \to (\mathbb{Z}, +)$ given by $\varphi(g^k) = k$ is a well defined bijection. Since

$$\varphi(g^{k_1} g^{k_2}) = \varphi(g^{k_1+k_2}) = k_1 + k_2 = \varphi(g^{k_1}) + \varphi(g^{k_2}),$$

φ is a homomorphism. Thus, part (1) follows. For part (2), Theorem 2.11 shows $g^{k_1} = g^{k_2}$, $k_1, k_2 \in \mathbb{Z}$, if and only if $k_1 \equiv k_2 \pmod{|g|}$. In particular, it follows that the map $\varphi : \langle g \rangle \to (\mathbb{Z}_{|g|}, +)$ given by $\varphi(g^k) = [k]$ is a well defined injection. As φ is also clearly surjective and a homomorphism (similar to the $\mathbb{Z}$ case), the result follows. □

Naturally, there are plenty of homomorphisms that are not isomorphisms. An insipid example is the map $\varphi : G \to H$ given by $\varphi(g) = e_H$, which is called the *trivial homomorphism*. For a more interesting example, the map $\varphi : \mathbb{Z} \to \mathbb{Z}_n$ given by $\varphi(k) = [k]$, $k \in \mathbb{Z}$, is a homomorphism since

$$\varphi(k_1 + k_2) = [k_1 + k_2] = [k_1] + [k_2] = \varphi(k_1) + \varphi(k_2).$$

It is clear that φ is not one-to-one and therefore not an isomorphism. Similarly, the map $\phi : \mathbb{Z} \to \mathbb{Z}$ given by $\phi(k) = 2k$ is a homomorphism since

$$\phi(k_1 + k_2) = 2(k_1 + k_2) = 2k_1 + 2k_2 = \phi(k_1) + \phi(k_2).$$

Yet ϕ is definitely not onto and so it is not an isomorphism.

Turning to automorphisms, a large family can often be constructed in a uniform manner. When $G = M_{n \times n}(\mathbb{R})$, the following definition is used in the context of diagonalizing matrices.

DEFINITION 2.22. Let G be a group and let $g, h \in G$. The *conjugation* map $c_g : G \to G$ is defined by

$$c_g(h) = ghg^{-1}.$$

THEOREM 2.23. *Let G be a group and let $g \in G$. Then c_g is an automorphism.*

PROOF. Let $h_1, h_2 \in G$. Calculate that

$$c_g(h_1 h_2) = gh_1 h_2 g^{-1} = gh_1 g^{-1} gh_2 g^{-1} = c_g(h_1) c_g(h_2)$$

so that c_g is a homomorphism. Since it is clear that $c_{g^{-1}}$ is an inverse to c_g, it follows that c_g is bijective and therefore it is an automorphism. □

The set of automorphisms on G actually forms a group under composition of functions and is called $\mathrm{Aut}(G)$ (Exercise 2.115). Automorphisms of the form c_g where $g \in G$ are called *inner automorphisms* and they form a subgroup of $\mathrm{Aut}(G)$ that is called $\mathrm{Inn}(G)$ (Exercise 2.115). Of course for abelian groups, only the identity map is inner.

4.3. Basic Properties.

DEFINITION 2.24. Let $\varphi : G \to H$ be a homomorphism between groups.

(1) The *image* of φ is

$$\mathrm{Im}\, \varphi = \{\varphi(g) \mid g \in G\}.$$

(2) The *kernel* of φ is

$$\ker \varphi = \{g \in G \mid \varphi(g) = e_H\}.$$

For example, if $\varphi : \mathbb{Z} \to \mathbb{Z}$ is the homomorphism $\varphi(k) = 2k$, then $\mathrm{Im}\, \varphi = \{2k \mid k \in \mathbb{Z}\} = 2\mathbb{Z}$ and $\ker \varphi = \{k \mid 2k = 0,\, k \in \mathbb{Z}\} = \{0\}$. As another example, consider the homomorphism $\psi : GL(n, \mathbb{R}) \to \mathbb{R}^\times$ given by $\psi(g) = \det(g)$. Then $\mathrm{Im}\, \psi = \{\det g \mid g \in GL(n, \mathbb{R})\}$. Looking at, say, matrices of the form $\mathrm{diag}(r, 1, \ldots, 1)$ for $r \in \mathbb{R}^\times$, it follows that any nonzero number can be achieved by a determinant. Thus $\mathrm{Im}\, \psi = \mathbb{R}^\times$. Turning to the kernel, $\ker \psi = \{g \in GL(n, \mathbb{R}) \mid \det g = 1\} = SL(n, \mathbb{R})$.

THEOREM 2.25. *Let $\varphi : G \to H$ be a homomorphism between groups and let $g \in G$.*

(1) *The composition of homomorphisms is a homomorphism.*
(2) *If φ is an isomorphism, then so is φ^{-1}.*
(3) $\varphi(e_G) = e_H$.
(4) $\varphi(g^{-1}) = \varphi(g)^{-1}$.
(5) $\varphi(g_1 g_2 \cdots g_N) = \varphi(g_1)\varphi(g_2) \cdots \varphi(g_N)$ *for* $g_i \in G$.
(6) $\mathrm{Im}\, \varphi$ *is a subgroup of* H.
(7) $\ker \varphi$ *is a subgroup of* G.
(8) φ *is one-to-one if and only if* $\ker \varphi = \{e_G\}$.

PROOF. For part (1), suppose $\tau : H \to K$ is a homomorphism of groups and $g_1, g_2 \in G$. Then

$$\begin{aligned}(\tau \circ \varphi)(g_1 g_2) &= \tau(\varphi(g_1 g_2)) = \tau(\varphi(g_1)\varphi(g_2)) \\ &= \tau(\varphi(g_1))\tau(\varphi(g_2)) = (\tau \circ \varphi)(g_1)(\tau \circ \varphi)(g_2)\end{aligned}$$

as desired. For part (2), first note that φ^{-1} is well defined since φ is bijective. For $h_1, h_2 \in H$, pick $g_1, g_2 \in G$ so that $\varphi(g_i) = h_i$. Observe that $\varphi(g_1 g_2) = \varphi(g_1)\varphi(g_2) = h_1 h_2$. Thus

$$\varphi^{-1}(h_1 h_2) = g_1 g_2 = \varphi^{-1}(h_1)\varphi^{-1}(h_2)$$

and so φ^{-1} is a homomorphism. As φ^{-1} is already bijective, it is an isomorphism.

For part (3), write

$$\varphi(e_G) = \varphi(e_G e_G) = \varphi(e_G)\varphi(e_G).$$

Multiply both sides by $\varphi(e_G)^{-1}$ to get $e_H = \varphi(e_G)$. For part (4), write

$$e_H = \varphi(e_G) = \varphi(gg^{-1}) = \varphi(g)\varphi(g^{-1})$$

so that the inverse to $\varphi(g)$ is $\varphi(g^{-1})$ as desired. Part (5) follows by induction on N.

For part (6), first note $e_H \in \operatorname{Im}\varphi$. Then observe that

$$\varphi(g_1)\varphi(g_2)^{-1} = \varphi(g_1)\varphi(g_2^{-1}) = \varphi(g_1 g_2^{-1}) \in \operatorname{Im}\varphi$$

so that $\operatorname{Im}\varphi$ is a subgroup. Similarly for part (7), first note that $e_G \in \ker\varphi$. Next check that if $g_1, g_2 \in \ker\varphi$, then

$$\varphi(g_1 g_2^{-1}) = \varphi(g_1)\varphi(g_2)^{-1} = e_H e_H^{-1} = e_H$$

so that $g_1 g_2^{-1} \in \ker\varphi$. Thus $\ker\varphi$ is also a subgroup.

For part (8), first suppose that φ is one-to-one and $g \in \ker\varphi$. Then $\varphi(g) = e_H$ and $\varphi(e_G) = e_H$. Since φ is one-to-one, $g = e_G$. Thus $\ker\varphi = \{e_G\}$. Conversely, suppose $\ker\varphi = \{e_G\}$ and suppose $\varphi(g_1) = \varphi(g_2)$. Multiplying by $\varphi(g_2)^{-1}$ on both sides shows $\varphi(g_1 g_2^{-1}) = e_H$ so that $g_1 g_2^{-1} \in \ker\varphi$. Therefore $g_1 g_2^{-1} = e_G$. Multiplying by g_2 on both sides shows $g_1 = g_2$. Hence φ is one-to-one. □

It should be noted that the conditions in parts (7) and (8) of Theorem 2.25 are used very frequently in group theory. For example, consider the homomorphisms $\varphi(k) = 2k$ and $\psi(g) = \det(g)$ discussed above. Though direct verification is not difficult in either case, the fact that $\ker\varphi = \{0\}$ shows φ is injective and the fact that $\ker\psi = SL(n,\mathbb{R})$ shows $SL(n,\mathbb{R})$ is a subgroup of $GL(n,\mathbb{R})$.

4.4. Exercises 2.92–2.118.

EXERCISE 2.92. Show the following maps are homomorphisms:
(a) For $r \in \mathbb{R}$, $\varphi : \mathbb{R} \to \mathbb{R}$ given by $\varphi(x) = rx$.
(b) For $m \in \mathbb{Z}$ and $n \in \mathbb{N}$, $\varphi : \mathbb{Z}_n \to \mathbb{Z}_n$ given by $\varphi([k]) = m\,[k]$.
(c) For $r \in \mathbb{R}$, $\varphi : \mathbb{R}^+ \to \mathbb{R}^+$ given by $\varphi(x) = x^r$.
(d) For $m \in \mathbb{Z}$ and $n \in \mathbb{N}$, $\varphi : U_n \to U_n$ given by $\varphi([k]) = [k]^m$.
(e) $\varphi : \mathbb{R} \to \mathbb{R}^+$ given by $\varphi(x) = e^x$.
(f) $\varphi : D_n \to \{\pm 1\}$ given by $\varphi(R_j) = 1$ and $\varphi(W_j) = -1$.
(g) $\theta : GL(n,\mathbb{F}) \to GL(n,\mathbb{F})$ given by $\theta(g) = \left(g^{-1}\right)^T$.
(h) For groups G, H, $\pi_1 : G \times H \to G$ given by $\pi_1(g,h) = g$.
(i) $\varphi : G \to G \times G$ given by $\varphi(g) = (g,g)$.

EXERCISE 2.93. Show the following maps are not homomorphisms:
(a) $\varphi : \mathbb{Z} \to \mathbb{Z}$ given by $\varphi(k) = k + 1$.
(b) $\varphi : \mathbb{R} \to \mathbb{R}$ given by $\varphi(x) = x^2$.
(c) $\varphi : \mathbb{R}^\times \to \mathbb{R}^\times$ given by $\varphi(x) = 2x$.
(d) $\varphi : S_4 \to S_4$ given by $\varphi(\sigma) = \sigma^{-1}$.

(e) $\varphi : GL(2,\mathbb{R}) \to GL(2,\mathbb{R})$ given by $\varphi(g) = g^T$.

EXERCISE 2.94. Show the following groups are isomorphic.

(a) $\mathbb{Z} \cong 2\mathbb{Z}$.
(b) $\mathbb{C} \cong \mathbb{R} \times \mathbb{R}$.
(c) For $n \in \mathbb{N}$, $\{z \in \mathbb{C} \mid z^n = 1\} \cong \mathbb{Z}_n$ (cf. Exercise 2.4).
(d) $\mathbb{R} \cong \mathbb{R}^+$. *Hint:* e^x.
(e) $U_7 \cong \mathbb{Z}_2 \times \mathbb{Z}_3$.
(f) $U_{12} \cong \mathbb{Z}_2 \times \mathbb{Z}_2$.
(g) $\mathbb{R} \cong U(2,\mathbb{R}) = \left\{ \begin{pmatrix} 1 & x \\ 0 & 1 \end{pmatrix} \mid x \in \mathbb{R} \right\}$.
(h) $\mathbb{R}^\times \times \mathbb{R}^\times \cong D(2,\mathbb{R}) = \left\{ \begin{pmatrix} x & 0 \\ 0 & y \end{pmatrix} \mid x, y \in \mathbb{R}^\times \right\}$.
(i) $\mathbb{C}^\times \cong \left\{ \begin{pmatrix} a & b \\ -b & a \end{pmatrix} \mid a, b \in \mathbb{R},\ a^2 + b^2 \neq 0 \right\}$.
(j) $GL(2,\mathbb{Z}_2) \cong D_3$.
(k) $D_n \cong \left\{ \begin{pmatrix} \pm[1] & [b] \\ [0] & [1] \end{pmatrix} \mid [b] \in \mathbb{Z}_p \right\}$.
(l) $Q_2 \cong \mathbb{Z}_4$ (cf. Exercise 2.60).
(m) $Q_4 \cong Q$ (cf. Exercises 2.13 and 2.60).

EXERCISE 2.95. Show the following groups are not isomorphic.

(a) $\mathbb{Z}_4 \ncong \mathbb{Z}_5$.
(b) $S_3 \ncong \mathbb{Z}_6$.
(c) $\mathbb{Z}_4 \ncong \mathbb{Z}_2 \times \mathbb{Z}_2$.
(d) $\mathbb{R}^\times \ncong \mathbb{R}$. *Hint:* Count the solutions to $x^2 = 1$ in $\mathbb{R}^\times$ and to $2x = 0$ in $\mathbb{R}$.
(e) $\mathbb{Z} \ncong \mathbb{Q}$.

EXERCISE 2.96. Calculate the kernels and images of the following homomorphisms.

(a) $\varphi : \mathbb{Z}_{10} \to \mathbb{Z}_{10}$ by $\varphi([k]) = 2[k]$.
(b) $\varphi : U_{10} \to U_{10}$ by $\varphi([k]) = [k]^2$.
(c) $\varphi : D_n \to \{\pm 1\}$ by $\varphi(R_j) = 1$ and $\varphi(W_j) = -1$.

EXERCISE 2.97. **(a)** Let $\varphi : G \to H$ be a homomorphism with $\varphi(g) = h$ for some $g \in G$ and $h \in H$. Show $\varphi^{-1}(h) = g \ker \varphi = \{gk \mid k \in \ker \varphi\}$.

(b) Suppose $\varphi : U_{10} \to H$ satisfies $\varphi([3]) = h$ with $\ker \varphi = \{[1], [9]\}$. Find all $x \in U_{10}$ so $\varphi(x) = h$.

EXERCISE 2.98. Let $\varphi : \mathbb{R} \to \mathbb{C}^\times$ be given by $\varphi(x) = e^{ix}$ and let $\psi : \mathbb{C}^\times \to \mathbb{R}^\times$ be given by $\psi(z) = |z|$, the norm.

(a) Show φ and ψ are homomorphisms.
(b) Calculate $\operatorname{Im} \varphi$ and $\operatorname{Im} \psi$.
(c) Calculate $\ker \varphi$ and $\ker \psi$.

EXERCISE 2.99 (Homomorphisms from $\mathbb{Z}$). **(a)** If G is a group and $\varphi : \mathbb{Z} \to G$ is a homomorphism, show $\varphi(k) = \varphi(1)^k$ for $k \in \mathbb{Z}$.

(b) If $\varphi' : \mathbb{Z} \to G$ is another homomorphism, show $\varphi = \varphi'$ if and only if $\varphi(1) = \varphi'(1)$.

(c) If $g \in G$, show the map $\varphi : \mathbb{Z} \to G$ given by $\varphi(k) = g^k$ is a homomorphism.

(d) Conclude that the set of homomorphisms from $\mathbb{Z}$ to G is in bijection with the set of elements of G.

(e) List all homomorphisms from $\mathbb{Z}$ to D_6.

EXERCISE 2.100 (Homomorphisms from $\mathbb{Z}_n$). **(a)** If G is a group, $n \in \mathbb{N}$, and $\varphi : \mathbb{Z}_n \to G$ is a homomorphism, show $\varphi([k]) = \varphi([1])^k$ for $k \in \mathbb{Z}$. Conclude $\varphi([1]) \mid n$.

(b) If $\varphi' : \mathbb{Z}_n \to G$ is another homomorphism, show $\varphi = \varphi'$ if and only if $\varphi(1) = \varphi'(1)$.

(c) If $g \in G$, show the map $\varphi : \mathbb{Z}_n \to G$ given by $\varphi([k]) = g^k$ is well defined if and only if $|g| \mid n$. In that case, show φ is a homomorphism.

(d) Conclude that the set of homomorphisms from $\mathbb{Z}_n$ to G is in bijection with $\{g \in G \mid |g| \mid n\}$.

(e) List all homomorphisms from $\mathbb{Z}_{20}$ to $\mathbb{Z}_{25}$.

EXERCISE 2.101. Let $n, m \in \mathbb{N}$.

(a) Show the map $\varphi : \mathbb{Z}_n \to \mathbb{Z}_m$ given by $\varphi([k]_n) = [k]_m$ is well defined if and only if $m \mid n$. In that case, show φ is a surjective homomorphism.

(b) Let p be a positive prime and now assume $n \leq m$. Show the map $\psi : \mathbb{Z}_{p^n} \to \mathbb{Z}_{p^m}$ given by $\psi([k]_{p^n}) = [p^{m-n}k]_{p^m}$ is a well defined injective homomorphism.

EXERCISE 2.102. If $\varphi : G \to H$ is a homomorphism and K is a subgroup of H, show $\varphi^{-1}(K) = \{g \in G \mid \pi(g) \in K\}$ is a subgroup of G.

EXERCISE 2.103. Let G be an abelian group and let $\varphi : G \to \mathbb{Z}$ be a surjective homomorphism. Show $G \cong \mathbb{Z} \times \ker \varphi$. *Hint:* Pick $t \in G$ so $\varphi(t) = 1$ and look at $\langle t \rangle$ (cf. Exercise 2.97).

EXERCISE 2.104. Suppose G is a group with $S \subseteq G$ and $G = \langle S \rangle$. If $\varphi, \varphi' : G \to H$ are homomorphisms satisfying $\varphi(s) = \varphi'(s)$ for all $s \in S$, show $\varphi = \varphi'$.

EXERCISE 2.105. Find all homomorphisms from $\mathbb{Z}_4$ to $\mathbb{Z}_2 \times \mathbb{Z}_2$.

EXERCISE 2.106 ($D_n \hookrightarrow S_n$). Let $n \geq 3$. Define $\iota : \mathbb{Z}_n \to \{1, \ldots, n\}$ by $\iota([k]) = k$ for $1 \leq k \leq n$. For $g \in D_n$, define $\sigma_g : \{1, \ldots, n\} \to \{1, \ldots, n\}$ by $\sigma_g(k) = \iota(g([k]))$.

(a) Show $\sigma_g \in S_n$.

(b) Show the map $\varphi : D_n \to S_n$ by $\varphi(g) = \sigma_g$ is an injective homomorphism.

(c) Show φ is an isomorphism if and only if $n = 3$.

(d) For $n = 4$, calculate $\operatorname{Im} \varphi$.

EXERCISE 2.107 ($D_n \hookrightarrow GL(2, \mathbb{C})$). For $n \geq 3$, let $r = \begin{pmatrix} e^{2\pi i/n} & 0 \\ 0 & e^{-2\pi i/n} \end{pmatrix}$ and $w = \begin{pmatrix} 0 & 1 \\ 1 & 0 \end{pmatrix}$ be elements of $GL(2, \mathbb{C})$. Show $\langle r, w \rangle \cong D_n$.

EXERCISE 2.108. Let G be the set of all functions from $\mathbb{R}\backslash\{0,1\}$ to $\mathbb{R}\backslash\{0,1\}$ with binary operation given by composition of functions. Let $H = \langle f, g \rangle \subseteq G$ where $f(x) = x^{-1}$ and $g(x) = 1 - x^{-1}$. Show $H \cong S_3$.

EXERCISE 2.109. **(a)** For groups G and H, show $G \times H \cong H \times G$.

(b) If $G \cong G'$ and $H \cong H'$, show $G \times H \cong G' \times H'$.

EXERCISE 2.110. Let G be a group. Show the map $\varphi(g) = g^{-1}$ is an isomorphism if and only if G is abelian.

EXERCISE 2.111. Let G be a group. For $g \in G$, let $l_g : G \to G$ be given by $l_g(h) = gh$ for $h \in G$. Let $G^* = \{l_g \mid g \in G\}$.

(a) With respect to composition of functions, show G^* is a group, called the *left regular representation.*

(b) Show $G \cong G^*$.

EXERCISE 2.112. Let $\varphi : G \to H$ be a homomorphism. If $g \in G$ has finite order, show $|\varphi(g)| \mid |g|$ with equality for isomorphisms.

EXERCISE 2.113 (Small Groups). **(a)** If G is a group of order 2, show $G \cong \mathbb{Z}_2$.

(b) If G is a group of order 3, show $G \cong \mathbb{Z}_3$.

(c) If G is a group of order 4, show G is isomorphic to either $\mathbb{Z}_4$ or $\mathbb{Z}_2 \times \mathbb{Z}_2$.

(d) If G is a group of order 5, show $G \cong \mathbb{Z}_5$.

(e) Conclude that the smallest nonabelian group has order 6 and give an example.

EXERCISE 2.114. Show $G \cong G^{\text{opp}}$ (cf. Exercise 2.52).

EXERCISE 2.115 (Aut(G) and Inn(G)). **(a)** Let G be a group and write Aut(G) for the set of automorphisms of G. Under composition of functions, show Aut(G) is a group.

(b) Write Inn(G) for the set of inner automorphisms of G. Show Inn(G) is a subgroup of Aut(G).

EXERCISE 2.116 (Aut($\mathbb{Z}_n$) $\cong U_n$). **(a)** Show Aut($\mathbb{Z}$) $\cong \mathbb{Z}_2$ (cf. Exercise 2.115).

(b) Show Aut($\mathbb{Z}_n$) $\cong U_n$ (cf. Exercise 2.100).

(c) Show Inn(D_4) $\cong \mathbb{Z}_2 \times \mathbb{Z}_2$.

EXERCISE 2.117 (Aut($\mathbb{Z}_n^k$)). For $n, k \in \mathbb{N}$ with $n \geq 2$, write $\mathbb{Z}_n^k = \overbrace{\mathbb{Z}_n \times \cdots \times \mathbb{Z}_n}^{k \text{ copies}}$. Show Aut($\mathbb{Z}_n^k$) $\cong GL(k, \mathbb{Z}_n)$ (cf. Exercise 2.115). *Hint:* Use ideas from linear algebra.

EXERCISE 2.118 (Generators for D_n). Let $n \in \mathbb{Z}$, $n \geq 3$. Suppose G is a group of order $2n$ containing elements r and w that satisfy (i) $|r| = n$ and $|w| = 2$ and (ii) $wrw = r^{-1}$. This exercise shows $G \cong D_n$.

(a) Show $r^{-1} \neq r$, $rw = wr^{-1}$, and $rw \neq wr$. Conclude that $w \notin \langle r \rangle$ so that $\langle r \rangle \cap \langle w \rangle = \{e\}$.

(b) Show $G = \{r^j, r^j w \mid 1 \leq j \leq n\}$.

(c) By examining $(wrw)^j$, show $wr^j w = w^{-j}$ and $wr^j = r^{-j} w$.

(d) Show $G \cong D_n$ by using $\varphi(r) = R_1$ and $\varphi(w) = W_1$.

5. Quotients

5.1. Definitions. As we have already commented many times, there are plenty of occasions where mathematicians want to view two technically different objects as essentially the same. We first ran into such ideas when dealing with $\mathbb{Z}_n$ for $n \in \mathbb{N}$. There we introduced the notion of a congruence class. Recall for $a \in \mathbb{Z}$,

$$\begin{aligned}[a] &= \{b \in \mathbb{Z} \mid b \equiv a \pmod{n}\} \\ &= \{a + nk \mid k \in \mathbb{Z}\} \\ &= a + n\mathbb{Z}.\end{aligned}$$

Here if $S \subseteq \mathbb{Z}$, we use the suggestive notation $a + S = \{a + s \mid s \in S\}$ and the notation $aS = \{as \mid s \in S\}$. With this notation, $\mathbb{Z}_n = \{a + n\mathbb{Z} \mid a \in \mathbb{Z}\}$. Our definition of addition on $\mathbb{Z}_n$ can then be written as

$$(a + n\mathbb{Z}) + (b + n\mathbb{Z}) = (a + b) + n\mathbb{Z}$$

for $a, b \in \mathbb{Z}$. Notice that the set $n\mathbb{Z}$ is not just any random subset of $\mathbb{Z}$ but actually a subgroup of $(\mathbb{Z}, +)$.

Situations analogous to the above discussion are very common in group theory. In such settings (writing the group multiplicatively now), one usually has a group G and a subgroup H. In these scenarios, the added twist is that one desires to count different elements of G as essentially the same when they are only off by an element of H. Taking inspiration from the $\mathbb{Z}_n$ case above, it therefore makes sense to look at subsets of the form

$$gH = \{gh \mid h \in H\}.$$

Just as with conjugacy classes, this construction lumps together all the elements we want to count as essentially the same.

If life were fair, we would go on to define multiplication on such sets by

$$(g_1 H)(g_2 H) = (g_1 g_2) H$$

just as we did with conjugacy classes in $\mathbb{Z}_n$ (though using multiplicative instead of additive notation here). In many practical cases, this idea works splendidly. However, similarly to Definition 1.17, g will be called a representative for gH and there is a possible problem stemming from the fact that representatives are usually not unique. As a result, an extra condition on H called *normality* (Definition 2.30 below) is needed in order to make the proposed binary operation well defined.

In any case, let us first see how far we can get just using the trick of looking at subsets of G of the form gH. We will address the possibility of a group structure in §5.3.

DEFINITION 2.26. Let G be a group and let H be a subgroup of G. Let $g \in G$.

(1) The *left coset* of g with respect to H is

$$gH = \{gh \mid h \in H\}.$$

(2) If C is a left coset with respect to H and $C = gH$, then g is called a *representative* of C.

(3) The set of left cosets is denoted G/H and is called the *quotient* of G by H.

(4) The *index* of H in G is $|G/H|$ and is denoted $[G : H]$.

It should be noted that it is fairly common for people to write $[g]$ or $\overline{g}$ for the left coset gH. Also, at the risk of death by repetition, observe that the above left coset notation is naturally written in its multiplicative form. If G happens to be written additively, then the left coset is usually written as

$$g + H = \{g + h \mid h \in H\}.$$

Before launching into examples, let us stress two points that have already come up in our study of $\mathbb{Z}_n$. The first is that representatives for left cosets are usually not unique (cf. Lemma 2.27). In other words, it is possible to have different $g_1, g_2 \in G$ so that $g_1 H = g_2 H$. The second point to be stressed is that G/H is a set of sets. In other words, elements of G/H are certain subsets of G.

Our first example, of course, is $\mathbb{Z}_n$. Here the notation is additive and we start with $G = (\mathbb{Z}, +)$ and the subgroup $H = n\mathbb{Z}$. Then $\mathbb{Z}/n\mathbb{Z}$ consists of all left cosets $k + n\mathbb{Z}$, $k \in \mathbb{Z}$. Since this is simply new notation for the conjugacy class $[k]$, we see $\mathbb{Z}/n\mathbb{Z} = \mathbb{Z}_n$ as sets.

As a second example, consider the group $(\mathbb{R}^2, +)$ and the subgroup $W = \{(x, 0) \mid x \in \mathbb{R}\}$ (clearly $W \cong \mathbb{R}$). Then $\mathbb{R}^2/W$ consists of the set of left cosets $\{(x, y) + W \mid (x, y) \in \mathbb{R}^2\}$. Geometrically speaking, the left coset $(x, y) + W$ is the horizontal line in the plane that passes through (x, y). Tracing the line back to the y-axis, it is clear that each left coset can be uniquely represented by an element of the form $(0, y) + W$. Algebraically, this is accomplished by observing

$$\begin{aligned}(x, y) + W &= \{(x, y) + (x', 0) \mid x' \in \mathbb{R}\} = \{(x + x', y) \mid x' \in \mathbb{R}\} = \{(x'', y) \mid x'' \in \mathbb{R}\} \\ &= \{(0, y) + (x'', 0) \mid x'' \in \mathbb{R}\} = (0, y) + W.\end{aligned}$$

To summarize, $\mathbb{R}^2/W$ consists of all the horizontal lines in the plane and the map $(0, y) + W \to y$ establishes a bijection between $\mathbb{R}^2/W$ and $\mathbb{R}$.

As a third example, consider $D_3 = \{I, R_1, R_1^2, W_1, W_1R_1, W_1R_1^2\}$ and its rotational subgroup $\mathcal{R}_3 = \{I, R_1, R_1^2\}$. By exhaustive enumeration, there are exactly two left cosets: $\mathcal{R}_3 = \{I, R_1, R_1^2\}$ and $W_1\mathcal{R}_3 = \{W_1, W_1R_1, W_1R_1^2\}$. Thus $|D_3/\mathcal{R}_3| = 2$. As usual, these cosets can be written with different representatives: $I\mathcal{R}_3 = R_1\mathcal{R}_3 = R_1^2\mathcal{R}_3$ and $W_1\mathcal{R}_3 = (W_1R_1)\,\mathcal{R}_3 = (W_1R_1^2)\,\mathcal{R}_3$.

Our first result tells us how to write the most general representative for a left coset gH.

LEMMA 2.27. *Let G be a group, let H be a subgroup of G, and let $g_1, g_2 \in G$. Then $g_1H = g_2H$ if and only if $g_2 = g_1h$ for some $h \in H$.*

PROOF. Suppose $g_1H = g_2H$. Since $e \in H$, $g_2 \in g_2H = g_1H$ so $g_2 = g_1h$ for some $h \in H$. Conversely, suppose $g_2 = g_1h$. As H is closed under multiplication, multiplying on the right by an element of H shows $g_2H \subseteq g_1H$. For the reverse inclusion, note that $g_1 = g_2h^{-1}$. Since H is also closed under inverses, the same argument shows $g_1H \subseteq g_2H$ as desired. □

5.2. Lagrange's Theorem. It turns out that the left cosets partition a group into very nice subsets. For finite groups, this places a strong condition on the possible order of a subgroup.

THEOREM 2.28 (Lagrange's Theorem). *Let G be a group and let H be a subgroup of G. Let $g, g_1, g_2 \in G$.*

(1) *There is a bijection between the left coset gH and H.*

(2) *The relation on G defined by $g_1 \sim g_2$ if $g_2 = g_1h$ for some $h \in H$ is an equivalence relation. The equivalence classes are the left cosets.*

(3) ***(Lagrange)** If G is finite,*

$$|G/H| = \frac{|G|}{|H|}.$$

In particular, $|H|$ divides $|G|$.

PROOF. For part (1), observe that left multiplication by g^{-1} gives a bijection $gH \to H$ (left multiplication by g is the inverse map). Part (2) follows trivially from the definition of a subgroup, an equivalence class, and a left coset (with the help of Lemma 2.27). For instance, to check transitivity, suppose $g_1 \sim g_2$ and

$g_2 \sim g_3$. Write $g_2 = g_1 h_1$ and $g_3 = g_2 h_2$ for $h_i \in H$. Then $g_3 = g_1 h_1 h_2$ so that $g_3 \sim g_1$ as desired. The rest of the verifications are even easier and are left to the reader.

For part (3), notice that part (2) coupled with Theorem 1.37 shows G is a disjoint union of its left cosets. In particular,

$$|G| = \sum_{gH \in G/H} |gH|.$$

However, part (1) shows $|gH| = |H|$. Thus

$$|G| = \sum_{gH \in G/H} |H| = |G/H|\,|H|. \qquad \square$$

Lagrange's Theorem is an exceptionally powerful result with many applications. The theorem places an extremely restrictive combinatorial constraint on the order of subgroups of finite groups. For instance, if G is a group with $|G| = 15$, any proper subgroup must have order 3 or 5. Similarly, it follows that any group G of prime order has only two subgroups: $\{e\}$ and G with nothing in-between (Exercise 2.123).

The following is another application of Lagrange's Theorem and is used frequently.

COROLLARY 2.29. *Let G be a finite group.*

(1) *If $g \in G$, then $|g|$ divides $|G|$.*
(2) *In particular, $g^{|G|} = e$.*
(3) *If $|G| = p$, p a prime, then $G \cong \mathbb{Z}_p$.*

PROOF. Since $\langle g \rangle$ is a subgroup and Theorem 2.21 shows $|\langle g \rangle| = |g|$, Lagrange's Theorem implies $|g|$ divides $|G|$, which finishes part (1). To see $g^{|G|} = e$ for part (2), simply write $|G| = |g|\,k$ for some $k \in \mathbb{N}$. Then

$$g^{|G|} = \left(g^{|g|}\right)^k = e^k = e.$$

For part (3), pick $g \in G$ so $g \neq e$. Then $|g|$ divides $|G|$ with $|g| > 1$. Since $|G|$ is prime, it follows that $|g| = |G|$ so that $\langle g \rangle = G$. Use Theorem 2.21 to see $G \cong \mathbb{Z}_p$. $\square$

Corollary 2.29 has many nice applications. For instance, consider $[a] \in U_n$ for $n \in \mathbb{N}$ with $n > 1$. Since $|U_n| = \varphi(n)$, Corollary 2.29 shows $[a]^{\varphi(n)} = [1]$. This, of course, is Euler's Theorem (and Fermat's Little Theorem when n is prime) proved completely effortlessly.

As a final remark, it can easily happen that n divides $|G|$, $n \in \mathbb{N}$, but that no subgroup of order n exists (Exercise 2.211). However as a partial converse, we will see in Section 9 that when $|G| = p^n m$ with p a prime and $(p, m) = 1$, there exist subgroups of order p^k for each $0 \leq k \leq n$.

5.3. Normality. Let G be a group and let H be a subgroup. As already mentioned, we want to define a group structure on the set of left cosets, G/H, by defining $(g_1 H)(g_2 H) = (g_1 g_2) H$ for $g_1, g_2 \in G$. Theorem 2.31 below will tell us exactly when this is a well defined operation.

To motivate the theorem and accompanying definitions, suppose we define $(g_1 H)(g_2 H) = (g_1 g_2) H$. The problem with this definition is that we are showing a bias in our choice of representatives g_1 and g_2 to describe the left cosets g_1H and g_2H. What if we instead use the representatives g_1h_1 and g_2h_2 to describe the left cosets $g_1H = g_1h_1H$ and $g_2H = g_2h_2H$ where $h_1, h_2 \in H$ (Lemma 2.27)? In that case, our definition requires that

$$(g_1g_2)\,H = (g_1H)\,(g_2H) = (g_1h_1H)\,(g_2h_2H) = (g_1h_1g_2h_2)\,H.$$

In particular, this means that $(g_1g_2)\,H = (g_1h_1g_2h_2)\,H$. In turn, this happens if and only if $(g_1g_2)^{-1}(g_1h_1g_2h_2) \in H$. Since $(g_1g_2)^{-1}(g_1h_1g_2h_2) = g_2^{-1}h_1g_2h_2$, we see that our definition of multiplication of cosets can only be well defined if

$$g_2^{-1}h_1g_2 \in H$$

for any $g_2 \in G$ and $h_1 \in H$. This condition on H will be called *normality* below and it will turn out that G/H inherits a group structure only when H is *normal.*

DEFINITION 2.30. Let G be a group, let H be a subgroup of G, and let $g \in G$.

(1) The *right coset* of g with respect to H is $Hg = \{hg \mid h \in H\}$.

(2) H is called *normal* if $gHg^{-1} \subseteq H$ for all $g \in G$ where $gHg^{-1} = \{ghg^{-1} \mid h \in H\}$.

It is common to write $H \lhd G$ to indicate that H is a normal subgroup of G and to write $H\backslash G$ for the set of right cosets. Unfortunately, the notation for the set of right cosets, $H\backslash G = \{Hg \mid g \in G\}$, conflicts with the set theory notation $A\backslash B = \{a \in A \mid a \notin B\}$ so the reader will have to judge from the context which definition is appropriate. If it is any consolation, we will almost never use right cosets outside of this section and its exercises.

THEOREM 2.31. *Let G be a group and let H be a subgroup of G. The following are equivalent.*

(1) *H is normal.*

(2) *$gHg^{-1} = H$ for all $g \in G$.*

(3) *Left cosets are right cosets; i.e., $gH = Hg$ for all $g \in G$.*

(4) *$gH \subseteq Hg$ for each $g \in G$.*

(5) *$Hg \subseteq gH$ for each $g \in G$.*

(6) *The equation*

$$(g_1H)\,(g_2H) = (g_1g_2)\,H$$

gives a well defined binary operation on G/H where $g_1, g_2 \in G$.

PROOF. For (1) $\Longrightarrow$ (2), assume $gHg^{-1} \subseteq H$ for all $g \in G$. In particular, $g^{-1} \in G$ so $g^{-1}Hg \subseteq H$. Conjugating both sides with c_g shows $H \subseteq gHg^{-1}$ as desired. (2) $\Longrightarrow$ (1) is clear. For (2) $\Longrightarrow$ (3), simply multiply by g on the right while (3) $\Longrightarrow$ (2) is achieved by multiplying by g^{-1} on the right. Similarly, (1) $\Longleftrightarrow$ (4) and (1) $\Longleftrightarrow$ (5) are done analogously.

For part (6), recall that the most general choice of representative for g_iH is g_ih_i for any $h_i \in H$. Thus the supposed definition

$$(g_1H)\,(g_2H) = (g_1g_2)\,H$$

is well defined if and only if we also have

$$(g_1h_1H)\,(g_2h_2H) = (g_1g_2)\,H$$

for all $h_i \in H$. However, the definition of multiplication of left cosets says

$$(g_1h_1H)\,(g_2h_2H) = (g_1h_1g_2h_2)\,H.$$

Therefore the definition of multiplication is well defined if and only if

$$(g_1h_1g_2h_2)\,H = (g_1g_2)\,H.$$

By Lemma 2.27, this happens if and only if $g_1h_1g_2h_2 = g_1g_2h$ for some $h \in H$. Multiplying by g_1^{-1} on the left, h_2^{-1} on the right, and writing $h' = hh_2^{-1}$, the definition in question is well defined if and only if $h_1g_2 = g_2h'$ for some $h' \in H$. Thus the equation in part (6) gives a well defined binary operation if and only if $Hg_2 \subseteq g_2H$. In particular, (6) $\Longleftrightarrow$ (4). □

Note that the equation $gH = Hg$ does not mean that $gh = hg$ for all h. Rather, it is an equality of sets so that $\{gh \mid h \in H\} = \{h'g \mid h' \in H\}$. In other words, it means that for each $h \in H$ there is some $h' \in H$ so that $gh = h'g$.

Theorem 2.31 allows us to put a group structure on G/H when H is normal. Associativity follows from the associativity of G, the identity element is clearly eH, and the inverse to gH is clearly $g^{-1}H$ where $g \in G$.

DEFINITION 2.32. Let G be a group and let H be a normal subgroup of G. Under the bilinear operation

$$(g_1H)\,(g_2H) = (g_1g_2)\,H$$

for $g_1, g_2 \in G$, G/H is a group and is called the *quotient group* or *factor group*.

Our first example, of course, is $\mathbb{Z}_n$. Here the notation is additive and we start with $G = (\mathbb{Z}, +)$ and $H = n\mathbb{Z}$. As $\mathbb{Z}$ is abelian, any subgroup is normal. In particular, $\mathbb{Z}/n\mathbb{Z}$ is therefore a group. By definition, of course, $\mathbb{Z}/n\mathbb{Z} = \mathbb{Z}_n$ as groups.

As a second example, look again at the abelian group $G = \left(\mathbb{R}^2, +\right)$ and the (automatically normal) subgroup $W = \{(x,0) \mid x \in \mathbb{R}\}$. Recall that we have seen in §5.1 that there is a bijection $\varphi : \mathbb{R}^2/W \to \mathbb{R}$ given by $\varphi\left((x,y) + W\right) = y$. In fact, the group structure on $\mathbb{R}^2/W$ makes φ into an isomorphism so that $\mathbb{R}^2/W \cong \mathbb{R}$:

$$\begin{aligned}\varphi\left[((x,y)+W) + ((x',y')+W)\right] &= \varphi\left[(x+x', y+y') + W\right] = y + y' \\ &= \varphi\left[(x,y)+W\right] + \varphi\left[(x',y')+W\right].\end{aligned}$$

As a third example, look at $D_3 = \langle R_1, W_1 \rangle$ and its rotational subgroup $\mathcal{R}_3$. We have seen that $D_3/\mathcal{R}_3$ consists of two elements: $\mathcal{R}_3$ and $W_1\mathcal{R}_3$. To see $\mathcal{R}_3$ is a normal subgroup, it suffices to verify the normality condition for the generators of D_3 (Exercise 2.133). Namely, it suffices to show that $R_1\mathcal{R}_3R_1^{-1} \subseteq \mathcal{R}_3$ (which is obvious) and that $W_1\mathcal{R}_3W_1^{-1} \subseteq \mathcal{R}_3$. However by Theorem 2.4, we already know that $W_1R_jW_1 = R_{-j}$. Thus $\mathcal{R}_3$ is normal and $D_3/\mathcal{R}_3$ is a group. Moreover, it is easy to explicitly calculate the group law from the definitions (note that $\mathcal{R}_3 = I\mathcal{R}_3$):

$$\begin{aligned}(\mathcal{R}_3)\,(\mathcal{R}_3) &= \left(I^2\right)\mathcal{R}_3 = \mathcal{R}_3, \\ (W_1\mathcal{R}_3)\,(W_1\mathcal{R}_3) &= \left(W_1^2\right)\mathcal{R}_3 = \mathcal{R}_3, \\ (\mathcal{R}_3)\,(W_1\mathcal{R}_3) &= (IW_1)\mathcal{R}_3 = W_1\mathcal{R}_3, \\ (W_1\mathcal{R}_3)\,(\mathcal{R}_3) &= (W_1I)\mathcal{R}_3 = W_1\mathcal{R}_3.\end{aligned}$$

As a result, it follows that the map $\varphi : D_3/\mathcal{R}_3 \to (\mathbb{Z}_2, +)$ given by $\varphi(\mathcal{R}_3) = [0]$ and $\varphi(W_1\mathcal{R}_3) = [1]$ is an isomorphism. Hence $D_3/\mathcal{R}_3 \cong \mathbb{Z}_2$.

5.4. Direct Products and the Correspondence Theorem. As can be seen from our discussion above, normal subgroups are clearly defined to be used in quotient group constructions. However, they also have other applications. A fairly useful one is the following theorem. A generalization will be given in Theorem 2.76.

THEOREM 2.33. *Suppose M and N are normal subgroups of G. Writing MN for $\{mn \mid m \in M \text{ and } n \in N\}$, suppose $G = MN$ and $M \cap N = \{e\}$. Then*

$$G \cong M \times N.$$

In particular, elements of M and N commute with each other.

PROOF. Consider the map $\varphi : M \times N \to G$ given by $\varphi(m,n) = mn$. By hypothesis, the map is surjective. Furthermore, if $mn = e$, then $m = n^{-1}$. Thus $m, n \in M \cap N$ so that $m = n = e$ and $\ker \varphi = \{(e,e)\}$. If we knew that φ were a homomorphism, this would show that φ is injective and therefore an isomorphism. It therefore remains only to show that φ is a homomorphism. Since

$$\varphi\left((m,n)\left(m',n'\right)\right) = \varphi\left(mm',nn'\right) = mm'nn'$$

and

$$\varphi\left(m,n\right)\varphi\left(m',n'\right) = mnm'n',$$

it is necessary to show that M and N commute, i.e., that $mn = nm$ for all $m \in M$ and $n \in N$. To this end, consider the element $mnm^{-1}n^{-1}$ (which we hope to show is the identity element). Since $mnm^{-1} \in N$ and $nm^{-1}n^{-1} \in M$ by normality, it follows that $\left(mnm^{-1}\right)n^{-1} = m\left(nm^{-1}n^{-1}\right) \in M \cap N$ so $mnm^{-1}n^{-1} = e$ as desired. □

As usual, when groups are written additively, it is a common practice to use the direct sum notation instead of the direct product notation. For instance instead of MN, the notation $M + N = \{m + n \mid m \in M,\ n \in N\}$ is employed. In that case, the above theorem says that when M and N are normal subgroups of G with $M \cap N = \{0\}$ and $G = M + N$, then $G \cong M \oplus N$.

As an example, consider the case of $G = \mathbb{Z}_6$ and

$$M = 2\mathbb{Z}_6 = \{2[x] \mid [x] \in \mathbb{Z}_6\} = \{[0],[2],[4]\},$$
$$N = 3\mathbb{Z}_6 = \{3[x] \mid [x] \in \mathbb{Z}_6\} = \{[0],[3]\}.$$

Both M and N are subgroups and automatically normal since $\mathbb{Z}_6$ is abelian. By inspection, $M \cap N = \{0\}$ and $\mathbb{Z}_6 = 2\mathbb{Z}_6 + 3\mathbb{Z}_6$. Thus $\mathbb{Z}_6 \cong 2\mathbb{Z}_6 \oplus 3\mathbb{Z}_6$. In order to make this more familiar, it is worth noting that $2\mathbb{Z}_6 \cong \mathbb{Z}_3$ by the map $[x]_6 \to \left[\frac{x}{2}\right]_3$ and that $3\mathbb{Z}_6 \cong \mathbb{Z}_2$ by the map $[x]_6 \to [\frac{x}{3}]_2$. The straightforward details are left to the reader (Exercise 2.156). We therefore get the isomorphism $\mathbb{Z}_6 \cong \mathbb{Z}_3 \oplus \mathbb{Z}_2$.

To end this subsection, we give a useful theorem for passing back and forth between a group and its quotient.

THEOREM 2.34 (Correspondence Theorem). *Let N be a normal subgroup of G. Write*

$$\mathcal{S}_{G,N} = \{\textit{subgroups of } G \textit{ containing } N\},$$
$$\mathcal{S}_{G/N} = \{\textit{subgroups of } G/N\},$$

and

$$\pi : G \to G/N$$

for the projection map *given by $\pi(g) = gN$, $g \in G$.*

(1) *The map* π *is a surjective homomorphism.*

(2) *There is a bijection from* $\mathcal{S}_{G,N}$ *to* $\mathcal{S}_{G/N}$ *mapping* $H \in \mathcal{S}_{G,N}$ *to* $\overline{H} \in \mathcal{S}_{G/N}$ *where* $\overline{H}$ *is given by*

$$\overline{H} = \pi(H) = \{hN \mid h \in H\}.$$

The inverse maps $\overline{H} \in \mathcal{S}_{G/N}$ *to* $H \in \mathcal{S}_{G,N}$ *where* H *is given by*

$$H = \pi^{-1}(\overline{H}) = \{g \in G \mid gN \in \overline{H}\}.$$

(3) *The above bijection preserves normality; i.e.,* $H \in \mathcal{S}_{G,N}$ *is normal in* G *if and only if the corresponding* $\overline{H} \in \mathcal{S}_{G/N}$ *is normal in* G/N.

PROOF. For part (1), calculate

$$\pi(gg') = (gg')N = (gN)\,(g'N) = \pi(g)\pi(g').$$

Surjectivity is clear by the definition of G/N.

For part (2), first notice that if $H \in \mathcal{S}_{G,N}$, then $\pi(H)$ is the image of a group under a homomorphism and so $\pi(H) \in \mathcal{S}_{G/N}$. Similarly, if $\overline{H} \in \mathcal{S}_{G/N}$, then $\pi^{-1}(\overline{H})$ is easily seen to be a subgroup (cf. Exercise 2.102). Moreover if $n \in N$, $\pi(n) = nN = e_G N = e_{G/N}$. Therefore $\pi^{-1}(\overline{H}) \supseteq \pi^{-1}(e_{G/H}) \supseteq N$ so that $\pi^{-1}(\overline{H}) \in \mathcal{S}_{G,N}$. It only remains to see these two operations invert each other. First calculate

$$\pi(\pi^{-1}(\overline{H})) = \{kN \mid k \in \pi^{-1}(\overline{H})\} = \{kN \mid kN \in \overline{H}\} = \overline{H}.$$

Next calculate

$$\begin{aligned}\pi^{-1}(\pi(H)) &= \{g \in G \mid gN \in \pi(H)\} = \{g \in G \mid gN \in \{hN \mid h \in H\}\} \\ &= \{g \in G \mid g = hn \text{ for some } h \in H, n \in N\} = H\end{aligned}$$

since $e \in N \subseteq H$.

For part (3), suppose $H \in \mathcal{S}_{G,N}$ is a normal subgroup of G. Then

$$(gN)\pi(H)(g^{-1}N) = \{ghg^{-1}N \mid h \in H\} = \pi(gHg^{-1}) = \pi(H)$$

for $g \in G$ so that $\pi(H)$ is normal in G/H. Conversely, suppose $\overline{H} \in \mathcal{S}_{G/N}$ is a normal subgroup of G/H with $k \in \pi^{-1}(\overline{H})$. Then

$$\pi(gkg^{-1}) = \pi(g)\pi(k)\pi(g)^{-1} \subseteq \pi(g)\overline{H}\pi(g)^{-1} \subseteq \overline{H}$$

so that $gkg^{-1} \in \pi^{-1}(\overline{H})$. It follows that $g\left(\pi^{-1}(\overline{H})\right)g^{-1} \subseteq \pi^{-1}(\overline{H})$ as desired (cf. Exercise 2.137). □

For example, recall that we have seen that $D_3/\mathcal{R}_3 \cong \mathbb{Z}_2$. Since there are obviously only two subgroups of $\mathbb{Z}_2$, namely $\{[0]\}$ and $\mathbb{Z}_2$, it follows that there are only two subgroups of D_3 containing $\mathcal{R}_3$, namely $\mathcal{R}_3$ and all of D_3 (of course this can also be easily deduced from Lagrange's Theorem).

5.5. First Isomorphism Theorem. One of the main uses for quotient groups stems from the following theorem.

THEOREM 2.35 (First Isomorphism Theorem). *Let* $\varphi : G \to H$ *be a homomorphism of groups.*

(1) $\ker\varphi$ *is a normal subgroup of* G.

(2) $G/\ker\varphi \cong \operatorname{Im}\varphi$. *In particular, if* φ *is surjective, then*

$$G/\ker\varphi \cong H.$$

PROOF. For part (1), suppose $k \in \ker\varphi$ and $g \in G$. Then, by definition of the kernel,

$$\varphi(gkg^{-1}) = \varphi(g)\varphi(k)\varphi(g)^{-1} = \varphi(g)e_H\varphi(g)^{-1} = \varphi(g)\varphi(g)^{-1} = e_H.$$

Thus $g(\ker\varphi)g^{-1} \subseteq \ker\varphi$ and $\ker\varphi$ is normal.

For part (2), attempt to define a map $\widetilde{\varphi} : G/\ker\varphi \to H$ by $\widetilde{\varphi}(g\ker\varphi) = \varphi(g)$. To see this map is well defined, we must show it is independent of the choice of representative for $g\ker\varphi$. Namely, we must show $\widetilde{\varphi}(g\ker\varphi) = \widetilde{\varphi}(gk\ker\varphi)$ for any $k \in \ker\varphi$. However, by definition,

$$\widetilde{\varphi}(gk\ker\varphi) = \varphi(gk) = \varphi(g)\varphi(k) = \varphi(g)e_H = \varphi(g) = \widetilde{\varphi}(g\ker\varphi)$$

as needed.

To see $\widetilde{\varphi}$ is a homomorphism, follow the definitions to see

$$\begin{aligned}\widetilde{\varphi}\left((g_1\ker\varphi)\,(g_2\ker\varphi)\right) &= \widetilde{\varphi}\left((g_1g_2)\ker\varphi\right) = \varphi(g_1g_2) = \varphi(g_1)\varphi(g_2)\\ &= \widetilde{\varphi}(g_1\ker\varphi)\widetilde{\varphi}(g_2\ker\varphi).\end{aligned}$$

By construction, φ and $\widetilde{\varphi}$ clearly have the same image. Thus we have a well defined homomorphism $\widetilde{\varphi} : G/\ker\varphi \to \operatorname{Im}\varphi$ that is onto.

It remains only to see $\widetilde{\varphi}$ is one-to-one. For this, it suffices to show $\ker\widetilde{\varphi} = \{e_G\ker\varphi\}$ by Theorem 2.25. To this end, observe that $g\ker\varphi \in \ker\widetilde{\varphi}$ if and only if $\widetilde{\varphi}(g\ker\varphi) = e_H$ if and only if $\varphi(g) = e_H$ if and only if $g \in \ker\varphi$ if and only if $g\ker\varphi = e_G\ker\varphi$ as desired. □

As a boringly repetitive example, let $n \in \mathbb{N}$ and consider the surjective homomorphism $\varphi : \mathbb{Z} \to \mathbb{Z}_n$ given by $\varphi(k) = [k]$, $k \in \mathbb{Z}$. Since $\ker\varphi = n\mathbb{Z}$, the First Isomorphism Theorem shows $\mathbb{Z}/n\mathbb{Z} \cong \mathbb{Z}_n$ (which we already knew and is essentially a tautology).

As another example, consider the surjective homomorphism $\varphi : \mathbb{Z}_6 \to \mathbb{Z}_3$ given by $\varphi([k]_6) = [k]_3$ (cf. Exercise 2.101). Clearly $\ker\varphi = 3\mathbb{Z}_6 = \{[0],[3]\}$ so that the First Isomorphism Theorem shows $\mathbb{Z}_6/3\mathbb{Z}_6 \cong \mathbb{Z}_3$.

As a final example, consider the surjective homomorphism $\det : GL(n,\mathbb{F}) \to (\mathbb{F}^\times,\cdot)$ given by the determinant. Since the identity element of $\mathbb{F}^\times$ is 1, it follows that

$$\ker(\det) = \{g \in GL(n,\mathbb{F}) \mid \det g = 1\} = SL(n,\mathbb{F}).$$

Thus $SL(n,\mathbb{F})$ is a normal subgroup of $GL(n,\mathbb{F})$ and the First Isomorphism Theorem shows $GL(n,\mathbb{F})/SL(n,\mathbb{F}) \cong \mathbb{F}^\times$.

5.6. Exercises 2.119–2.169.

5.6.1. *Cosets.*

EXERCISE 2.119. Calculate the index of the following subgroups and find a representative for each left coset:

(a) $\langle[5]\rangle$ in $\mathbb{Z}_{15}$,
(b) $\langle[5]\rangle$ in U_{21},
(c) $\mathbb{R}$ in $\mathbb{C}$,
(d) $\{(x,-x) \mid x \in \mathbb{R}\}$ in $\mathbb{R}^2$,
(e) $\left\langle \begin{pmatrix} 1 & 2 & 3 \\ 3 & 2 & 1 \end{pmatrix} \right\rangle$ in S_3,
(f) $\{\sigma \in S_n \mid \sigma(n) = n\}$ in S_n,
(g) $\langle W_1\rangle$ in D_n,

(h) $SL(n, \mathbb{R})$ in $GL(n, \mathbb{R})$,

(i) $\left\langle \begin{pmatrix} 1 & 1 \\ 0 & 1 \end{pmatrix} \right\rangle$ in $GL(2, \mathbb{Z}_2)$.

EXERCISE 2.120. Let G be a group and let H be a subgroup. Show $|G/H| = |H \backslash G|$ (here $H \backslash G$ stands for the set of right cosets). *Hint:* Map gH to Hg^{-1}.

EXERCISE 2.121. **(a)** For $n \in \mathbb{N}$, show there exits a subgroup H of $(\mathbb{Q}^+, \cdot)$ so that $|\mathbb{Q}^+/H| = n$. *Hint:* Consider, say, $\langle \{2^n, 3, 5, 7, 11, ...\} \rangle$.

(b) Suppose $H \neq \mathbb{Q}$ is a subgroup of $\mathbb{Q}$. Show $|\mathbb{Q}/H| = \infty$. *Hint:* If $|\mathbb{Q}/H| = n < \infty$, show $nq \in H$ for $q \in \mathbb{Q}$. Now look at $n\frac{q}{n}$.

(c) Conclude that $\mathbb{Q}^+ \not\cong \mathbb{Q}$.

5.6.2. *Lagrange's Theorem.*

EXERCISE 2.122. What are the possible orders of subgroups of:

(a) $\mathbb{Z}_{20}$,

(b) D_6,

(c) S_5,

(d) $\mathbb{Z}_{12} \times S_3$.

EXERCISE 2.123. Let $p \in \mathbb{N}$ be a prime and let G be a group with $|G| = p$. Show G has exactly two subgroups: $\{e\}$ and G with nothing in-between.

EXERCISE 2.124. **(a)** If H and K are finite subgroups of G, show $|H \cap K|$ divides $(|H|, |K|)$.

(b) If $(|H|, |K|) = 1$, show $H \cap K = \{e\}$.

(c) If $|H|$ is prime, show $H \cap K$ is either $\{e\}$ or H.

EXERCISE 2.125. Suppose $K \subseteq H \subseteq G$ is a chain of subgroups. Show $|G/K| = |G/H|\,|H/K|$.

EXERCISE 2.126 (Cauchy's Theorem for Finite Abelian Groups). Let G be a finite abelian group and let p be a positive prime dividing the order of G. This exercise shows G has an element of order p (cf. Corollary 2.70).

(a) Write $G = \{g_1, \ldots, g_n\}$ with $|g_i| = d_i$. Let $H = \prod_{i=1}^n \mathbb{Z}_{d_i}$ and define $\varphi : H \to G$ by $\varphi(k_1, \ldots, k_n) = g_1^{k_1} \cdots g_n^{k_n}$. Show φ is a well defined surjective homomorphism.

(b) Using $|H| = |G|\,|\ker \varphi|$, show there exists i, $1 \leq i \leq n$, so $p \mid d_i$.

(c) Show G has an element of order p. *Hint:* $g_i^{d_i/p}$.

EXERCISE 2.127. Let $p \in \mathbb{N}$ be a prime with $p > 2$ and let $p_1, \ldots, p_k$ be the distinct prime factors of $p - 1$. Suppose $[c] \in U_p$ satisfies $[c]^{(p-1)/p_i} \neq [1]$ for $1 \leq i \leq k$. Show that $[c]$ generates U_p.

EXERCISE 2.128. **(a)** Suppose H is a subgroup of G. Show $H = G$ if and only if $|G/H| = 1$.

(b) Suppose H_i, $i \in \mathbb{N}$, is a subgroup if G with $H_i \supsetneq H_{i+1}$. Show $|G/\bigcap_{i=1}^\infty H_i| = \infty$. *Hint:* Generalize $|G/\bigcap_{i=1}^\infty H_i| = |G/H_1|\,|H_1/H_2|\,|H_2/\bigcap_{i=1}^\infty H_i|$.

EXERCISE 2.129. **(a)** Suppose H and K are subgroups of G. Define $i : K/(H \cap K) \to G/H$ by $i(k(H \cap K)) = kH$ for $k \in K$. Show i is well defined and injective.

(b) Suppose H_i, $i \in \mathbb{N}$, is a subgroup if G. Show

$$\left|G/\left(\bigcap_{i=1}^N H_i\right)\right| \leq \prod_{i=1}^{N} |G/H_i|.$$

Hint: Generalize the equality

$$\left|G/\left(\bigcap_{i=1}^N H_i\right)\right| = |G/H_1|\ |H_1/(H_1 \cap H_2)|\ \left|(H_1 \cap H_2)/\left(\bigcap_{i=1}^N H_i\right)\right|$$

(e.g., Exercise 2.125) and use part (a).

EXERCISE 2.130 (Groups of Order $2p$). Suppose $p \geq 3$ is prime and that $|G| = 2p$. This exercise shows that either $G \cong \mathbb{Z}_{2p}$ or that $G \cong D_p$.

(a) If $g \in G$, $g \neq e$, show $|g| \in \{2, p, 2p\}$.
(b) If there is $g \in G$ so $|g| = 2p$, show $G \cong \mathbb{Z}_{2p}$.
(c) Assume now $G \ncong \mathbb{Z}_{2p}$. If no g has order p, use Exercises 2.45 and 2.126 to get a contradiction.
(d) Choose $r \in G$ so $|r| = p$. If H is a subgroup of G, show $H \cap \langle r \rangle$ is either $\{e\}$ or $\langle r \rangle$.
(e) Let $w \in G \backslash \{\langle r \rangle \cup \{e\}\}$. Show $|w| = 2$.
(f) Show $\langle r \rangle$ is normal and that $wxw = x^k$, $1 \leq k < p$.
(g) Show $k^2 \equiv 1 \bmod(p)$ and conclude that $k \equiv \pm 1$.
(h) If $k \equiv 1$, show $xy = yx$ so that $|xy| = 2p$, a contradiction. If $k = -1$, use Exercise 2.118 to show $G \cong D_p$.

5.6.3. *Normality.*

EXERCISE 2.131. Show the following subgroups are normal:
(a) $\{R_{3j} \mid j \in \mathbb{Z}\}$ in D_{21},
(b) $SL(n, \mathbb{F})$ in $GL(n, \mathbb{F})$,
(c) $\{\begin{pmatrix} 1 & x \\ 0 & 1 \end{pmatrix} \mid x \in \mathbb{R}\}$ in $\{\begin{pmatrix} a & x \\ 0 & a^{-1} \end{pmatrix} \mid a, x \in \mathbb{R},\ a \neq 0\}$,
(d) $\{(g, e_H) \mid g \in G\}$ in $G \times H$,
(e) for $n \in \mathbb{N}$, $\{g \in G \mid g^n = e\}$,
(f) $\{R_{2j}, W_{2j} \mid j \in \mathbb{Z}\}$ in D_{2n}.

EXERCISE 2.132. Show the following subgroups are not normal:
(a) $\{\mathrm{diag}(c_1, c_2) \mid c_i \neq 0\}$ in $GL(2, \mathbb{R})$,
(b) $\{\begin{pmatrix} 1 & x \\ 0 & 1 \end{pmatrix} \mid x \in \mathbb{R}\}$ in $GL(2, \mathbb{R})$,
(c) $\left\langle \begin{pmatrix} 1 & 2 & 3 \\ 3 & 2 & 1 \end{pmatrix} \right\rangle$ in S_3.

EXERCISE 2.133 (Normality and Generators). If $G = \langle S \rangle$ is a group and H is a subgroup, show H is normal if and only if $sHs^{-1} \subseteq H$ for all $s \in S \cup S^{-1}$.

EXERCISE 2.134. Let G be a group and let H_α be a normal subgroup, α in some index set $\mathcal{A}$. Show $H = \bigcap_{\alpha \in \mathcal{A}} H_\alpha$ is a normal subgroup of G.

EXERCISE 2.135. If N and N' are normal subgroups of G and G', respectively, show $N \times N'$ is a normal subgroup of $G \times G'$.

EXERCISE 2.136. If H and N are subgroups of G with N normal, show $N \cap H$ is a normal subgroup of H.

EXERCISE 2.137. Suppose $\varphi : G \to H$ is a homomorphism between groups and that N is a normal subgroup of H. Show $\varphi^{-1}(N)$ is a normal subgroup of G (cf. Exercise 2.102).

EXERCISE 2.138. **(a)** If K is a normal subgroup of H and if H is a normal subgroup of G, show by example that K need not be normal in G.

(b) Show the center of a group is normal.

(c) If Z_H is the center of a normal subgroup H of G, show Z_H is normal in G.

EXERCISE 2.139. Let H be a subgroup of G. The *normalizer* of H in G is $N(H) = \{g \in G \mid gHg^{-1} = H\}$ (cf. Exercise 2.72).

(a) Show H is normal in $N(H)$.

(b) Show $N(H)$ is the largest subgroup of G containing H in which H is normal.

EXERCISE 2.140 (Internal Product). If H and N are subgroups of G, let $NH = \{nh \mid n \in N \text{ and } h \in H\}$.

(a) Show by example that NH is not always a subgroup. *Hint:* Look at a nonabelian example.

(b) If N is normal, show NH is a subgroup.

(c) If N and H are normal, show NH is normal.

EXERCISE 2.141 (Index 2 Subgroups Are Normal). If H is a subgroup of G of index 2, show H is normal. *Hint:* Write $G = H \coprod (g_0H)$ for $g_0 \notin H$ and consider what happens if $g_0Hg_0^{-1} \cap g_0H \neq \varnothing$.

EXERCISE 2.142. Suppose G is a group and that H is the only subgroup of order $|H|$. Show H is normal. *Hint:* Only if you absolutely need it: Exercise 2.71.

EXERCISE 2.143. Let G be a group and let $\Delta = \{(g, g) \mid g \in G\} \subseteq G \times G$, the *diagonal subgroup*. Show Δ is normal in $G \times G$ if and only if G is abelian.

EXERCISE 2.144 (Normal subgroups of D_n). **(a)** Show $\mathcal{R}_n \cong \mathbb{Z}_n$ and conclude that subgroups of $\mathcal{R}_n$ are of the form $\{R_{k_0j} \mid j \in \mathbb{Z}\}$ for k_0 a positive divisor of n (cf. Exercise 2.63).

(b) Show that any subgroup of $\mathcal{R}_n$ is a normal subgroup of D_n.

(c) If N is a normal subgroup of D_n containing W_{j_0}, show N contains $\{W_{j_0+2j} \mid j \in \mathbb{Z}\}$ and $\{R_{2j} \mid j \in \mathbb{Z}\}$.

(d) If n is odd, show that the only normal subgroup not contained in $\mathcal{R}_n$ is D_n.

(e) If n is even, show there are three normal subgroups not contained in $\mathcal{R}_n$: D_n, $\{R_{2j}, W_{2j} \mid j \in \mathbb{Z}\}$, and $\{R_{2j}, W_{2j+1} \mid j \in \mathbb{Z}\}$.

EXERCISE 2.145. Show all subgroups of Q are normal (cf. Exercises 2.13 and 2.90).

EXERCISE 2.146. Find all normal subgroups of S_4.

EXERCISE 2.147. If G is a group, show $\text{Inn}(G)$ is a normal subgroup of $\text{Aut}(G)$ (cf. Exercise 2.115).

EXERCISE 2.148. Suppose G is a cyclic group with $|G| = mn$ for $m, n \in \mathbb{N}$ with $(m, n) = 1$ (cf. Exercise 2.153). Show $G \cong \mathbb{Z}_m \times \mathbb{Z}_n$. *Hint:* If $G = \langle g \rangle$, look at $M = \langle g^n \rangle$ and $N = \langle g^m \rangle$.

EXERCISE 2.149 (Core). Let H be a subgroup of G. The *core* of H (also called the *normal interior*) is

$$\operatorname{cor}(H) = \bigcap_{g \in G} c_g H.$$

Show $\operatorname{cor}(H)$ is the largest normal subgroup of G contained in H.

EXERCISE 2.150 (Semidirect Products). Let N and H be groups along with a homomorphism $\varphi : H \to \operatorname{Aut}(N)$ (cf. Exercise 2.115). For $n \in N$ and $h \in H$, write $\varphi_h(n)$ for $(\varphi(h))(n)$. Define the *semidirect product* (cf. Definition 2.75) of N and H as

$$N \rtimes H = \{(n,h) \mid n \in N,\, h \in H\}$$

with group law given by

$$(n_1, h_1)(n_2, h_2) = (n_1\varphi_{h_1}(n_2),\, h_1h_2).$$

Write $N' = \{(n, e_H) \mid n \in N\} \cong N$ and $H' = \{(e_N, h) \mid h \in H\} \cong H$.

(a) Show $N \rtimes H$ is a group.

(b) Show (i) N' is normal, (ii) $N'H' = N \rtimes H$ (cf. Exercise 2.140), and (iii) $N' \cap H' = \{e_{N\rtimes H}\}$.

(c) If $\varphi_h = \operatorname{Id}$ for all $h \in H$, show $N \rtimes H = N \times H$.

Conversely, suppose N and H are subgroups of G satisfying (i) N is normal, (ii) $NH = G$, and (iii) $N \cap H = \{e_G\}$. Define $\varphi : H \to \operatorname{Aut}(N)$ by $\varphi_h(n) = c_h(n)$.

(d) Show φ is a well defined homomorphism.

(e) Show the map $\psi : N \rtimes H \to G$ by $\psi(n,h) = nh$ establishes an isomorphism $G \cong N \rtimes H$.

Establish the following isomorphisms:

(f) $D_n \cong \mathbb{Z}_n \rtimes \mathbb{Z}_2$.

(g) $\{\text{affine motions of } \mathbb{R}^n\} \cong \mathbb{R}^n \rtimes GL(n,\mathbb{R})$ (cf. Exercise 2.29) with $\varphi_g(x) = gx$ for $g \in GL(n,\mathbb{R})$ and $x \in \mathbb{R}^n$.

(h) $O(n) \cong SO(n) \rtimes \mathbb{Z}_2$ (in fact, $O(n) \cong SO(n) \times \mathbb{Z}_2$ exactly when n is odd).

5.6.4. *Quotient Groups.*

EXERCISE 2.151. Establish the following isomorphisms:

(a) $\mathbb{Z}_{12}/\langle[4]\rangle \cong \mathbb{Z}_4$.

(b) $U_{16}/\langle[-1]\rangle \cong \mathbb{Z}_4$.

(c) $U_{16}/\langle[9]\rangle \cong \mathbb{Z}_2 \times \mathbb{Z}_2$.

(d) $\{\begin{pmatrix} a & x \\ 0 & a^{-1}\end{pmatrix} \mid a, x \in \mathbb{R},\, a \neq 0\}/\{\begin{pmatrix} 1 & x \\ 0 & 1\end{pmatrix} \mid x \in \mathbb{R}\} \cong \mathbb{R}^\times$.

(e) $D_8/\langle R_2\rangle \cong \mathbb{Z}_2 \times \mathbb{Z}_2$.

(f) $D_8/\langle R_4\rangle \cong D_4$.

(g) $\mathbb{R}^\times/\{\pm 1\} \cong \mathbb{R}^+$.

(h) $(G \times H)/\{(g, e_H) \mid g \in G\} \cong H$.

(i) $(\mathbb{Z} \times \mathbb{Z})/\langle(2,2)\rangle \cong \mathbb{Z} \times \mathbb{Z}_2$.

EXERCISE 2.152 ($G/Z(G)$ Cyclic Implies Abelian). Suppose G is a group and $G/Z(G)$ is cyclic.

(a) Show G is abelian. *Hint:* If $G/Z(G) = \langle gZ(G)\rangle$, first show $G = \{g^n z \mid z \in Z(G),\, n \in \mathbb{Z}\}$.

(b) Show the center of a nonabelian group cannot have prime index.

EXERCISE 2.153. Let p, q be distinct positive primes and suppose G is a group with $|G| = pq$.

(a) Show that either G has trivial center or that G is abelian. *Hint:* Exercise 2.152.

(b) If G is abelian, show G is isomorphic to $\mathbb{Z}_p \times \mathbb{Z}_q$. *Hint:* Exercise 2.126.

(c) Give an example of a nonabelian group of order $2 \cdot 3$.

EXERCISE 2.154. **(a)** If G is cyclic, show $|G/H| < \infty$ for any subgroup $H \neq \{e\}$.

Show the following groups are not cyclic:

(b) $\mathbb{Q}$,

(c) $\mathbb{Q}^+$,

(d) $\mathbb{Z} \oplus \mathbb{Z}$,

(e) $\{z \in \mathbb{C}^\times \mid z^n = 1,\, n \in \mathbb{N}\}$.

EXERCISE 2.155. Suppose G is a group and let G' be the subgroup generated by $\{g_1 g_2 g_1^{-1} g_2^{-1} \mid g_i \in G\}$, the *commutator subgroup* of G (cf. Exercise 2.73).

(a) Show G' is normal.

(b) Show G/G' is an abelian group.

(c) If H is a normal subgroup of G so that G/H is abelian, show $G' \subseteq H$.

(d) Conclude that G' is the unique smallest normal subgroup whose quotient with G is abelian.

5.6.5. *First Isomorphism Theorem.*

EXERCISE 2.156. Use the First Isomorphism Theorem to establish the following group isomorphisms:

(a) $\mathbb{Z}_n / \langle [m] \rangle \cong \mathbb{Z}_m$ for $m \mid n$.

(b) $(m\mathbb{Z}) / (n\mathbb{Z}) \cong \mathbb{Z}_{\frac{n}{m}}$ for $m \mid n$.

(c) $D_n / \langle R_1 \rangle \cong \mathbb{Z}_2$.

(d) $D_{2n} / \langle R_2 \rangle \cong \mathbb{Z}_2 \times \mathbb{Z}_2$.

(e) $D_n / \langle R_m \rangle \cong D_m$ for $m \mid n$ and $m \geq 3$.

(f) $\mathbb{R}/\mathbb{Z} \cong S^1$.

EXERCISE 2.157. Suppose N is a normal subgroup of G and let $g \in G$.

(a) As an element of G/N, show $|gN| = \min\{k \in \mathbb{N} \mid g^k \in N\}$.

(b) If g has finite order, show $|gN| \mid |g|$. *Hint:* Show $\langle gN \rangle \cong \langle g \rangle / (\langle g \rangle \cap N)$.

EXERCISE 2.158. If N is a normal subgroup of G and all elements of G/N and N have finite order, show every element of G has finite order. *Hint:* Exercise 2.157.

EXERCISE 2.159. Let G be an abelian group, $n \in \mathbb{Z}_{\geq 0}$, $U = \{g \in G \mid g^n = e\}$, and $P = \{g^n \mid g \in G\}$.

(a) Show $P \cong G/U$.

(b) If G is finite, show $|G| = |P|\,|U|$.

EXERCISE 2.160. For $a, b \in \mathbb{Z}^\times$, set $d = (a, b)$. This exercise shows

$$(\mathbb{Z} \times \mathbb{Z}) / \langle (a, b) \rangle \cong \mathbb{Z}_d \times \mathbb{Z}.$$

(a) Let $\varphi : \mathbb{Z} \times \mathbb{Z} \to \mathbb{Z}$ be given by $\varphi(x, y) = \frac{b}{d}x - \frac{a}{d}y$. Show φ is a surjective homomorphism with $\ker \varphi = \left\langle \left(\frac{a}{d}, \frac{b}{d}\right) \right\rangle$.

(b) In the special case of $(a,b)=1$, conclude that $(\mathbb{Z}\times\mathbb{Z})\,/\,\langle(a,b)\rangle\cong\mathbb{Z}$.

(c) In general, choose $x_0,y_0\in\mathbb{Z}$ so $\frac{b}{d}x_0-\frac{a}{d}y_0=1$. For $k\in\mathbb{Z}$, show $\varphi^{-1}(k)=\{k(x_0,y_0)+j(\frac{a}{d},\frac{b}{d})\mid j\in\mathbb{Z}\}$. Conclude that each element of $\mathbb{Z}\times\mathbb{Z}$ can be uniquely written as $k(x_0,y_0)+j(\frac{a}{d},\frac{b}{d})$ for $k,j\in\mathbb{Z}$.

(d) Define $\psi:\mathbb{Z}\times\mathbb{Z}\to\mathbb{Z}_d\times\mathbb{Z}$ by $\psi(k(x_0,y_0)+j(\frac{a}{d},\frac{b}{d}))=([j],k)$. Show ψ is a surjective homomorphism with $\ker\psi=\langle(a,b)\rangle$. Conclude that $(\mathbb{Z}\times\mathbb{Z})\,/\,\langle(a,b)\rangle\cong\mathbb{Z}_d\times\mathbb{Z}$.

EXERCISE 2.161. Let $m,n\in\mathbb{N}$ with $m\mid n$. This exercise shows

$$U_n/\{[k]_n\in U_n\mid k\equiv 1 \bmod m\}\cong U_m.$$

(a) Let $\varphi_m:U_n\to U_m$ be given by $\varphi([k]_n)=[k]_m$. Show φ_m is a well defined homomorphism with $\ker\varphi_m=\{[k]_n\in U_n\mid [k]_m=[1]_m\}$.

(b) Suppose p is a positive prime dividing n. If $p^2\mid n$, show $(k,n)=1$ if and only if $(k,\frac{n}{p})=1$. If $p^2\nmid n$ and $p\nmid k$, show $(k,n)=1$ if and only if $(k,\frac{n}{p})=1$.

(c) Show $\varphi_{\frac{n}{p}}:U_n\to U_{\frac{n}{p}}$ is surjective. *Hint:* To hit $[k]_{\frac{n}{p}}\in U_{\frac{n}{p}}$ with $\varphi_{\frac{n}{p}}$, look at either $[k]_n$ or $[k+\frac{n}{p}]_n$.

(d) Use successive composition of maps to show $\varphi_m:U_n\to U_m$ is surjective. Conclude $U_n/\{[k]_n\in U_n\mid k\equiv 1\bmod(m)\}\cong U_m$.

EXERCISE 2.162 ($G/Z(G)\cong\mathrm{Inn}(G)$). Let G be a group. Show $G/Z(G)\cong\mathrm{Inn}(G)$ (cf. Exercise 2.115).

EXERCISE 2.163. Show a group G has a nontrivial homomorphism to $\mathbb{Z}_2$ if and only if G has a subgroup of index 2 (cf. Exercise 2.141).

EXERCISE 2.164. Up to isomorphism, write down all homomorphic images of $\mathbb{Z}_{24}$ (cf. Exercise 2.63).

EXERCISE 2.165. Show that a subgroup is normal if and only if it is the kernel of a homomorphism.

EXERCISE 2.166. Let G be an abelian group and let T be the set of all elements of G with finite order, the *torsion subgroup* (cf. Exercise 2.74). Show G/T is *torsion free*; i.e., show the torsion subgroup of G/T is trivial.

EXERCISE 2.167 (Second Isomorphism Theorem). Let N and H be subgroups of G with N normal.

(a) Show N is a normal subgroup of NH (cf. Exercise 2.140).

(b) Define $\varphi:H\to(NH)\,/N$ by $\varphi(h)=hN$ for $h\in H$. Show φ is a surjective homomorphism with $\ker\varphi=N\cap H$.

(c) Conclude that

$$(NH)\,/N\cong H/\,(N\cap H)\,.$$

EXERCISE 2.168 (Third Isomorphism Theorem). Let N and K be normal subgroups of G with $N\subseteq K$.

(a) Show K/N is a normal subgroup of G/N.

(b) Define $\varphi:G\to(G/N)\,/\,(K/N)$ by $\varphi(g)=(gN)\,(K/N)$ for $g\in G$. Show φ is a surjective homomorphism with $\ker\varphi=K$.

(c) Conclude that

$$(G/N)\,/\,(K/N)\cong G/K.$$

EXERCISE 2.169 ($PGL(n,\mathbb{F})$ and $PSL(n,\mathbb{F})$). Let $\mathbb{F}$ be one of $\mathbb{Q}$, $\mathbb{R}$, $\mathbb{C}$, or $\mathbb{Z}_p$ for $p \in \mathbb{N}$ prime and let $Z(n,\mathbb{F}) = Z(GL(n,\mathbb{F})) = \{cI_n \mid c \in \mathbb{F}^\times\}$ (cf. Exercise 2.85). Define the *Projective Linear Group* as

$$PGL(n,\mathbb{F}) = GL(n,\mathbb{F})/Z(n,\mathbb{F})$$

and the *Projective Special Linear Group* as

$$PSL(n,\mathbb{F}) = SL(n,\mathbb{F})/\left(Z(n,\mathbb{F}) \cap SL(n,\mathbb{F})\right).$$

If $\mathbb{F} = \mathbb{Z}_p$, $PSL(n,\mathbb{Z}_p)$ is sometimes written as $L_n(p)$.

(a) Show $Z(n,\mathbb{F}) \cong \mathbb{F}^\times$ and $Z(n,\mathbb{F}) \cap SL(n,\mathbb{F}) \cong \{\zeta \in \mathbb{F}^\times \mid \zeta^n = 1\}$.

(b) Show $PSL(n,\mathbb{F}) \cong \left(Z(n,\mathbb{F})SL(n,\mathbb{F})\right)/Z(n,\mathbb{F})$, a subgroup of $PGL(n,\mathbb{F})$. *Hint:* Exercise 2.167.

(c) Under this identification, show $PSL(n,\mathbb{F}) = PGL(n,\mathbb{F})$ if and only if every element of $\mathbb{F}$ is an nth power.

6. Fundamental Theorem of Finite Abelian Groups

When looking for finite abelian groups, surely the group $(\mathbb{Z}_n, +)$ comes immediately to mind. Obviously, we can also toss in a few direct products to mix things up a bit. The stunning and incredibly useful fact of this section is that these are the only finite abelian groups. Before working on this result, let us take a moment to look at how $\mathbb{Z}_n$ itself can sometimes be written as a direct product.

It should be noted that even though $\mathbb{Z}_n$ is a group written additively, it is slightly more common for people to use the direct product notation in this setting instead of the direct sum notation, though either choice is acceptable. Therefore, instead of writing the slightly more natural $\mathbb{Z}_6 \cong \mathbb{Z}_2 \oplus \mathbb{Z}_3$ (§5.3), we will write

$$\mathbb{Z}_6 \cong \mathbb{Z}_2 \times \mathbb{Z}_3 = \left\{([a]_2, [b]_3) \;\middle|\; [a]_2 \in \mathbb{Z}_2, [b]_3 \in \mathbb{Z}_3\right\}.$$

If the reader is disturbed at this blatant inconsistency, feel free to replace all the $\prod_{i=1}^k$ symbols by $\bigoplus_{i=1}^k$.

THEOREM 2.36. *Let $n, m \in \mathbb{N}$.*

(1) *Then $\mathbb{Z}_{nm} \cong \mathbb{Z}_n \times \mathbb{Z}_m$ if and only if $(n,m) = 1$.*

(2) *In particular, if $n = \prod_{i=1}^k p_i^{m_i}$ is the prime decomposition of n, then $\mathbb{Z}_n \cong \prod_{i=1}^k \mathbb{Z}_{p_i^{m_i}}$.*

PROOF. As part (2) clearly follows from part (1) via induction, consider only part (1). Look at necessity first. Write $d = (n,m)$ with $n = dn_1$ and $m = dm_1$ for $n_1, m_1 \in \mathbb{N}$. If $([a]_n, [b]_m) \in \mathbb{Z}_n \times \mathbb{Z}_m$, then $n[a]_n = [0]_n$ and $m[b]_m = [0]_m$. Thus

$$\frac{nm}{d}([a]_n, [b]_m) = (m_1 n[a]_n, n_1 m[b]_m) = ([0]_n, [0]_m).$$

In particular, the order of any element of $\mathbb{Z}_n \times \mathbb{Z}_m$ is at most $\frac{nm}{d}$. On the other hand, the element $[1]_{nm} \in \mathbb{Z}_{nm}$ has order nm. Thus if $\mathbb{Z}_{nm} \cong \mathbb{Z}_n \times \mathbb{Z}_m$, then, at the very least, $\frac{nm}{d} \geq nm$ so that $d = 1$.

Conversely, suppose $d = 1$ and consider the map $\varphi : \mathbb{Z}_{nm} \to \mathbb{Z}_n \times \mathbb{Z}_m$ given by $\varphi([k]_{nm}) = ([k]_n, [k]_m)$. This map is well defined as multiples of nm are clearly multiples of n and multiples of m. The map is trivially a homomorphism:

$$\begin{aligned}\varphi([k_1]_{nm} + [k_2]_{nm}) &= \varphi([k_1 + k_2]_{nm}) = ([k_1 + k_2]_n, [k_1 + k_2]_m)\\ &= ([k_1]_n, [k_1]_m) + ([k_2]_n, [k_2]_m) = \varphi([k_1]_{nm}) + \varphi([k_2]_{nm}).\end{aligned}$$

Next, if $[k]_{nm} \in \ker\varphi$, then $n \mid k$ and $m \mid k$. Since $(n, m) = 1$, this implies $(nm) \mid k$ (Exercise 1.18) so that the kernel is trivial. As φ is now injective and since the range and codomain have the same number of elements, it follows that φ is surjective as well and thus is an isomorphism. $\square$

For example, $\mathbb{Z}_6 \cong \mathbb{Z}_2 \times \mathbb{Z}_3$, but $\mathbb{Z}_4 \ncong \mathbb{Z}_2 \times \mathbb{Z}_2$. In any case, the decomposition $\mathbb{Z}_n \cong \prod_{i=1}^k \mathbb{Z}_{p_i^{m_i}}$ from Theorem 2.36 motivates the following definition in an effort to pick out pieces such as $\mathbb{Z}_{p_i^{m_i}}$.

DEFINITION 2.37. Let G be an abelian group and let $m \in \mathbb{N}$. Write

$$G(m) = \{g \in G \mid g^m = e\}.$$

The following lemma provides the muscle to classify all finite abelian groups. Part (3) is the hard part—probably the trickiest result we have yet undertaken.

LEMMA 2.38. *Let G be an abelian group.*

(1) *For $m \in \mathbb{N}$, $G(m)$ is a subgroup of G.*

(2) *If G is finite and $|G| = mn$ with $(m, n) = 1$, then $G \cong G(m) \times G(n)$.*

(3) *Suppose G is finite with $G = G(p^n)$ for a positive prime p and $n \in \mathbb{N}$. Choose $g \in G$ with maximal order. Then there exists a subgroup H of G so $G \cong \langle g \rangle \times H$. If n is minimal so that $G = G(p^n)$, then $\langle g \rangle \cong \mathbb{Z}_{p^n}$.*

PROOF. Part (1) is straightforward and is left to Exercise 2.174. For part (2), use Theorem 2.33. Since G is abelian, it only remains to show $G = G(m)G(n)$ and $G(m) \cap G(n) = \{e\}$. For the first condition, let $g \in G$. Pick $u, v \in \mathbb{Z}$ so $mu + nv = 1$ so that $g = g^{nv}g^{mu}$. It suffices to show $g^{nv} \in G(m)$ and $g^{mu} \in G(n)$. To see this, use Corollary 2.29 to calculate

$$(g^{nv})^m = (g^v)^{mn} = e \quad \text{and} \quad (g^{mu})^n = (g^u)^{nm} = e.$$

For the second condition, let $g \in G(m) \cap G(n)$. It follows that $|g| \mid m$ and $|g| \mid n$. But $(m, n) = 1$ so that $|g| = 1$ and thus $g = e$ as desired.

Part (3) is where we earn our stripes. To begin with, we might as well choose $n \in \mathbb{N}$ as small as possible with $G = G(p^n)$. In particular, the order of any element of G divides p^n and $|g| = p^n$. Next, let H be a maximal subgroup of G (with respect to inclusion) such that $H \cap \langle g \rangle = \{e\}$. Since G is finite, a moment's thought will show such a subgroup always exists. By Theorem 2.33, the whole problem reduces to showing $G = \langle g \rangle H$. Via proof by contradiction, suppose $G \backslash (\langle g \rangle H) \neq \varnothing$ and seek a contradiction.

In this scenario, we claim there exists $y \in G \backslash (\langle g \rangle H)$ with $h = y^p \in H$. Temporarily assuming this, we can finish the proof by considering the subgroup $K = \langle y, H \rangle$. By the maximality of H, $\langle y, H \rangle \cap \langle g \rangle \neq \{e\}$. Thus there is $i, j \in \mathbb{N}$, $1 \leq j < p^n$ and $1 \leq i < p$, and $h_1 \in H$ so $y^i h_1 = g^j$. Choose $u, v \in \mathbb{Z}$ so $iu + pv = 1$. Then

$$y = \left(y^i\right)^u \left(y^p\right)^v = \left(g^j h_1^{-1}\right)^u h^v \in \langle g \rangle H,$$

which is a contradiction.

It remains only to verify the claim that there exists $y \in G \backslash (\langle g \rangle H)$ with $y^p \in H$. To see this, choose $x \in G \backslash (\langle g \rangle H)$ with minimal order and write $|x| = p^k$, $k \in \mathbb{N}$ with $1 \leq k \leq n$. If $k = 1$, we are done. Otherwise, $|x^p| = p^{k-1} < |x|$. By minimality, there exists $m \in \mathbb{Z}$ and $h \in H$ so that $x^p = g^m h$. We would like to set $y = xg^{-\frac{m}{p}}$, but it is not yet clear that $p \mid m$ and so y might not be well defined.

To rectify this situation, take the p^{k-1} power of both sides of $x^p = g^m h$. This shows that $e = g^{mp^{k-1}} h^{p^{k-1}}$ so that $g^{mp^{k-1}} = h^{-p^{k-1}}$. Since $H \cap \langle g \rangle = \{e\}$, it follows that $g^{mp^{k-1}} = e$ (and $e = h^{p^{k-1}}$, though we do not care). As $|g| = p^n$, we must have $p \mid m$. Therefore we can legally define $y = xg^{-\frac{m}{p}}$. Since $x \notin \langle g \rangle H$, $y \notin \langle g \rangle H$ as well. However, $y^p = x^p g^{-m} = h$ as desired. □

If we write $G \cong G(m) \times G(n)$ as in part (3) of Lemma 2.38, we expect that $|G(m)| = m$ and $|G(n)| = n$. To see this, we need the following result (which will be significantly generalized in Corollary 2.70).

LEMMA 2.39 (Cauchy's Theorem for Finite Abelian Groups). *Let G be a finite abelian group and let p be a positive prime with $p \mid |G|$. Then G has an element of order p.*

PROOF. Write $G = \{g_1, \ldots, g_n\}$ with $|g_i| = d_i$. Let $H = \prod_{i=1}^n \mathbb{Z}_{d_i}$ and define $\varphi : H \to G$ by $\varphi(k_1, \ldots, k_n) = g_1^{k_1} \cdots g_n^{k_n}$. Since G is abelian, it is obvious that φ is a well defined surjective homomorphism. Therefore the First Isomorphism Theorem shows that $|H| = |G| \, |\ker \varphi|$. Since $p \mid |G|$, it follows that $p \mid |H|$. As $|H| = \prod_{i=1}^n d_i$, $p \mid d_i$ for some i, $1 \le i \le n$. Thus $g = g_i^{d_i/p}$ is well defined and, by Theorem 2.11, part (3), $|g| = p$ as desired. □

COROLLARY 2.40. *Let G be a finite abelian group. If $|G| = mn$ with $m, n \in \mathbb{N}$ and $(m, n) = 1$, then $|G(m)| = m$ and $|G(n)| = n$.*

PROOF. From Lemma 2.38, $G \cong G(m) \times G(n)$. Particularly, $mn = |G(m)| \, |G(n)|$. If p is a positive prime with $p^r \mid m$ for some $r \in \mathbb{N}$, we claim that $p^r \mid |G(m)|$. If not, we would necessarily have, at the very least, $p \mid |G(n)|$. By Cauchy's Theorem, this would imply that there exists a $g \in G(n)$ so that $|g| = p$. But this would mean that $g \in G(m)$, which would violate the fact that $G(m) \cap G(n) = \{e\}$. Thus $p^r \mid |G(m)|$. Looking at the prime decomposition of m, it follows that $|G(m)| \ge m$. Similarly, $|G(n)| \ge n$. As $|G(m)| \, |G(n)| = mn$, we must have $|G(m)| = m$ and $|G(n)| = n$. □

THEOREM 2.41 (Fundamental Theorem of Finite Abelian Groups). **(1)** *Each nontrivial finite abelian group is isomorphic to exactly one group of the form*

$$\left(\mathbb{Z}_{p_1^{n_{1,1}}} \times \cdots \times \mathbb{Z}_{p_1^{n_{1,k_1}}}\right) \times \left(\mathbb{Z}_{p_2^{n_{2,1}}} \times \cdots \times \mathbb{Z}_{p_2^{n_{2,k_2}}}\right) \times \cdots \times \left(\mathbb{Z}_{p_m^{n_{m,1}}} \times \cdots \times \mathbb{Z}_{p_m^{n_{m,k_m}}}\right)$$

where $m, k_i, n_{i,j} \in \mathbb{N}$, $p_1 < p_2 < \cdots < p_m$ are positive primes, and $n_{i,1} \ge n_{i,2} \ge \cdots \ge n_{i,k_i}$. The numbers $p_i^{n_{i,j}}$ are called the elementary divisors *of the group.*

(2) *Each nontrivial finite abelian group is isomorphic to exactly one group of the form*

$$\mathbb{Z}_{n_1} \times \mathbb{Z}_{n_2} \times \cdots \times \mathbb{Z}_{n_s}$$

where $s, n_i \in \mathbb{N}$, $n_i \ge 2$, and $n_1 \mid n_2 \mid \cdots \mid n_s$. The numbers n_i are called the invariant factors *of the group.*

PROOF. Suppose G is a finite abelian group and look at part (1) first. If $\prod_{i=1}^m p_i^{r_i}$ is the prime decomposition of $|G|$ with $p_1 < p_2 < \cdots < p_m$ positive primes, it follows from Lemma 2.38, part (2), Corollary 2.40, and induction that $G \cong \prod_{i=1}^m G(p_i^{r_i})$. It then follows from Lemma 2.38, part (3), and induction that $G(p_i^{r_i}) \cong \prod_{j=1}^{k_i} \mathbb{Z}_{p_i^{n_{i,j}}}$ for some $r_i \ge n_{i,1} \ge n_{i,2} \ge \cdots \ge n_{i,k_i}$. Thus, up to isomorphism, every finite abelian group appears on our list.

It remains only to see that isomorphic copies of the same group cannot appear in different positions on our list. In other words, we must show that isomorphic finite abelian groups have the same elementary divisors. Now if G and H are isomorphic copies, then, at the very least, $|G| = \prod_{i=1}^{m} p_i^{r_i} = |H|$. Since the order of an element is preserved under an isomorphism, it follows that $G(p_i^{r_i}) \cong H(p_i^{r_i})$. It therefore suffices to show that $G(p_i^{r_i})$ and $H(p_i^{r_i})$ have the same elementary divisors for each p_i.

Rephrasing the problem, it suffices to look at the special case of $G = G(p^N)$, p a positive prime and $N \in \mathbb{N}$, and to show that the elementary divisors of G are determined by quantities that are left unchanged by isomorphisms. Our approach will be to show that the elementary divisors are completely determined by the numbers $|G(p^n)|$, $n \in \mathbb{Z}_{\geq 0}$, which will suffice since, as already remarked, these numbers are left invariant under any isomorphism.

To this end, write the elementary divisors of G as

$$\overbrace{p^{n_1}, \ldots, p^{n_1}}^{k_1 \text{ copies}}, \overbrace{p^{n_2}, \ldots, p^{n_2}}^{k_2 \text{ copies}}, \ldots, \overbrace{p^{n_r}, \ldots, p^{n_r}}^{k_r \text{ copies}}$$

with $N \geq n_1 > n_2 > \cdots > n_r > 0$ so that we may assume

$$G = \prod_{i=1}^{r} \overbrace{\mathbb{Z}_{p^{n_i}} \times \cdots \times \mathbb{Z}_{p^{n_i}}}^{k_i \text{ copies}}.$$

Observe that for $m, n \in \mathbb{N}$, $m \leq n$, $\mathbb{Z}_{p^m} \cong \mathbb{Z}_{p^n}(p^m)$ under the map $[x]_{p^m} \to [p^{n-m}x]_{p^n}$ (Exercise 2.175) so that $|\mathbb{Z}_{p^n}(p^m)| = p^m$. If $m \geq n$, clearly $|\mathbb{Z}_{p^n}(p^m)| = p^n$. In particular, if $m < n$, then $m + 1 \leq n$ so that $|\mathbb{Z}_{p^n}(p^m)| = p^m$ and $\left|\mathbb{Z}_{p^n}(p^{m+1})\right| = p^{m+1}$. If $m \geq n$, then $m + 1 \geq n$ so that $|\mathbb{Z}_{p^n}(p^m)| = p^n$ and $\left|\mathbb{Z}_{p^n}(p^{m+1})\right| = p^n$. In summary,

$$\frac{\left|\mathbb{Z}_{p^n}(p^{m+1})\right|}{|\mathbb{Z}_{p^n}(p^m)|} = \begin{cases} p, & \text{if } m < n, \\ 1, & \text{if } m \geq n. \end{cases}$$

Next, for $n \in \mathbb{N}$, notice that

$$G(p^n) = \prod_{i=1}^{r} \mathbb{Z}_{p^{n_i}}(p^n) \times \cdots \times \mathbb{Z}_{p^{n_i}}(p^n).$$

Thus if $n_{j-1} > n \geq n_j$, it follows that

$$\frac{\left|G(p^{n+1})\right|}{|G(p^n)|} = p^{k_1 + k_2 + \cdots + k_{j-1}}.$$

In particular, the function $n \to \left|G(p^{n+1})\right| / |G(p^n)|$ stays constant (with a value of $p^{k_1+k_2+\cdots+k_{j-1}}$) when $n_{j-1} > n \geq n_j$ and then decreases (to a value of $p^{k_1+k_2+\cdots+k_{j-2}}$) when $n = n_{j-1}$.

Therefore the set $\{n_1, \ldots, n_r\}$ is determined by looking at the values of $n \in \mathbb{N}$ for which the function $n \to \left|G(p^{n+1})\right| / |G(p^n)|$ decreases. Once $\{n_1, \ldots, n_r\}$ is determined, the values of $\{k_1, \ldots, k_r\}$ can be read off inductively from the ratios $\left|G(p^{n_i+1})\right| / |G(p^{n_i})|$. This finishes part (1).

Though "proving" a theorem by looking at an example is one of the cardinal sins of mathematics, I won't tell if you won't. More seriously, the connection between part (1) and part (2) really is best seen by way of example (via Theorem

2.36). For instance, if we start with the group $(\mathbb{Z}_{2^3} \times \mathbb{Z}_{2^2} \times \mathbb{Z}_2) \times (\mathbb{Z}_5 \times \mathbb{Z}_5)$ written in the form of part (1), we can rewrite it in the form of part (2) as

$$\begin{aligned}(\mathbb{Z}_{2^3} \times \mathbb{Z}_{2^2} \times \mathbb{Z}_2) \times (\mathbb{Z}_5 \times \mathbb{Z}_5) &\cong (\mathbb{Z}_2) \times (\mathbb{Z}_{2^2} \times \mathbb{Z}_5) \times (\mathbb{Z}_{2^3} \times \mathbb{Z}_5) \\ &\cong \mathbb{Z}_2 \times \mathbb{Z}_{2^2 5} \times \mathbb{Z}_{2^3 5}.\end{aligned}$$

Going from the format of part (2) to part (1) obviously just reverses this process. Here is a messier example:

$$\begin{aligned}&(\mathbb{Z}_{2^7} \times \mathbb{Z}_{2^7} \times \mathbb{Z}_{2^3} \times \mathbb{Z}_2 \times \mathbb{Z}_2 \times \mathbb{Z}_2) \times (\mathbb{Z}_{3^5} \times \mathbb{Z}_{3^2}) \times (\mathbb{Z}_{5^2} \times \mathbb{Z}_{5^2} \times \mathbb{Z}_{5^2} \times \mathbb{Z}_{5^2}) \\ &\cong (\mathbb{Z}_2) \times (\mathbb{Z}_2) \times (\mathbb{Z}_2 \times \mathbb{Z}_{5^2}) \times (\mathbb{Z}_{2^3} \times \mathbb{Z}_{5^2}) \times (\mathbb{Z}_{2^7} \times \mathbb{Z}_{3^2} \times \mathbb{Z}_{5^2}) \times (\mathbb{Z}_{2^7} \times \mathbb{Z}_{3^5} \times \mathbb{Z}_{5^2}) \\ &\cong \mathbb{Z}_2 \times \mathbb{Z}_2 \times \mathbb{Z}_{2^1 5^2} \times \mathbb{Z}_{2^3 5^2} \times \mathbb{Z}_{2^7 3^2 5^2} \times \mathbb{Z}_{2^7 3^5 5^2}.\end{aligned}$$

In general, the equivalence of part (1) and part (2) follows in a fairly straightforward way and the official details are relegated to Exercise 2.182. □

6.1. Exercises 2.170–2.188.

EXERCISE 2.170. Up to isomorphism, list all abelian groups of the following orders:

(a) 10,
(b) 11,
(c) 72,
(d) 100,
(e) 360.

EXERCISE 2.171. Find the elementary divisors and invariant factors of the following groups:

(a) $\mathbb{Z}_{12}$,
(b) $\mathbb{Z}_{36}$,
(c) $\mathbb{Z}_{10} \times \mathbb{Z}_{28}$,
(d) $\mathbb{Z}_2 \times \mathbb{Z}_{130}$.

EXERCISE 2.172. Decompose the following groups as a direct product of cyclic groups:

(a) U_8,
(b) U_{25},
(c) U_{60}.

EXERCISE 2.173. Up to permutation of factors, write down every possible way to write the following groups as a product of cyclic groups:

(a) $\mathbb{Z}_{12}$,
(b) $\mathbb{Z}_{60}$.

EXERCISE 2.174. Let G be an abelian group and let $m \in \mathbb{N}$.

(a) Show $G(m)$ is a subgroup.
(b) If $G = \mathbb{Q}/\mathbb{Z}$, determine $G(m)$.

EXERCISE 2.175. Let $m, n \in \mathbb{N}$ with $m \leq n$. Show the map $[x]_{p^m} \to [p^{n-m}x]_{p^n}$ gives rise to the isomorphism $(\mathbb{Z}_{p^m}, +) \cong (\mathbb{Z}_{p^n}(p^m), +)$.

EXERCISE 2.176. Let G be a finite abelian group. Show the following are equivalent:

(a) G is cyclic.

(b) $G \cong \mathbb{Z}_{p_1^{n_1}} \times \mathbb{Z}_{p_2^{n_2}} \times \cdots \times \mathbb{Z}_{p_m^{n_m}}$ where $m, k_i, n_i \in \mathbb{N}$, $p_1 < p_2 < \cdots < p_m$ are positive primes.

(c) $\operatorname{Exp}(G) = |G|$ where $\operatorname{Exp}(G) = \min\{n \in \mathbb{N} \mid g^n = e,\ g \in G\}$, called the *exponent* of G.

(d) For each prime p with $p \mid |G|$, there are exactly p solutions to $x^p = e$ in G.

EXERCISE 2.177 (Roots of Unity in a Finite Cyclic Group). Let G be a cyclic group of order N and let $k \in \mathbb{N}$. Show $|\{g \in G \mid g^n = e\}| = (N, n)$.

EXERCISE 2.178. **(a)** Let $n, k \in \mathbb{N}$ with $k \leq n$ and let p be a positive prime. Show $\mathbb{Z}_{p^n}$ has a subgroup of order p^k.

(b) If G is a finite abelian group and $p^n \mid |G|$, show G has a subgroup of order p^n (cf. Exercise 2.126 and Theorem 2.69).

EXERCISE 2.179 (Structure of U_p). Let p be a positive prime. This exercise shows that U_p is cyclic and thus $U_p \cong \mathbb{Z}_{p-1}$ (cf. Corollary 3.31).

(a) Let m be a positive divisor of $|U_p|$. Show $|G(m)| \leq m$. *Hint:* Exercise 2.7.

(b) Let $\psi(m) = |\{g \in U_p \mid |g| = m\}|$. If $\psi(m) \neq 0$, show that $|G(m)| = m$ and that $G(m) \cong \mathbb{Z}_m$.

(c) Show that either $\psi(m) = 0$ or $\psi(m) = \varphi(m)$, the Euler φ-function. *Hint:* Use Theorem 2.11 on $[k = k \cdot 1] \in \mathbb{Z}_m$, $1 \leq k \leq m$.

(d) Show $|U_p| = \sum_{m \mid |U_p|} \psi(m)$. Conclude that $\psi(m) = \varphi(m)$ for all m dividing $|U_p|$. *Hint:* Exercise 2.47.

(e) In particular, show that U_p is cyclic and thus $U_p \cong \mathbb{Z}_{p-1}$. *Hint:* $m = |U_p|$.

(f) Show part (e) can be false if p is replaced by a nonprime.

EXERCISE 2.180. Let G, H, and K be finite abelian groups.

(a) If $G \times K \cong H \times K$, show $G \cong H$.

(b) If $G \times G \cong H \times H$, show $G \cong H$.

(c) Show parts (a) and (b) can fail if the groups are infinite. *Hint:* Try $K = \prod_{n=1}^{\infty} ((\mathbb{Z}_2 \times \mathbb{Z}_2) \times \mathbb{Z}_4)$ (cf. Exercise 2.91).

EXERCISE 2.181. Let G be a finite abelian group and let $\widehat{G} = \{$homomorphisms from G to $\mathbb{C}^\times\}$, called the *dual group* of G. For $\varphi, \psi \in \widehat{G}$, define a binary operation $(\varphi\psi)(g) = \varphi(g)\psi(g)$.

(a) Show $\widehat{G}$ is a group.

(b) For $n \in \mathbb{N}$, show $\widehat{\mathbb{Z}_n} = \{\varphi_j : \mathbb{Z}_n \to \mathbb{C}^\times$ given by $\varphi_j([k]) = e^{\frac{2\pi ijk}{n}} \mid 0 \leq j < n\}$. Conclude $\widehat{\mathbb{Z}_n} \cong \mathbb{Z}_n$.

(c) If H is another finite abelian group, show $\widehat{G \times H} \cong \widehat{G} \times \widehat{H}$. Conclude that $\widehat{G} \cong G$.

EXERCISE 2.182. Prove part (2) of the Fundamental Theorem of Finite Abelian Groups, Theorem 2.41.

EXERCISE 2.183 (Generators and Relations). Let $x_1, \ldots, x_n$ be formal independent variables (called *generators*) and let

$$\mathbb{Z}[x_1, \ldots, x_n] = \{a_1 x_1 + \cdots + a_n x_n \mid a_i \in \mathbb{Z},\ 1 \leq i \leq n\}$$

equipped with the binary operation of polynomial addition. Fix $R \subseteq \mathbb{Z}[x_1, \ldots, x_n]$ (R for *relations*) and consider the abelian group

$$\mathbb{Z}[x_1, \ldots, x_n] / \langle R \rangle .$$

(a) Show $\mathbb{Z}[x_1, \ldots, x_n] \cong \mathbb{Z}^n$.

(b) If $a_i \in \mathbb{Z}_{\geq 0}$, $1 \leq i \leq n$, show $\mathbb{Z}[x_1, \ldots, x_n] / \langle \{a_1 x_1, \ldots, a_n x_n\} \rangle \cong \mathbb{Z}_{a_1} \times \cdots \times \mathbb{Z}_{a_n}$ (where $\mathbb{Z}_0$ is defined as $\mathbb{Z}$).

(c) Show $\mathbb{Z}[x_1, x_2, x_3] / \langle \{2x_1 + x_2 + 2x_3,\ 4x_1 + 3x_2\} \rangle \cong \mathbb{Z} \times \mathbb{Z}_2$. *Hint:* Show every element of $\mathbb{Z}[x_1, x_2, x_3]$ can be uniquely written as a $\mathbb{Z}$-linear combination of $2x_1 + x_2 + 2x_3$, $x_1 + x_2 - x_3$, and x_3. Note that $(4x_1 + 3x_2) - (2x_1 + x_2 + 2x_3) = 2x_1 + 2x_2 - 2x_3$.

(d) A group G is called *finitely generated* if there is a finite set $X \subseteq G$ so that $G = \langle X \rangle$. Up to isomorphism, show any finitely generated abelian group is of the form $\mathbb{Z}[x_1, \ldots, x_n] / \langle R \rangle$ for some $R \subseteq \mathbb{Z}[x_1, \ldots, x_n]$.

EXERCISE 2.184. To any matrix $M \in M_{k \times n}(\mathbb{Z})$, $k \in \mathbb{Z}_{\geq 0}$, let

$$\iota(M) = \langle \{M_{1,1}x_1 + \cdots + M_{1,n}x_n,\ M_{2,1}x_1 + \cdots + M_{2,n}x_n,\ \ldots\} \rangle$$

in $\mathbb{Z}[x_1, \ldots, x_n]$ (cf. Exercise 2.183) and define

$$\mathbb{Z}[x_1, \ldots, x_n]/M = \mathbb{Z}[x_1, \ldots, x_n]/\iota(M).$$

Recall that an *elementary row operation* on $M_{k \times n}(\mathbb{Z})$ consists of **(i)** multiplying a row by ± 1, **(ii)** interchanging two rows, or **(iii)** adding an integer multiple of one row to another row. Similar language is used for *elementary column operations.* For $N \in M_{k \times n}(\mathbb{Z})$, we say M and N are *equivalent* if M can be transformed to N by a finite sequence of elementary row and column operations.

(a) If $\mathcal{R}$ is an elementary row operation, show $\iota(\mathcal{R}M) = \iota(M)$. Conclude $\mathbb{Z}[x_1, \ldots, x_n]/M = \mathbb{Z}[x_1, \ldots, x_n]/\mathcal{R}M$.

(b) If $\mathcal{C}$ is an elementary row operation, show

$$\mathbb{Z}[x_1, \ldots, x_n]/M \cong \mathbb{Z}[x_1, \ldots, x_n]/\mathcal{C}M.$$

Hint: Use the following change of variables to construct isomorphisms: $x_i \to -x_i$, $x_i \longleftrightarrow x_j$, $x_j \to x_j - m x_i$.

(c) Conclude that if M and N are equivalent, then

$$\mathbb{Z}[x_1, \ldots, x_n]/M \cong \mathbb{Z}[x_1, \ldots, x_n]/N.$$

EXERCISE 2.185. Continue the notation from Exercise 2.184. We show that any $M \in M_{k \times n}(\mathbb{Z})$ is equivalent to a diagonal matrix.

(a) For $M \neq 0$, show M is equivalent to some N with $N_{1,1} \geq 1$.

(b) For $M \neq 0$, show M is equivalent to some N with $N_{1,1} \geq 1$ and $0 \leq N_{i,1} < N_{1,1}$ for $i \geq 2$.

(c) If some $N_{i,1} \neq 0$ in part (b), show M is equivalent to some N' with $N'_{1,1} = N_{i,1} < N_{1,1}$ and $0 \leq N'_{i,1} < N'_{1,1}$ for $i \geq 2$.

(d) For $M \neq 0$, show M is equivalent to some N with $N_{1,1} \geq 1$ and $N_{i,1} = 0$ for $i \geq 2$.

(e) For $M \neq 0$, show M is equivalent to some N with $N_{1,1} \geq 1$, $N_{i,1} = 0$ for $i \geq 2$, and $0 \leq N_{1,j} < N_{1,1}$ for $j \geq 2$.

(f) If some $N_{1,j} \neq 0$ in part (e), show M is equivalent to some N' with $N'_{1,1} = N_{1,j} < N_{1,1}$.

(g) For $M \neq 0$, show M is equivalent to some N with $N_{1,1} \geq 1$, $N_{i,1} = 0$ for $i \geq 2$, and $N_{1,j} = 0$ for $j \geq 2$.

(h) Conclude M is equivalent to a matrix of the form

$$\begin{pmatrix} \operatorname{diag}(a_1, \ldots, a_m) & 0_{m \times r} \\ 0_{(k-m) \times m} & 0_{(k-m) \times r} \end{pmatrix}$$

for some $a_i \in \mathbb{N}$ and $m \in \mathbb{Z}_{\geq 0}$ where $r = n - m$.

EXERCISE 2.186. Let $M \in M_{\infty \times n}(\mathbb{Z})$. Fix a row of M and let $\mathcal{R}_k$ be an elementary row operation that adds a multiple of the fixed row to row k. Define $\prod_{k=1}^{\infty} \mathcal{R}_k M$ to be the matrix whose k^{th} row is the same as the kth row of $\mathcal{R}_k M$.

(a) Show $\iota\left(\prod_{k=1}^{\infty} \mathcal{R}_k M\right) = \iota(M)$. Conclude

$$\mathbb{Z}[x_1, \ldots, x_n]/M = \mathbb{Z}[x_1, \ldots, x_n]/\prod_{k=1}^{\infty} \mathcal{R}_k M.$$

(b) Show the techniques from Exercise 2.185 can be extended to the case of $M \in M_{\infty \times n}(\mathbb{Z})$ and that M is equivalent to a matrix of the form

$$\begin{pmatrix} \operatorname{diag}(a_1, \ldots, a_m) & 0_{m \times r} \\ 0_{\infty \times m} & 0_{\infty \times r} \end{pmatrix}$$

for some $a_i \in \mathbb{N}$ and $m \in \mathbb{Z}_{\geq 0}$ where $r = n - m$.

(c) Show subgroups of finitely generated abelian groups are finitely generated.

EXERCISE 2.187. Let $M \in M_{k \times n}(\mathbb{Z})$, $k \in \mathbb{Z}_{\geq 0} \cup \{\infty\}$ (cf. Exercise 2.186).

(a) Show

$$\mathbb{Z}[x_1, \ldots, x_n]/M \cong \mathbb{Z}_{a_1} \times \cdots \times \mathbb{Z}_{a_m} \times \mathbb{Z}^r$$

for some $a_i \in \mathbb{N}$ and $m \in \mathbb{Z}_{\geq 0}$ where $r = n - m$. The number r is called the *free rank* of $\mathbb{Z}[x_1, \ldots, x_n]/M$.

(b) For $a, b \in \mathbb{N}$, show $\mathbb{Z}^a \cong \mathbb{Z}^b$ if and only if $a = b$.

(c) Show r and $\mathbb{Z}_{a_1} \times \cdots \times \mathbb{Z}_{a_m}$ (up to isomorphism) are uniquely determined by M. *Hint:* The torsion subgroup T of $\mathbb{Z}[x_1, \ldots, x_n]/M$ (cf. Exercise 2.74) is uniquely determined by M as is $\left(\mathbb{Z}[x_1, \ldots, x_n]/M\right)/T$.

(d) Write $\mathbb{Z}[x, y, z, w]/\langle\{2x + 3y,\ 7x + 4y + z + w,\ 3y + 2z + 5w\}\rangle$ as a product of cyclic groups.

EXERCISE 2.188 (Fundamental Theorem of Finitely Generated Abelian Groups).

(a) Using the preceding five exercises, show that every nontrivial finitely generated abelian group is isomorphic to exactly one of the following groups:

$$\mathbb{Z}^r \times \left(\mathbb{Z}_{p_1^{n_{1,1}}} \times \cdots \times \mathbb{Z}_{p_1^{n_{1,k_1}}}\right) \times \left(\mathbb{Z}_{p_2^{n_{2,1}}} \times \cdots \times \mathbb{Z}_{p_2^{n_{2,k_2}}}\right) \times \cdots \times \left(\mathbb{Z}_{p_m^{n_{m,1}}} \times \cdots \times \mathbb{Z}_{p_m^{n_{m,k_m}}}\right)$$

where $r, m \in \mathbb{Z}_{\geq 0}$ (not both zero), $k_i, n_{i,j} \in \mathbb{N}$, $p_1 < p_2 < \cdots < p_m$ are positive primes, and $n_{i,1} \geq n_{i,2} \geq \cdots \geq n_{i,k_i}$. The numbers $p_i^{n_{i,j}}$ are called the *elementary divisors* of the group and r is called the *free rank*.

(b) Equivalently, show that every nontrivial finitely generated abelian group is isomorphic to exactly one of the following groups:

$$\mathbb{Z}^r \times \mathbb{Z}_{n_1} \times \mathbb{Z}_{n_2} \times \cdots \times \mathbb{Z}_{n_s}$$

where $r, s \in \mathbb{Z}_{\geq 0}$ (not both zero), $n_i \in \mathbb{N}$, $n_i \geq 2$, and $n_1 \mid n_2 \mid \cdots \mid n_s$. The numbers n_i are called the *invariant factors* of the group and r is called the *free rank* (and is the same r as above).

7. The Symmetric Group

7.1. Cayley's Theorem. Recall that the symmetric group, S_n, consists of all permutations of the set $\{1, 2, \ldots, n\}$. More precisely, S_n is the set of bijections $\sigma : \{1, 2, \ldots, n\} \to \{1, 2, \ldots, n\}$. We have already seen $|S_n| = n!$. It turns out that the symmetric group plays a very important role in many mathematical and physical constructions. One small example of this is the following theorem showing that the symmetric groups contain all possible finite groups.

Though the following proof is technically simple, it can be mentally painful if this is your first real brush with functions mapping into function spaces. In particular, S_n is a set of *functions.* With this in mind, suppose we want to map some set G into S_n, $\varphi : G \to S_n$. For $g \in G$, it follows that $\varphi(g)$ must be a bijective function from $\{1, 2, \ldots, n\}$ to $\{1, 2, \ldots, n\}$. In order to specify $\varphi(g)$, it is necessary to spell out where $\varphi(g)$ maps each element of the domain, i.e., it is necessary to evaluate the function $\varphi(g)$ on each $k \in \{1, 2, \ldots, n\}$. Though somewhat choked by parenthesis, the most natural way to write the value of $\varphi(g)$ on k is by writing

$$(\varphi(g))(k).$$

Hence, defining a function $\varphi : G \to S_n$ amounts to defining $(\varphi(g))(k)$ for each $g \in G$ and $k \in \{1, 2, \ldots, n\}$ and then showing that the map $k \to (\varphi(g))(k)$ is a bijection of $\{1, 2, \ldots, n\}$.

THEOREM 2.42 (Cayley's Theorem). *Every finite group G is isomorphic to a subgroup of S_n where $n = |G|$.*

PROOF. It suffices to construct an injective homomorphism $\varphi : G \to S_n$ since this shows that G is isomorphic to the subgroup $\operatorname{Im} \varphi$ of S_n (Theorem 2.25). To this end, first write $l_g : G \to G$ for the left multiplication map $l_g(h) = gh$, $g, h \in G$. Notice the definition shows $l_g \circ l_h = l_{gh}$. In particular, the inverse to l_g is $l_{g^{-1}}$ so that l_g is bijective.

Next label the elements of G as $\{g_1, g_2, \ldots, g_n\}$. For each fixed g_i, $l_{g_i}(g_j)$ is some unique g_k (k depending on j). As j varies over $\{1, 2, \ldots, n\}$, it will turn out that the values of k form a rearrangement, i.e., permutation, of $\{1, 2, \ldots, n\}$. We therefore want to define $\varphi(g_i)$ to be the permutation of $\{1, 2, \ldots, n\}$ that maps j to k when $l_{g_i}(g_j) = g_k$.

To be a bit more precise, we want to define $\varphi : G \to S_n$ by letting $(\varphi(g_i))(j) = k$ where k is the unique element of $\{1, 2, \ldots, n\}$ satisfying

$$l_{g_i}(g_j) = g_k.$$

Technically, k depends on both i and on j so we ought to write something like $k(i, j)$. But since $(\varphi(g_i))(j) = k(i, j)$, we will save on notation and not bother with the k's at all. We officially define $\varphi : G \to S_n$ by letting $(\varphi(g_i))(j)$ be the unique element of $\{1, 2, \ldots, n\}$ satisfying

$$l_{g_i}(g_j) = g_{(\varphi(g_i))(j)}.$$

To finish the proof, we must show that (a) the resulting map $j \to (\varphi(g_i))(j)$ is bijective (so that $\varphi(g_i) \in S_n$) for each g_i; (b) the map $g_i \to \varphi(g_i)$ is injective; and (c) the map $g_i \to \varphi(g_i)$ is a homomorphism.

For part (a), suppose $(\varphi(g_i))(j) = (\varphi(g_i))(j')$. Then by definition,

$$l_{g_i}(g_j) = g_{(\varphi(g_i))(j)} = g_{(\varphi(g_i))(j')} = l_{g_i}(g_{j'}).$$

Since l_{g_i} is injective, it follows that $g_j = g_{j'}$. Thus $j = j'$ and the map $j \to (\varphi(g_i))(j)$ is injective. For surjectivity, the equation $(\varphi(g_i))(j) = j'$ can be solved for j if and only if the equation $g_{(\varphi(g_i))(j)} = g_{j'}$ can be solved if and only if $l_{g_i}(g_j) = g_{j'}$ can be solved if and only if $g_j = l_{g_i}^{-1}(g_{j'})$ can be solved for j. Since it is automatic that there is some j so that $g_j = l_{g_i}^{-1}(g_{j'})$ (after all, $l_{g_i}^{-1}(g_{j'}) \in \{g_1, g_2, \ldots, g_n\}$), surjectivity is finished.

For part (b), suppose $\varphi(g_i) = \varphi(g_j)$. In particular, $(\varphi(g_i))(1) = (\varphi(g_j))(1)$ so that

$$g_i g_1 = l_{g_i}(g_1) = g_{(\varphi(g_i))(1)} = g_{(\varphi(g_j))(1)} = l_{g_j}(g_1) = g_j g_1.$$

Multiplying on the right by g_1^{-1} shows $g_i = g_j$ so that φ is injective.

For (c), recall $\varphi(g_i)\varphi(g_j) = \varphi(g_i) \circ \varphi(g_j)$ and compare $(\varphi(g_i)\varphi(g_j))(k) = (\varphi(g_i))[(\varphi(g_j))(k)]$ to $(\varphi(g_i g_j))(k)$. It suffices to show $g_{(\varphi(g_i))[(\varphi(g_j))(k)]}$ is the same as $g_{(\varphi(g_i g_j))(k)}$. For this, simply calculate

$$g_{(\varphi(g_i))[(\varphi(g_j))(k)]} = l_{g_i}(g_{(\varphi(g_j))(k)}) = l_{g_i}(l_{g_j}(g_k)) = l_{g_i g_j}(g_k) = g_{(\varphi(g_i g_j))(k)}. \qquad \square$$

While Cayley's Theorem is aesthetically pleasing, its practical use is limited by the fact that $n!$ rapidly grows into an enormous number. For instance, looking for groups of only order 20 in S_{20} requires sifting through a group of size 2,432,902,008,176,640,000—which might take a while.

7.2. Cyclic Decomposition. Recall that we have associated to any $\sigma \in S_n$ the array

$$\begin{pmatrix} 1 & 2 & \cdots & n \\ \sigma(1) & \sigma(2) & \cdots & \sigma(n) \end{pmatrix}.$$

While this notation is explicit and handy in its own way, it quickly becomes much too cumbersome in practice. One of the most common and useful ways to condense this notation is known as *cyclic notation.*

DEFINITION 2.43. Let $k_1, \ldots, k_r$ be distinct elements of $\{1, \ldots, n\}$.

(1) Write $\sigma = (k_1, \ldots, k_r)$ for the element of S_n defined by

$$\begin{aligned} \sigma(k_i) &= k_{i+1} \quad \text{for } 1 \leq i \leq r-1, \\ \sigma(k_r) &= k_1, \\ \sigma(k) &= k \quad \text{if } k \notin \{k_1, \ldots, k_r\}. \end{aligned}$$

The element $(k_1, \ldots, k_r) \in S_n$ is called a *cycle* or, more specifically, an *r-cycle* or a *cycle of length r*.

(2) Let $\operatorname{dom}(k_1, \ldots, k_r) = \{k_1, \ldots, k_r\}$.

(3) Two cycles $(k_1, \ldots, k_r)$ and $(k_1', \ldots, k_s')$ are called *disjoint* if $\operatorname{dom}(k_1, \ldots, k_r) \cap \operatorname{dom}(k_1', \ldots, k_s') = \varnothing$.

(4) The cycle $(k_1, \ldots, k_r)$ is called *trivial* if $r = 1$ and is called *nontrivial* if $r \geq 2$.

(5) The cycle $(k_1, \ldots, k_r)$ is called a *transposition* if $r = 2$.

Cycles are often thought of as acting in a circular or cyclical fashion. For instance, $(1, 2, 3) \in S_3$ can be represented by the picture

$$\begin{array}{ccccc} & & 2 & & \\ & \nearrow & & \searrow & \\ 1 & & \longleftarrow & & 3 \end{array}$$

In terms of our old array notation, the element $\begin{pmatrix} 1 & 2 & 3 \\ 2 & 3 & 1 \end{pmatrix} \in S_3$ can be written as the 3-cycle $(1,2,3)$ while the element $\begin{pmatrix} 1 & 2 & 3 \\ 2 & 1 & 3 \end{pmatrix}$ can be written as the transposition $(1,2)$. In fact, it turns out that every element of S_3 is actually a cycle (Exercise 2.193). However, this is not true for S_n when $n \geq 4$. For instance, $\begin{pmatrix} 1 & 2 & 3 & 4 \\ 2 & 1 & 4 & 3 \end{pmatrix} \in S_4$ is not a cycle. Nevertheless, $\begin{pmatrix} 1 & 2 & 3 & 4 \\ 2 & 1 & 4 & 3 \end{pmatrix}$ can be written as the product of the cycles $(1,2)$ and $(3,4)$ and it is this observation that generalizes to any element of S_n.

We begin with some pretty obvious properties of cycles.

LEMMA 2.44. *Let $k_1, \ldots, k_r$ be distinct elements of $\{1, \ldots, n\}$.*

(1) *$(k_1, \ldots, k_r)$ is trivial if and only if it is the identity element of S_n.*

(2) *$|(k_1, \ldots, k_r)| = r$.*

(3) *$(k_1, \ldots, k_{r-1}, k_r)^{-1} = (k_r, k_{r-1}, \ldots, k_1)$.*

(4) *Two nontrivial cycles $(k_1, \ldots, k_r)$ and $(k'_1, \ldots, k'_s)$ are equal (as elements of S_n) if and only if $r = s$ and there is a t, $1 \leq t \leq r$, so that*

$$(k'_1, \ldots, k'_r) = (k_t, k_{t+1}, \ldots, k_r, k_1, k_2, \ldots, k_{t-1}),$$

i.e., so that

$$\begin{aligned} k'_i &= k_{i-1+t} && \text{for } 1 \leq i \leq r+1-t, \\ k'_i &= k_{i-1+t-r} && \text{for } r+2-t \leq i \leq r. \end{aligned}$$

(5) *Two disjoint cycles commute.*

PROOF. This lemma follows immediately from the definitions and is left to Exercise 2.192. □

The following relation and associated notation will be the key to decomposing an element of S_n as products of cycles.

DEFINITION 2.45. Let $\sigma \in S_n$.

(1) Define a relation $\sim_\sigma$ on $\{1, \ldots, n\}$ given by $i \sim_\sigma j$ if there is a $k \in \mathbb{Z}$ so $\sigma^k(i) = j$.

(2) Let $[i]_\sigma = \{\sigma^j(i) \mid j \in \mathbb{Z}\}$.

(3) The *fixed point set* of σ is written $\{1, \ldots, n\}^\sigma$ and is defined as the set of all $i \in \{1, \ldots, n\}$ satisfying $\sigma(i) = i$.

Of course, $\sim_\sigma$ will actually be an equivalence relation. Moreover, $\sim_\sigma$ is straightforward to calculate in practice. For example, if we start with a product of cycles such as $\sigma = (1,2)(3,4,5) \in S_6$, we see that $[1]_\sigma = [2]_\sigma = \{1,2\}$, $[3]_\sigma = [4]_\sigma = [5]_\sigma = \{3,4,5\}$, and $[6]_\sigma = \{6\}$ (in fact, $\{1,2,3,4,5,6\}^\sigma = \{6\}$). In particular, the decomposition of σ as a product of the cycles $(1,2)$ and $(3,4,5)$ appears to be closely linked to the *nonsingleton* equivalence classes of $\sim_\sigma$, i.e., the equivalence classes consisting of more than a single element. As we see next, this is no accident.

THEOREM 2.46. *Let $\sigma \in S_n$.*

(1) *The relation $\sim_\sigma$ on $\{1, \ldots, n\}$ is an equivalence relation and $[i]_\sigma$ is the equivalence class of i, $i \in \{1, \ldots, n\}$.*

(2) *For each i, there is a maximal $m \in \mathbb{N}$ so that $i, \sigma(i), \sigma^2(i), \ldots, \sigma^{m-1}(i)$ are distinct. In that case, $\sigma^m(i) = i$ and the equivalence class $[i]_\sigma$ is exactly*

$$\{i, \sigma(i), \sigma^2(i), \ldots, \sigma^{m-1}(i)\}.$$

(3) *If $\sigma = c_1 c_2 \cdots c_k$ is a product of nontrivial disjoint cycles, then the set of equivalence classes of $\sim_\sigma$ on $\{1, \ldots, n\}$ is*

$$\{\operatorname{dom} c_j \mid 1 \leq j \leq k\} \bigcup \{\{i\} \mid i \in \{1, \ldots, n\}^\sigma\}$$

and

$$\sigma|_{\operatorname{dom} c_j} = c_j|_{\operatorname{dom} c_j} \quad \textit{and} \quad \sigma|_{\{1,\ldots,n\}^\sigma} = \operatorname{Id}$$

for $1 \leq j \leq k$.

PROOF. Part (1) follows immediately from the definitions. For part (2), first observe that m exists since $\{1, \ldots, n\}$ is finite. For the next statement in (2), proceed by way of contradiction and suppose $\sigma^m(i) \neq i$. By construction, this means there is a j, $1 \leq j \leq m-1$, so that $\sigma^m(i) = \sigma^j(i)$ and so $\sigma^{m-j}(i) = i$. However, this contradicts the fact that i is distinct from $\sigma(i), \sigma^2(i), \ldots, \sigma^{m-1}(i)$. Thus $\sigma^m(i) = i$. To finish the final statement of (2), first apply σ^{-m} to get $i = \sigma^{-m}(i)$. Iteration of the relations $\sigma^{\pm m}(i) = i$ shows $\sigma^{km}(i) = i$ for any $k \in \mathbb{Z}$. Thus $\sigma^{j+km}(i) = \sigma^j(i)$. Modular arithmetic then shows $\{\sigma^j(i) \mid j \in \mathbb{Z}\} = \{i, \sigma(i), \sigma^2(i), \ldots, \sigma^{m-1}(i)\}$ as desired.

For the proof of part (3), first observe that $[i]_\sigma = \{i\}$ if and only if i is a fixed point of σ, i.e., if and only if $\sigma(i) = i$. Thus $\{\{i\} \mid i \in \{1, \ldots, n\}^\sigma\}$ captures all the single point equivalence classes. Obviously $\sigma|_{\{1,\ldots,n\}^\sigma} = \operatorname{Id}$.

Turn now to $\operatorname{dom} c_j$. Of course c_j preserves $\operatorname{dom} c_j$ by construction. On the other hand, the definition of disjointedness and of a cycle shows $c_{j'}$ acts by Id on $\operatorname{dom} c_j$ for all $j' \neq j$, $1 \leq j' \leq k$. Since $\sigma = c_1 c_2 \cdots c_k$, it follows that $\sigma|_{\operatorname{dom} c_j} = c_j|_{\operatorname{dom} c_j}$.

If we pick $l_j \in \operatorname{dom} c_j$, we claim that (i) $[l_j]_\sigma = \operatorname{dom} c_j$ and (ii) every nonsingleton equivalence class is of this form. If so, the proof will be complete. To verify the first claim, note that $[l_j]_\sigma$ consists of all $\sigma^k(l_j)$ where $k \in \mathbb{Z}$. However, $l_j \in \operatorname{dom} c_j$ so that $\sigma^k(l_j) = c_j^k(l_j)$. By definition of a cycle, it immediately follows that $\{c_j^k(l_j) \mid k \in \mathbb{Z}\} = \operatorname{dom} c_j$ and we are done. To verify the second claim, suppose $[l]_\sigma$ is not a single point equivalence class. In particular, we must have $\sigma(l) \neq l$. However, if $l \notin \bigcup_j \operatorname{dom} c_j$, then the fact that $\sigma = c_1 c_2 \cdots c_k$ would show that $\sigma(l) = l$, a contradiction. Therefore $l \in \operatorname{dom} c_j = [l_j]$ for some j and we are done. □

We are finally in a position to prove that every element of S_n is a product of cycles.

THEOREM 2.47 (Cyclic Decomposition Theorem). *Any nontrivial element of S_n can be written as a product of disjoint, nontrivial cycles. Up to rearrangement of the cycles, this decomposition is unique.*

PROOF. For existence, fix $\sigma \in S_n$. Using Theorem 1.37, write $\{1, \ldots, n\}$ as the disjoint union of equivalence classes $[i_1]_\sigma, [i_2]_\sigma, \ldots, [i_k]_\sigma$. Using Theorem 2.46, part (2), pick maximal $m_j \in \mathbb{N}$ so that $i_j, \sigma(i_j), \sigma^2(i_j), \ldots, \sigma^{m_j-1}(i_j)$ are distinct and let c_j be the m_j-cycle defined by

$$c_j = \left(i_j, \sigma(i_j), \sigma^2(i_j), \ldots, \sigma^{m_j-1}(i_j)\right).$$

By construction, it follows that $[i_j]_\sigma = \operatorname{dom} c_j$ and that $\sigma|_{[i_j]_\sigma} = c_j|_{\operatorname{dom} c_j}$. Since the c_j are disjoint from each other, Theorem 2.46, part (3), shows that $(c_1c_2\cdots c_k)\,|_{\operatorname{dom} c_j} = c_j|_{\operatorname{dom} c_j}$. Writing $\{1,\ldots,n\} = \coprod_{j=1}^k [i_j]_\sigma$ and noting that $\sigma|_{[i_j]_\sigma} = (c_1c_2\cdots c_k)\,|_{[i_j]_\sigma}$, we see immediately that

$$\sigma = c_1c_2\cdots c_k. \tag{2.48}$$

To finish existence, notice $c_j = \mathrm{Id}$ if and only if $m_j = 1$ (Lemma 2.44) and that the $\{c_j\}$ are disjoint cycles by construction. In particular for $\sigma \neq \mathrm{Id}$, at least one c_j must be nontrivial. Moreover, removing any of the trivial c_j does not affect the product in (2.48) and so finishes the proof of existence.

Turn now to the question of uniqueness. Suppose there are exactly r nontrivial cycles appearing in (2.48). Renumbering the c_j and i_j, if necessary, to put the trivial cycles at the end, our above discussion allows us to write $\sigma = c_1\cdots c_r$ as a product of nontrivial disjoint cycles with $\operatorname{dom} c_j = [i_j]_\sigma$, $1 \leq j \leq r$ (and $[i_j]_\sigma = \{i_j\}$, $r+1 \leq j \leq k$, not that we care anymore). Suppose also that $\sigma = c_1'c_2'\cdots c_{r'}'$ where $\{c_j'\}$ are disjoint nontrivial cycles. Using Theorem 2.46, part (3), we see that r and r' both count the number of nonsingleton equivalence classes of $\sim_\sigma$. In particular, $r = r'$. Moreover, those equivalence classes are given by $\{\operatorname{dom} c_j\}$ and also by $\{\operatorname{dom} c_j'\}$. Therefore we may renumber the c_j' so that $\operatorname{dom} c_j' = \operatorname{dom} c_j$, $1 \leq j \leq r$. Since we have seen $\sigma|_{\operatorname{dom} c_j} = c_j|_{\operatorname{dom} c_j}$ and similarly $\sigma|_{\operatorname{dom} c_j'} = c_j|_{\operatorname{dom} c_j'}$, it follows that $c_j|_{\operatorname{dom} c_j} = c_j'|_{\operatorname{dom} c_j}$. Since c_j and c_j' act by Id outside $\operatorname{dom} c_j$, we get $c_j = c_j'$ as desired. □

One nice thing about the proof of the Cyclic Decomposition Theorem is that it gives an algorithm for calculating the cyclic decomposition of any $\sigma \in S_n$. Basically, the decomposition comes down to determining the nonsingleton equivalence classes of $\sim_\sigma$. In turn, this reduces to picking various $i \in \{1,\ldots,n\}$ and applying successive iterations of σ to get distinct $i, \sigma(i), \sigma^2(i), \ldots, \sigma^{m-1}(i)$ until, eventually, $\sigma^m(i) = i$.

For example, consider the following element of S_{10}:

$$\sigma = \begin{pmatrix} 1 & 2 & 3 & 4 & 5 & 6 & 7 & 8 & 9 & 10 \\ 5 & 4 & 8 & 1 & 2 & 6 & 10 & 7 & 9 & 3 \end{pmatrix}.$$

We calculate that $\sigma(1) = 5$, $\sigma^2(1) = 2$, $\sigma^3(1) = 4$, and $\sigma^4(1) = 1$. Therefore one of the cycles appearing in the cyclic decomposition of σ is $(1,5,2,4)$. Next we see that $\sigma(3) = 8$, $\sigma^2(3) = 7$, $\sigma^3(3) = 10$, and $\sigma^4(3) = 3$, which gives rise to the cycle $(3,8,7,10)$. Finally we see that $\sigma(6) = 6$ and $\sigma(9) = 9$, which are singleton equivalence classes and are not needed for the cyclic decomposition of σ. As we have now exhausted all of $\{1,\ldots,10\}$, it follows that the cyclic decomposition of σ is $\sigma = (1,5,2,4)(3,8,7,10)$.

7.3. Conjugacy Classes. If $T \in M_{n\times n}(\mathbb{R})$ is a matrix representing the linear map $x \to Tx$, $x \in \mathbb{R}^n$, with respect to the standard basis, it is known from linear algebra that the matrix PTP^{-1}, $P \in GL(n,\mathbb{R})$, represents the same linear map with respect to another basis. It is also known that it is possible to choose P so that PTP^{-1} has a very simple form (e.g., Jordan Canonical Form). Once this is done, many difficult questions about T can be answered easily by examining PTP^{-1}.

Similarly, if G is a group with $g \in G$, certain difficult questions about g can sometimes be answered by examining an element of the form hgh^{-1} for some $h \in H$, called a *conjugate* of g. For instance, if $\varphi : G \to \mathbb{C}^\times$ is a homomorphism, then calculating $\varphi(g)$ is the same as calculating $\varphi(hgh^{-1})$ since $\varphi(hgh^{-1}) = \varphi(h)\varphi(g)\varphi(h)^{-1} = \varphi(g)$ ($\mathbb{C}^\times$ is abelian). As a result, it is often possible to make a judicious choice of h that greatly simplifies the calculation.

In this section we study the notion of conjugacy in S_n. This is especially important for an area of mathematics called *representation theory*, though we will not have the opportunity to explore such ideas in this text. In any event, we begin with definitions.

DEFINITION 2.49. **(1)** For $n \in \mathbb{N}$, a *partition* of n is the k-tuple $(n_1, n_2, \ldots, n_k)$, $1 \le k \le n$, with $n_j \in \mathbb{N}$, $n_1 \ge n_2 \ge \cdots \ge n_k$ so that $n = n_1 + n_2 + \cdots + n_k$.

(2) For $\sigma \in S_n$ and the equivalence relation $\sim_\sigma$, write $\{1, \ldots, n\}$ as the disjoint union of equivalence classes $[i_1], [i_2], \ldots, [i_k]$ and let $n_j = |[i_j]|$. Re-indexing if necessary, assume $n_1 \ge n_2 \ge \cdots \ge n_k$. Then the *associated partition* to σ (also called the *cycle structure*) is $(n_1, n_2, \ldots, n_k)$.

(3) If G is a group with $g_1, g_2 \in G$, then g_1 and g_2 are said to be *conjugate* if there exists an $h \in G$ so that $g_2 = hg_1h^{-1}$. The set of all elements in G conjugate to some $g \in G$ is called a *conjugacy class*. The map $c_h : G \to G$ given by $c_h(g) = hgh^{-1}$ is called *conjugation* by h.

Partitions are often represented graphically by drawing a left justified array with n_1 boxes in the first row, n_2 boxes in the second row, and so forth. For instance, the partition $(3, 2, 2, 1)$ of 8 can be drawn as

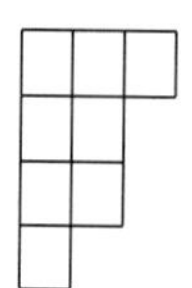

By definition, the above partition is the one associated to, say, the element $\sigma = (1, 2, 3)(4, 5)(6, 7)(8) \in S_8$. If we conjugate σ by, say, $(6, 8)$, we get

$$\begin{aligned}(6,8)\sigma(6,8)^{-1} &= (6,8)(1,2,3)(4,5)(6,7)(8)(6,8)\\ &= (1,2,3)(4,5)(6,8)(6,7)(8)(6,8).\end{aligned}$$

However, it is easy to see that $(6, 8)(6, 7)(8)(6, 8) = (8, 7)(6)$ so that

$$(6,8)\sigma(6,8)^{-1} = (1,2,3)(4,5)(7,8)(6),$$

which has the same associated partition as σ. This is no accident—we will see below that the associated partition only depends on the conjugacy class.

COROLLARY 2.50 (Conjugacy Classes of S_n). *Two elements of S_n are conjugate if and only if they have the same associated partition. In particular, there is a bijective correspondence between the set of conjugacy classes of S_n and partitions of n.*

PROOF. Let $\sigma \in S_n$ and let $(k_1, \ldots, k_r)$ be an r-cycle. By observation, $\sigma(k_1, \ldots, k_r)\sigma^{-1}$ maps $\sigma(k_j) \to \sigma(k_{j+1})$ for $1 \le j < r$, maps $\sigma(k_r) \to \sigma(k_1)$, and fixes the remaining elements of $\{1, 2, \ldots, n\}$. Thus

$$\sigma(k_1,\ldots,k_r)\sigma^{-1} = (\sigma(k_1), \sigma(k_2), \ldots, \sigma(k_r)). \tag{2.51}$$

Next observe that

$$\sigma\left(c_1c_2\cdots c_k\right)\sigma^{-1}=\left(\sigma c_1\sigma^{-1}\right)\left(\sigma c_2\sigma^{-1}\right)\cdots\left(\sigma c_k\sigma^{-1}\right)$$

for any $c_j \in S_n$. From these two facts and the Cyclic Decomposition Theorem, it immediately follows that conjugation preserves the associated partition.

Conversely, suppose two elements of S_n have the same associated partition. Using the above two facts, it will be easy to verify that they are conjugate. At the risk of being kicked out of the math club, I believe this result is most clearly seen by looking at an example. For instance, to conjugate $(1,2,3)(4,5)(6,7)(8) \in S_8$ to $(3,2,5)(1,8)(4,7)(6)$, simply use

$$\sigma=\begin{pmatrix}1&2&3&4&5&6&7&8\\3&2&5&1&8&4&7&6\end{pmatrix}.$$

The formal write-up is easy and is left to the reader. □

7.4. Parity and the Alternating Subgroup. For $k \in \mathbb{Z}\backslash\{0\}$, recall that the *sign* function is given by $\operatorname{sgn}(k)=\frac{k}{|k|}$ (so that $\operatorname{sgn}(k)$ is 1 when $k>0$ and is -1 when $k<0$). There is a generalization of this function appropriate for S_n.

DEFINITION 2.52. Let $P=\{(i,j)\mid 1\le i<j\le n\}$ and for $\sigma\in S_n$, define $\operatorname{sgn}: S_n\to\{\pm1\}$ by

$$\operatorname{sgn}(\sigma)=\prod_{(i,j)\in P}\operatorname{sgn}\left(\sigma(j)-\sigma(i)\right).$$

The first result shows that the sgn function is well behaved with respect to transpositions, i.e., 2-cycles.

LEMMA 2.53. *If $\sigma,\tau\in S_n$ with τ a transposition, then* $\operatorname{sgn}(\tau\sigma)=-\operatorname{sgn}(\sigma)$.

PROOF. Let $P_\sigma=\{(i,j)\in P\mid \sigma(i)>\sigma(j)\}$ (elements of P_σ are often called *inversions* of σ). Decomposing P as the disjoint union of P_σ and $P\backslash P_\sigma$, noting that $\operatorname{sgn}\left(\sigma(j)-\sigma(i)\right)$ is -1 when $(i,j)\in P_\sigma$ and is $+1$ when $(i,j)\in P\backslash P_\sigma$, it follows that $\operatorname{sgn}(\sigma)=(-1)^{|P_\sigma|}$.

Write $r_k=(k,k+1)$ for the transposition of two adjacent letters, called a *simple reflection*. We will show below that $|P_\sigma|=|P_{r_k\sigma}|\pm1$. Temporarily assuming this, it follows that $\operatorname{sgn}(r_k\sigma)=-\operatorname{sgn}(\sigma)$. To soup this result up to an arbitrary transposition, let $i<j$ and observe that (i,j) can be written as a product of an odd number of simple reflections,

$$(i,j)=r_ir_{i+1}\cdots r_{j-2}\left(r_{j-1}\right)r_{j-2}\cdots r_{i+1}r_i$$

(this can be easily verified by looking at what both sides do to each element of $\{i,i+1,\ldots,j\}$). Pulling off one r_k at a time at the cost of a minus sign each time, it follows that

$$\begin{aligned}\operatorname{sgn}((i,j)\sigma)&=\operatorname{sgn}(r_ir_{i+1}\cdots r_{j-2}\left(r_{j-1}\right)r_{j-2}\cdots r_{i+1}r_i\sigma)\\&=(-1)^{1+2(j-i)}\operatorname{sgn}(\sigma)=-\operatorname{sgn}(\sigma)\end{aligned}$$

as desired.

It remains to verify that $|P_\sigma|=|P_{r_k\sigma}|\pm1$. Let $1\le i_0<j_0\le n$ be chosen so that $\{\sigma(i_0),\sigma(j_0)\}=\{k,k+1\}$ (as unordered sets). Looking at the action of r_k on $\{k,k+1\}$, clearly (i_0,j_0) is in either P_σ or $P_{r_k\sigma}$, but not both. To finish the proof, it therefore suffices to show that $P_\sigma\backslash\{(i_0,j_0)\}=P_{r_k\sigma}\backslash\{(i_0,j_0)\}$.

This breaks into three cases. The first is the case $(i,j) \in P$ with $i \neq i_0$ and $j \neq j_0$. In this setting, $\sigma(i) = r_k\sigma(i)$ and $\sigma(j) = r_k\sigma(j)$. Therefore $\sigma(i) > \sigma(j)$ if and only if $r_k\sigma(i) > r_k\sigma(j)$ so that $(i,j) \in P_\sigma$ if and only if $(i,j) \in P_{r_k\sigma}$.

The second case occurs when $i = i_0$ and $j \neq j_0$ and the third case occurs when $i \neq i_0$ and $j = j_0$. Since the argument for these two cases is so similar, we only write out the details for the case of $i = i_0$ and $j \neq j_0$. Under this hypothesis, there are two further possibilities: either $\sigma(i) = k$ or $\sigma(i) = k+1$. Again, the argument in either situation is nearly identical so we only give the details for the case of $\sigma(i) = k$. In this setting, $r_k\sigma(i) = k+1$ and $r_k\sigma(j) = \sigma(j)$. Then the condition $\sigma(i) > \sigma(j)$ clearly implies $r_k\sigma(i) > r_k\sigma(j)$. Conversely, since $\sigma(j) \neq k+1$, the condition $\sigma(i) < \sigma(j)$ implies $\sigma(i) + 1 < \sigma(j)$ so that $r_k\sigma(i) < r_k\sigma(i)$. Again we see that $(i,j) \in P_\sigma$ if and only if $(i,j) \in P_{r_k\sigma}$. □

One of the frequently used facts about S_n is that there is an intrinsic notion of *parity* associated to a permutation. This concept is based upon the ability to express every element of S_n as products of transpositions.

THEOREM 2.54 (Parity). *Let $\sigma \in S_n$.*

(1) *Every element of S_n can be written (nonuniquely) as a product of transpositions.*

(2) *If $\sigma = \tau_1 \cdots \tau_r$ and $\sigma = \tau_1' \cdots \tau_s'$ are two such products of transpositions, then $r \equiv s \bmod(2)$.*

(3) *The function $\operatorname{sgn} : S_n \to \{\pm 1\}$ is a homomorphism that maps transpositions to -1.*

PROOF. We begin with part (1). The Cyclic Decomposition Theorem shows it is enough to write a cycle as a product of transpositions. To this end, the following formula does the trick (for $r \geq 2$):

$$(k_1, k_2, k_3, \ldots, k_{r-1}, k_r) = (k_1, k_r)(k_1, k_{r-1}) \cdots (k_1, k_4)(k_1, k_3)(k_1, k_2). \tag{2.55}$$

As this formula is easily verified by looking at what both sides do to each element of $\{k_1, k_2, \ldots, k_r\}$, the proof of existence is finished. For $r = 1$, use, say, $(1,2)^2 = \mathrm{Id}$.

For parts (2) and (3), use Lemma 2.53 and the fact that $\operatorname{sgn}(\mathrm{Id}) = 1$ to see that

$$\operatorname{sgn}(\tau_1 \cdots \tau_r) = (-1)^r. \tag{2.56}$$

This is enough to finish part (2) since the equation

$$(-1)^r = \operatorname{sgn}(\tau_1 \cdots \tau_r) = \operatorname{sgn}(\sigma) = \operatorname{sgn}(\tau_1' \cdots \tau_s') = (-1)^s$$

implies that $r \equiv s \bmod(2)$. We also get the rest of part (3) from (2.56) by writing $\sigma_i \in S_n$ as $\sigma_i = \tau_{1,i} \cdots \tau_{k_i,i}$ for some transpositions $\tau_{j,i}$ and calculating that

$$\begin{aligned}\operatorname{sgn}(\sigma_1\sigma_2) &= \operatorname{sgn}(\tau_{1,1} \cdots \tau_{k_1,1}\, \tau_{1,2} \cdots \tau_{k_2,2}) = (-1)^{k_1+k_2} = (-1)^{k_1}(-1)^{k_2} \\ &= \operatorname{sgn}(\tau_{1,1} \cdots \tau_{k_1,1}) \operatorname{sgn}(\tau_{1,2} \cdots \tau_{k_2,2}) = \operatorname{sgn}(\sigma_1) \operatorname{sgn}(\sigma_2).\end{aligned}$$

□

From (2.56), we now have an alternate way of defining the homomorphism $\operatorname{sgn} : S_n \to \{\pm 1\}$. Namely

$$\operatorname{sgn}(\sigma) = \begin{cases} +1 & \text{if } \sigma \text{ is the product of an } even \text{ number of transpositions,} \\ -1 & \text{if } \sigma \text{ is the product of an } odd \text{ number of transpositions.} \end{cases}$$

In actuality, this is the definition we really wanted to start with anyway (cf. Exercise 2.212). Now that we know the above definition is actually well defined, we use it freely. It should be noted that the sgn function is extremely important in mathematics. Among other places, it appears in the construction of the determinant function on matrices (Exercise 2.229). Moreover, it allows us to pick out a very important subgroup of S_n.

DEFINITION 2.57. **(1)** The *parity* of an element $\sigma \in S_n$ is *even* if $\text{sgn}(\sigma) = 1$ and is *odd* otherwise.

(2) The *alternating group* on n-letters, A_n, is $\{\sigma \in S_n \mid \sigma \text{ is } even\}$.

As an example, note that (2.55) shows an r-cycle is in A_n if and only if r is *odd.*

COROLLARY 2.58. **(1)** *A_n is a normal subgroup of S_n.*

(2) $|A_n| = \frac{n!}{2}$.

(3) *For $n \geq 3$, A_n is generated by all 3-cycles.*

PROOF. For part (1), observe that $A_n = \ker(\text{sgn})$ by definition and is therefore a normal subgroup of S_n. For part (2), since $\text{sgn} : S_n \to \{\pm 1\}$ is surjective, there is an isomorphism $S_n/A_n \cong \{\pm 1\}$. In particular, $\frac{n!}{|A_n|} = 2$ as desired. For part (3), first note that since 3 is odd, A_n contains all 3-cycles. Therefore it suffices to show that the product of any two transpositions can be written as a product of 3-cycles. There are really only two cases that need examination. Namely, choose distinct $i, j, k, l \in \{1, 2, \ldots, n\}$ and consider the cases of $(i, j)(k, l)$ (when $n \geq 4$) and $(i, j)(i, k)$. We finish by observing $(i, j)(k, l) = (i, l, k)(i, j, k)$ and $(i, j)(i, k) = (i, k, j)$. □

In fact for $n \geq 5$, we will see that A_n is the only proper normal subgroup of S_n (Exercise 2.287).

7.5. Exercises 2.189–2.229.

EXERCISE 2.189. **(a)** Use the ideas of Cayley's Theorem to write down an explicit subgroup of S_8 that is isomorphic to D_4.

(b) Use geometrical ideas to write down an explicit subgroup of S_4 that is isomorphic to D_4.

EXERCISE 2.190. Write the following elements of S_5 in the form

$$\begin{pmatrix} 1 & 2 & 3 & 4 & 5 \\ \sigma(1) & \sigma(2) & \sigma(3) & \sigma(4) & \sigma(5) \end{pmatrix}:$$

(a) $(1, 2, 3)(3, 4, 5)$,

(b) $(1, 5)(2, 1, 3, 5, 4)(2, 3, 5)$,

(c) $(2, 4, 5, 1)(3, 5)(5, 2)$.

EXERCISE 2.191. Write the following elements of S_5 as a product of disjoint, nontrivial cycles and draw the associated partition:

(a) $\begin{pmatrix} 1 & 2 & 3 & 4 & 5 \\ 4 & 2 & 1 & 3 & 5 \end{pmatrix}$,

(b) $\begin{pmatrix} 1 & 2 & 3 & 4 & 5 \\ 3 & 5 & 1 & 2 & 4 \end{pmatrix}$,

(c) $\begin{pmatrix} 1 & 2 & 3 & 4 & 5 \\ 3 & 5 & 1 & 4 & 2 \end{pmatrix}$,

(d) $(1, 2, 5)(2, 3)$,

(e) $(1, 2)(1, 3)(3, 4, 5)$.

EXERCISE 2.192. Prove Lemma 2.44.

EXERCISE 2.193. **(a)** Show that every element of S_3 is a cycle.
(b) Show part (a) is not true for S_n when $n \geq 4$.

EXERCISE 2.194. **(a)** For $\sigma = (1,3,5,4)(6,7) \in S_8$, write down the equivalence classes $[i]_\sigma$ for the equivalence relation $\sim_\sigma$.
(b) What is the fixed point set for σ?

EXERCISE 2.195. Determine which of the following elements of S_{10} are conjugate:
(a) $(1,3,7,8)(6,10)$,
(b) $(3,6,9)(2,5,7)$,
(c) $(1,2)(10,9,8,7)$.

EXERCISE 2.196. In terms of the associated partition, determine all $\sigma \in S_n$ satisfying $\sigma^2 = \mathrm{Id}$.

EXERCISE 2.197 (Order and Partitions). **(a)** Suppose the partition associated to $\sigma \in S_n$ is $(n_1, \ldots, n_k)$. Show $|\sigma| = \mathrm{lcm}(n_1, n_2, \ldots, n_k)$ (cf. Exercises 1.28 and 2.81).
(b) Determine the order of the elements of S_5 given in Exercise 2.191.
(c) A priori, the order of any $\sigma \in S_6$ is bounded by $6! = 720$. In fact, this is overkill. What is the largest order of an element of S_6?

EXERCISE 2.198. Suppose $\sigma \in S_n$ is a nontrivial r-cycle. Show σ^k, $k \in \mathbb{Z}$, remains an r-cycle if and only if $(k, r) = 1$.

EXERCISE 2.199. Show the number of r-cycles in S_n, $1 < r \leq n$, is $\frac{n!}{r(n-r)!}$ (cf. Exercise 2.239).

EXERCISE 2.200 (Generators of S_n). Show $(1,2)$ and $(1,2,\ldots,n)$ generate S_n.

EXERCISE 2.201. Suppose $\sigma \in S_n$ has prime order p.
(a) Show σ is a product of disjoint p-cycles.
(b) If the fixed point set of σ is empty, show $p \mid n$.

EXERCISE 2.202. For the elements of S_5 given in Exercise 2.191, determine the parity and use this to determine which are in A_5.

EXERCISE 2.203. **(a)** Show that the product of an even element of S_n with an odd element is odd.
(b) Show that the product of two odd elements is even.

EXERCISE 2.204. For $\sigma, \tau \in S_n$ show that both σ^{-1} and $c_\tau(\sigma)$ have the same parity as σ.

EXERCISE 2.205. Write the elements of S_5 given in Exercise 2.191 as a product of transpositions.

EXERCISE 2.206 ($|H \cap A_n|$). Let H be a subgroup of S_n. Prove that $|H \cap A_n|$ is either $|H|$ (so $H \subseteq A_n$) or $\frac{1}{2}|H|$. *Hint:* Try to define a map $H/(H \cap A_n) \to S_n/A_n$.

EXERCISE 2.207 (Order $\frac{n!}{2}$ Subgroups of S_n). Let H be a subgroup of S_n with $|H| = n!/2$.

(a) Show H is normal and conclude $\sigma^2 \in H$ for all $\sigma \in S_n$. *Hint:* Exercise 2.141 and note that $S_n/H \cong \mathbb{Z}_2$.

(b) Show H contains all 3-cycles. Conclude $H = A_n$.

EXERCISE 2.208. If $\sigma \in S_n$ has odd order, show $\sigma \in A_n$. *Hint:* Exercise 2.197.

EXERCISE 2.209. For $n \geq 5$, show A_n is generated by $\{\tau_1\tau_2 \mid \tau_1, \tau_2 \in S_n$ are nontrivial disjoint transpositions$\}$.

EXERCISE 2.210. Show that $\{\mathrm{Id}, (1,2)(3,4), (1,3)(2,4), (1,4)(2,3)\} \cong \mathbb{Z}_2 \times \mathbb{Z}_2$ is a normal subgroup of A_4.

EXERCISE 2.211. Though $|A_4| = 12$ and $6 \mid 12$, show A_4 has no subgroups of order 6.

EXERCISE 2.212 (Alternate Definitions of sgn). Let $\sigma \in S_n$. The following are three alternate ways of working with the sgn function. For two more, see Exercises 2.214, part (e), and 2.223, part (c).

(a) Show $\mathrm{sgn}(\sigma) = \prod_{1 \leq i < j \leq n} (\frac{\sigma(j) - \sigma(i)}{j - i})$.

(b) Consider the polynomial $p(x_1, \ldots, x_n) = \prod_{1 \leq i < j \leq n} (x_j - x_i)$. For $\sigma \in S_n$, show $\mathrm{sgn}(\sigma) = \frac{p(x_{\sigma(1)}, \ldots, x_{\sigma(n)})}{p(x_1, \ldots, x_n)}$.

(c) Let N_σ be the number of orbits of σ acting on $\{1, \ldots, n\}$. Show $\mathrm{sgn}(\sigma) = (-1)^{n - N_\sigma}$.

EXERCISE 2.213 (Left Regular Representation Partition). Let G be a finite group and let $g \in G$ have order m. Let $\varphi : G \hookrightarrow S_{|G|}$ be the injective homomorphism from Cayley's Theorem.

(a) Show that $\varphi(g)$ is a product of $\frac{|G|}{m}$ m-cycles so that the associated partition is $(m, m, \ldots, m)$. *Hint:* Look at the right coset space $\langle g \rangle \backslash G$.

(b) Conclude $\varphi(g) \in A_{|G|}$ if and only if m is odd or if $\frac{|G|}{m}$ is even.

EXERCISE 2.214 (Permutation Matrices). The set of *permutation matrices*, Perm_n, in $M_{n,n}(\mathbb{R})$ consists of all matrices having exactly a single 1 in each row and each column and 0's elsewhere. Write $e_i = (\overbrace{0, \ldots, 0, 1}^{i}, 0, \ldots, 0)$ for the *standard basis vectors* of $\mathbb{R}^n$.

(a) Show $|\mathrm{Perm}_n| = n!$.

(b) For $\sigma \in S_n$, let $p(\sigma)$ be the matrix satisfying $(p(\sigma))e_i = e_{\sigma(i)}$, $1 \leq i \leq n$. In other words, $p(\sigma)$ is the matrix whose i^{th} column is $e_{\sigma(i)}$ (transposed). Calculate $p(\sigma)$ for σ equal to the following elements of S_4: $(1,2,3,4)$, $(1,2)(3,4)$, and $(1,4,2)$.

(c) Show $p : S_n \to M_{n,n}(\mathbb{R})$ is a homomorphism.

(d) Show p is injective and that the image lies in Perm_n. Conclude p may be viewed as an isomorphism $p : S_n \xrightarrow{\cong} \mathrm{Perm}_n$.

(e) Show $\mathrm{sgn}(\sigma) = \det(p(\sigma))$.

(f) Show $\mathrm{Perm}_n \subseteq O(n, \mathbb{R})$; i.e., show the inverse of a permutation matrix is its transpose.

EXERCISE 2.215 (Permutation Representations). Suppose H is a subgroup of G so that $|G/H| = n$ is finite. Index the left cosets of G/H as $g_1H, \ldots, g_nH$.

(a) For $g \in G$ and $1 \leq i \leq n$, show there is a unique $1 \leq j \leq n$ so that $gg_iH = g_jH$. Define $\sigma(g) : \{1, \ldots, n\} \to \{1, \ldots, n\}$ by

$$gg_iH = g_{(\sigma(g))(i)}H.$$

(b) Show that $\sigma(g)$ is bijective so that the map $g \to \sigma(g)$ gives a map $\sigma : G \to S_n$.

(c) Show $\sigma : G \to S_n$ is a homomorphism (sometimes this is called the *Extended Cayley Theorem* since $H = \{e\}$ is Cayley's Theorem).

(d) Show $\ker \sigma = \bigcap_{g \in G} c_g H$ (cf. Exercise 2.149) so that $\sigma : G \to S_n$ is injective if and only if $\bigcap_{i=1}^{n} c_{g_i} H = \{e\}$.

(e) For any $i, j \in \{1, \ldots, n\}$, show there exists $g \in G$ so that $(\sigma(g))(i) = j$. In such settings, G is said to act *transitively* on $\{1, \ldots, n\}$ (cf. Definition 2.63).

(f) Suppose $\pi : G \to S_n$ is a homomorphism for which G acts transitively. Let $H = \{g \in G \mid (\pi(g))(1) = 1\}$. Show that $|G/H| = n$ and that the left cosets g_iH can be labeled so that π is the same as σ from part (a).

EXERCISE 2.216. Let $G = GL(2, \mathbb{Z}_2)$ and let U be the subgroup of G consisting of upper triangular matrices with [1]'s on the diagonal. Show $|G| = 6$ and $|U| = 2$. Use the results of Exercise 2.215 to show

$$GL(2, \mathbb{Z}_2) \cong S_3.$$

EXERCISE 2.217 (Existence of Certain Normal Subgroups). Let H be a subgroup of G, $H \neq G$, so that $|G/H| = n$ is finite.

(a) If $|G| \nmid n!$, show G contains a normal subgroup $N \neq \{e\}$ that is contained in H. *Hint:* Look at the kernel of the permutation representation $G \to S_n$ from Exercise 2.215.

(b) If $|G| \geq 3$ and $|G| \nmid (n!/2)$, then G contains a proper normal subgroup. *Hint:* Reduce to the case were the kernel of $G \to S_n$ is trivial and apply Exercises 2.215 and 2.206.

EXERCISE 2.218. Suppose p is the minimal positive prime dividing $|G|$ and that H is a subgroup of G so that $|G/H| = p$. Let $\varphi : G \to S_p$ be the homomorphism from Exercise 2.215.

(a) Show $|G/\ker \varphi| \mid p!$.

(b) Show $\ker \varphi = H$. Conclude H is normal.

EXERCISE 2.219. **(a)** Although $(1, 2, 3, 4, 5)$ and $(2, 1, 3, 4, 5)$ are conjugate in S_5, show they are not conjugate in A_5.

(b) For $n \geq 5$, show the set of 3-cycles is a single conjugacy class in A_n.

EXERCISE 2.220 (Conjugacy Classes of A_n). Let $(n_1, \ldots, n_k)$ be the partition associated to $\sigma \in S_n$.

(a) Show $\sigma \in A_n$ if and only if there are an even number of n_j's that are even.

(b) Suppose $\sigma \in A_n$ and that at least one n_j is even. If $\sigma' \in A_n$ is conjugate to σ by an element of S_n, show σ' is conjugate to σ by an element of A_n

(c) Suppose $\sigma \in A_n$ with all n_j odd and $n_j = n_{j+1}$ for some j, $1 \leq j < k$. If $\sigma' \in A_n$ is conjugate to σ by an element of S_n, show σ' is conjugate to σ by an element of A_n

(d) Suppose $\sigma \in A_n$ with all n_j odd and $n_j > n_{j+1}$. Show that the conjugacy class of σ under S_n breaks into exactly two conjugacy classes under A_n.

(e) Conclude that the conjugacy classes of A_n are again indexed by the associated partition $(n_1, \ldots, n_k)$ except in the case where all n_j are odd and strictly decreasing. In that case there are exactly two conjugacy classes associated to the partition.

EXERCISE 2.221. View S_n as the subgroup of S_{n+1} fixing the element $n+1$. Under this convention, show $A_n = S_n \cap A_{n+1}$.

EXERCISE 2.222. Show that the symmetry group of the tetrahedron is isomorphic to A_4.

EXERCISE 2.223 (Simple Reflections). Let $r_i = (i, i+1) \in S_n$, $1 \leq i < n$. (In the setting of the theory of the Lie algebra $\mathfrak{sl}_{n+1}$, r_i is called a *simple reflection.*)

(a) Show $\{r_1, \ldots, r_{n-1}\}$ generates S_n.

(b) Fix $\sigma \in S_n$. For $\sigma \neq \text{Id}$, write $\sigma = r_{i_1} \cdots r_{i_N}$ with N as small as possible. Then $r_{i_1} \cdots r_{i_N}$ is called a *reduced* expression for σ and the *length* of σ is defined by $l(\sigma) = N$. For $\sigma = \text{Id}$, let $l(\text{Id}) = 0$. Calculate the length of $(1,2,3) \in S_3$.

(c) Show $\text{sgn}(\sigma) = (-1)^{l(\sigma)}$.

EXERCISE 2.224. Let $\Delta = \{\text{ordered pairs } (i,j) \mid 1 \leq i,j \leq n,\ i \neq j\}$, $\Delta^+ = \{(i,j) \in \Delta \mid i < j\}$, and $\Delta^- = \{(i,j) \in \Delta \mid i > j\}$. For $\sigma \in S_n$ and $(i,j) \in \Delta$, define $\sigma \cdot (i,j) = (\sigma(i), \sigma(j))$. Let $n(\sigma) = |\{(i,j) \in \Delta^+ \mid \sigma \cdot (i,j) \in \Delta^-\}|$.

(a) For $(1,2,3,4,5) \in S_5$, calculate $n(1,2,3,4,5)$.

(b) Using the notation of Exercise 2.223, show that $r_i \cdot (i, i+1) \in \Delta^-$ and that $r_i \cdot (\{\Delta^+ \backslash \{(i,i+1)\}\}) = \Delta^+ \backslash \{(i,i+1)\}$.

(c) Show

$$n(\sigma r_i) = \begin{cases} n(\sigma) - 1 & \text{if } \sigma \cdot (i,j) \in \Delta^-, \\ n(\sigma) + 1 & \text{if } \sigma \cdot (i,j) \in \Delta^+. \end{cases}$$

(d) Show $n(w) \leq l(w)$.

(e) Use induction on the length to show $n(\sigma) = l(\sigma)$.

(f) Calculate the length of $(1, 2, \ldots, n) \in S_n$.

EXERCISE 2.225. Fix an automorphism $\pi : S_n \to S_n$ that maps transpositions to transpositions. For $1 \leq i \leq n$, let $T_i = \{(k,i) \mid 1 \leq k \leq n,\ k \neq i\}$. View S_{n-1} as the subgroup of S_n fixing the element n.

(a) Use (non)commutivity relations to show $\pi T_i = T_j$ for some $1 \leq j \leq n$.

(b) Show there exists $\sigma_n \in S_n$ so that $c_{\sigma_n} \pi T_n = T_n$.

(c) Conclude that $c_{\sigma_n} \pi$ is an automorphism mapping transpositions to transpositions that preserves $S_{n-1} \subseteq S_n$.

(d) Use induction on n to show there exists $\sigma \in S_n$ so that $\pi = c_\sigma$. In other words, conclude that π is an inner automorphism.

EXERCISE 2.226 (Automorphisms of S_n, $n \neq 6$). Fix an automorphism $\pi : S_n \to S_n$.

(a) Show π maps conjugacy classes to conjugacy classes.

(b) Show that the only conjugacy classes of elements of order 2 have associated partition of the form $P_{n,k} = (\overbrace{2, \ldots, 2}^{k}, 1, \ldots, 1)$ for $n \geq 2k$. Show that the number of elements in $P_{n,k}$ is

$$\frac{n!}{2^k k!(n-2k)!}.$$

(c) Show $|P_{n,1}| = |P_{n,k}|$ if and only if $k = 1$ or $(n,k) = (6,3)$. *Hint:* If $k \geq 2$, reduce to the equation

$$\begin{aligned} k &= \frac{(n-2)(n-3)}{2\,(k-1)} \cdots \frac{(n-2k+2j)(n-2k+2j-1)}{2j} \\ &\quad \cdots \frac{(n-2k+2)(n-2k+1)}{2} \\ &\geq \prod_{j=1}^{k-1} (2j-1) \end{aligned}$$

to first see $k \leq 3$. Then look at $k = 2, 3$ individually.

(d) Use Exercise 2.225 to conclude that when $n \neq 6$, all automorphisms of S_n are inner.

EXERCISE 2.227 (Outer Automorphism of S_6). **(a)** Identify the set $\{1, \ldots, 5\}$ with $\{[1], \ldots, [5]\} = \mathbb{Z}_5$. For any $a, b \in \mathbb{Z}_5$, $a \neq 0$, show the map $x \to ax + b$ is a bijection of $\{1, \ldots, 5\}$.

(b) Show $H = \{\text{maps of the form } x \to ax + b \mid a \neq 0\}$ is a subgroup of order 20 in S_5.

(c) Use Exercise 2.215 to show there is an injective homomorphism $\sigma : S_5 \to S_6$ for which S_5 acts transitively on $\{1, \ldots, 6\}$.

(d) Use the subgroup $\sigma(S_5) \cong S_5$ and Exercise 2.215 to show there is an automorphism $\pi : S_6 \to S_6$ satisfying $\{g \in S_6 \mid (\pi(g))\,(1) = 1\} = \sigma(S_5)$.

(e) Show that if π were inner, then $\{g \in S_6 \mid (\pi(g))\,(1) = 1\}$ would not act transitively on $\{1, \ldots, 6\}$. Conclude that π is an outer automorphism.

(f) Recalling Exercise 2.226, show that π flips the conjugacy classes $P_{6,1}$ and $P_{6,3}$ and that π^2 is inner. Show that the set of all automorphisms is generated by π and the inner automorphisms.

EXERCISE 2.228 (Automorphisms of A_n). **(a)** Show that the automorphism group of A_2 is trivial.

(b) Show that the automorphism group of A_3 is isomorphic to $\mathbb{Z}_2$ with trivial inner automorphism group.

(c) Using the results of Exercise 2.220, modify the techniques of Exercise 2.226 to apply to elements of order 3 in A_n. For $n \geq 4$, $n \neq 6$, show the automorphism group of A_n is given by S_n acting by conjugation.

(d) For A_6, use Exercise 2.227 to show the automorphism group is generated by S_6 and an additional element of order 2 that interchanges the conjugacy class associated to the partition $(3,1,1,1)$ and the conjugacy class associated to the partition $(3,3)$.

EXERCISE 2.229 (Determinants). An *alternating n-form* on $\mathbb{R}^n$ is a function $d : \overbrace{\mathbb{R}^n \times \cdots \times \mathbb{R}^n}^{n \text{ copies}} \to \mathbb{R}$ that is linear in each coordinate, i.e.

$$\begin{aligned} d(v_1, &\ldots, v_{i-1}, \mathbf{v}_i + \mathbf{c}\mathbf{v}'_i, v_{i+1}, \ldots, v_n) \\ &= d(v_1, \ldots, v_{i-1}, \mathbf{v}_i, v_{i+1}, \ldots, v_n) + \mathbf{c}d(v_1, \ldots, v_{i-1}, \mathbf{v}'_i, v_{i+1}, \ldots, v_n), \end{aligned}$$

and satisfies

$$(2.59)\quad d(v_1, \ldots, v_{i-1}, \mathbf{v}_i, v_{i+1}, \ldots, v_{j-1}, \mathbf{v}_j, v_{j+1}, \ldots, v_n) = -d(v_1, \ldots, v_{i-1}, \mathbf{v}_j, v_{i+1}, \ldots, v_{j-1}, \mathbf{v}_i, v_{j+1}, \ldots, v_n)$$

for $v_i, v_i' \in \mathbb{R}^n$, $c \in \mathbb{R}$, and $1 \leq i, j \leq n$, $i \neq j$. Write e_i for a standard basis vector of $\mathbb{R}^n$.

(a) Show the function $d_2 : \mathbb{R}^2 \times \mathbb{R}^2 \to \mathbb{R}$ given by $d_2((a,b),(c,d)) = ad - bc$ is an alternating 2-form on $\mathbb{R}^2$ that satisfies $d_2(e_1, e_2) = 1$.

(b) Show (2.59) is equivalent to requiring

$$d(v_1, \ldots, v_{i-1}, \mathbf{v}, v_{i+1}, \ldots, v_{j-1}, \mathbf{v}, v_{j+1}, \ldots, v_n) = 0$$

for $v \in \mathbb{R}^n$.

(c) If $\sigma \in S_n$ and d is an alternating n-form, show

$$d(v_{\sigma(1)}, \ldots, v_{\sigma(n)}) = \operatorname{sgn}(\sigma)\, d(v_1, \ldots, v_n).$$

(d) By writing vectors in terms of the basis $\{e_1, \ldots, e_n\}$, show that an alternating n-form is uniquely determined by its value on $(e_1, \ldots, e_n)$.

(e) Let $d_n : \mathbb{R}^n \times \cdots \times \mathbb{R}^n \to \mathbb{R}$ be defined by

$$d_n((v_{1,1}, \ldots, v_{1,n}), \ldots, (v_{n,1}, \ldots, v_{n,n})) = \sum_{\sigma \in S_n} \operatorname{sgn}(\sigma)\, v_{1,\sigma(1)} \cdots v_{n,\sigma(n)}$$

for $v_{i,j} \in \mathbb{R}$. Show d_n is the unique alternating n-form on $\mathbb{R}^n$ satisfying $d_n(e_1, \ldots, e_n) = 1$.

(f) By mapping vectors to rows, show a function $d : \mathbb{R}^n \times \cdots \times \mathbb{R}^n \to \mathbb{R}$ can be viewed as a function $d : M_{n,n}(\mathbb{R}) \to \mathbb{R}$. When this is done, the function d_n defined above is the *determinant*.

8. Group Actions

Sometimes groups arise very abstractly in a vacuum. More commonly, however, groups arise because they are operating on another space in some special way. For example, think of the set of rotations of $\mathbb{R}^2$ about the origin. Recall from linear algebra that such a rotation can be accomplished by multiplying the (transpose of the) vector (x, y) on the left by the matrix

$$R_\theta = \begin{pmatrix} \cos\theta & -\sin\theta \\ \sin\theta & \cos\theta \end{pmatrix}.$$

Since a counterclockwise rotation through an angle θ can be naturally identified with the complex number $e^{2\pi i\theta} \in S^1$ (Exercise 2.8), we can think of the group S^1 as acting on $\mathbb{R}^2$ by rotations. In fancier language, given $e^{2\pi i\theta} \in S^1$ and $(x,y) \in \mathbb{R}^2$, we will soon write $e^{2\pi i\theta} \cdot (x, y)$ to indicate the act of rotating the vector (x, y) through an angle of θ (note that the "$\cdot$" does not represent vector multiplication here). This action has two important properties. The first is that the identity element, $1 = e^{2\pi i 0}$, acts trivially, i.e., $1 \cdot (x,y) = (x,y)$, and the second is that composition of the action respects the group operation, i.e., $e^{2\pi i\theta_1} \cdot \left(e^{2\pi i\theta_2} \cdot (x,y)\right) = e^{2\pi i(\theta_1+\theta_2)} \cdot (x,y)$. These two properties are the basis for a general definition of a group action.

DEFINITION 2.60. Let G be a group and let X be a set. A map $\pi : G \times X \to X$ is called an *action* and X a *G-set* if:

(1) $\pi(e, x) = x$ for all $x \in X$.

(2) $\pi(g_1, \pi(g_2, x)) = \pi(g_1 g_2, x)$ for all $g_1, g_2 \in G$ and $x \in X$.

When π is understood, $\pi(g, x)$ is most often written as $g \cdot x$. In that case, condition (1) becomes

$$e \cdot x = x$$

and condition (2) takes the form

$$g_1 \cdot (g_2 \cdot x) = (g_1 g_2) \cdot x.$$

In fact, $g \cdot x$ is frequently further abbreviated by simply writing gx.

As a first example, of course, we have the action of S^1 on $\mathbb{R}^2$ from above that is given by

$$e^{2\pi i\theta} \cdot \begin{pmatrix} x \\ y \end{pmatrix} = \begin{pmatrix} \cos\theta & -\sin\theta \\ \sin\theta & \cos\theta \end{pmatrix} \begin{pmatrix} x \\ y \end{pmatrix} = \begin{pmatrix} x\cos\theta - y\sin\theta \\ x\sin\theta + y\cos\theta \end{pmatrix}.$$

As already discussed, this action rotates points in the plane about the origin. It is geometrically clear that it satisfies the axioms of an action since clearly (i) rotation by $\theta = 0$ is the identity map and (ii) rotation by θ_1 followed by rotation by θ_2 is the same as rotation by $\theta_1 + \theta_2$. Of course these facts can be verified directly using some trigonometric identities (Exercise 2.231).

As another example, consider $GL(n, \mathbb{R})$ and $\mathbb{R}^n$. If $g \in GL(n, \mathbb{R})$ and $v \in \mathbb{R}^n$, define an action by $g \cdot v = gv$ where gv is given by standard matrix multiplication. It is well known that $I_n v = v$ so the first axiom is satisfied. It is also well known that matrix multiplication is associative so that $g_1(g_2 v) = (g_1 g_2)v$, where $g_1, g_2 \in GL(n, \mathbb{R})$, and the second axiom is satisfied.

As another example, consider S_n and $X = \{1, 2, \ldots, n\}$. If $\sigma \in S_n$ and $k \in X$, define an action by $\sigma \cdot k = \sigma(k)$. Of course $\mathrm{Id} \cdot k = k$ and

$$\sigma_1 \cdot (\sigma_2 \cdot k) = \sigma_1(\sigma_2(k)) = (\sigma_1 \circ \sigma_2)(k) = (\sigma_1 \sigma_2) \cdot k$$

for $\sigma_1, \sigma_2 \in S_n$. Thus $\cdot$ is an action.

In general, a group comes equipped with a few intrinsic actions on itself.

DEFINITION 2.61. Let G be a group and let $g, h \in H$. Define *left multiplication* $l_g : G \to G$ and *right multiplication* $r_g : G \to G$ by

$$l_g(h) = gh,$$
$$r_g(h) = hg^{-1}.$$

THEOREM 2.62. *Let G be a group and let $g, h \in H$. The three maps $g \cdot h = l_g(h)$, $g \cdot h = r_g(h)$, and $g \cdot h = c_g(h)$ all define actions of G on G.*

PROOF. The fact that l_g is an action is simply the identity and associative law of a group. The fact that r_g is an action follows by checking

$$r_{g_1} \cdot (r_{g_2} \cdot h) = hg_2^{-1}g_1^{-1} = h(g_1 g_2)^{-1} = r_{g_1 g_2} \cdot h$$

for $g_1, g_2 \in G$. The fact that c_g is an action follows easily from the observation that $c_g = l_g \circ r_{g^{-1}} = r_{g^{-1}} \circ l_g$. □

Attached to any action are a number of standard constructions. We already have run into most of these in the context of the classification of the conjugacy classes of S_n.

DEFINITION 2.63. Let G be a group with an action on a set X. Let $g \in G$ and $x \in X$.

(1) The *orbit* of x is $\mathcal{O}_x = \{g \cdot x \mid g \in G\}$.

(2) The action is *transitive* if for each $x, y \in X$ there exists $g \in G$ so that $g \cdot x = y$.

(3) The *stabilizer* of x is $G_x = \{g \in G \mid g \cdot x = x\}$.

(4) The *kernel* of the action is $\{g \in G \mid g \cdot x = x \text{ for all } x \in X\} = \bigcap_{x \in X} G_x$ and the action is called *faithful* (sometimes called *effective*) if the kernel is $\{e\}$.

(5) If $H \subseteq G$, the *fixed point set* of H on X is $X^H = \{x \in X \mid h \cdot x = x \text{ for all } h \in H\}$.

As an example, consider the action of S^1 on $\mathbb{R}^2$ given by rotation about the origin. The orbit through $x \in \mathbb{R}^2$ is $\mathcal{O}_x = \{g \cdot x \mid g \in S^1\}$. From obvious geometrical considerations, it follows that $\mathcal{O}_x = \{y \in \mathbb{R}^2 \mid \|y\| = \|x\|\}$, the circle of radius $\|x\|$. This action is clearly not transitive since we cannot get from, say, $(1, 0)$ to $(2, 0)$ by a rotation. In general, an action is transitive if and only if the orbit of each point is the entire space. Of course we could obtain a transitive action of S^1 by restricting our attention to a single circle.

Turning to stabilizers, the stabilizer of the origin is $\left(S^1\right)_0 = \{g \in S^1 \mid g \cdot 0 = 0\}$ and so $\left(S^1\right)_0 = S^1$ since any rotation preserves the origin. Similarly, the stabilizer of any nonzero $x \in \mathbb{R}^2$ is $\left(S^1\right)_x = \{g \in S^1 \mid g \cdot x = x\}$ so that $\left(S^1\right)_x = \{1\}$ since the only rotation fixing x is the trivial rotation (here $e = 1 = e^{2\pi i 0}$). As the kernel of the action is $\bigcap_{x \in \mathbb{R}^2} \left(S^1\right)_x$, it follows that the kernel is $\{1\}$ and that the action is faithful.

Fixed point set calculations are not very interesting here. For instance, consider $H = \{e^{\frac{2\pi i k}{4}} \mid k \in \mathbb{Z}\} \subseteq S^1$ which is isomorphic to $\mathbb{Z}_4$. As H acts on $\mathbb{R}^2$ via rotation by multiples of $\frac{\pi}{4}$, the only point fixed by all of H is the origin. Thus $\left(\mathbb{R}^2\right)^H = \{0\}$.

As another example, consider the action of $GL(n, \mathbb{R})$ on $\mathbb{R}^n$ given by matrix multiplication on the left. The orbit of 0 is $\mathcal{O}_0 = \{g \cdot 0 \mid g \in GL(n, \mathbb{R})\}$ so clearly $\mathcal{O}_0 = \{0\}$. On the other hand, if $x, y \in \mathbb{R}^n$ are nonzero, it is known that there exists a $g \in GL(n, \mathbb{R})$ so that $gx = y$ (in fact it is known that given any two bases, there is some $g \in GL(n, \mathbb{R})$ taking one base to the other). Therefore the orbit of x is $\mathcal{O}_x = \{g \cdot x \mid g \in GL(n, \mathbb{R})\} = \mathbb{R}^n \backslash \{0\}$. Thus the action of $GL(n, \mathbb{R})$ on $\mathbb{R}^n$ is not transitive (since we cannot take nonzero vectors to zero vectors). However, if we were to restrict our attention to an action of $GL(n, \mathbb{R})$ on $\mathbb{R}^n \backslash \{0\}$, we would obtain a transitive action.

Turning to stabilizers, we calculate the stabilizer of e_1, the first standard basis vector. Since $GL(n, \mathbb{R})_{e_1} = \{g \in GL(n, \mathbb{R}) \mid ge_1 = e_1\}$, it trivially follows that

$$GL(n, \mathbb{R})_{e_1} = \{\begin{pmatrix} 1 & b \\ 0_{1 \times (n-1)} & d \end{pmatrix} \mid b \in \mathbb{R}^{n-1} \text{ and } d \in GL(n-1, \mathbb{R})\}$$

where we have written the matrix with respect to $1 \times (n-1)$ blocks. Other stabilizer calculations are similar. Turning to the related concept of the kernel, the kernel of the action is $\{g \in GL(n, \mathbb{R}) \mid gx = x, \text{ all } x \in \mathbb{R}^n\}$. Of course this condition is the same as saying that $g = I_n$ (e.g., look at $x = e_1, \ldots, e_n$). Thus the kernel is $\{I_n\}$ and the action is faithful.

Turning to fixed point sets, we calculate the fixed point set of the subgroup

$$H = \{\begin{pmatrix} 1 & b \\ 0_{1\times(n-1)} & d \end{pmatrix} \mid b \in \mathbb{R}^{n-1} \text{ and } d \in GL(n-1,\mathbb{R})\}.$$

Then $(\mathbb{R}^n)^H = \{x \in \mathbb{R}^n \mid hx = x, \text{ all } h \in H\}$. Note that if $x \in \mathbb{R}^n$ is nonzero, the condition $hx = x$ is equivalent to saying that x is an eigenvector of h with eigenvalue 1. In any case, obviously every $x \in \mathbb{R}e_1$ satisfies $hx = x$ for all $h \in H$. Therefore $\mathbb{R}e_1 \subseteq (\mathbb{R}^n)^H$. In fact, we claim that $\mathbb{R}e_1 = (\mathbb{R}^n)^H$. To see this, pick $x \notin \mathbb{R}e_1$ and write $x = (x_1, \ldots, x_n)$. Then $x_i \neq 0$ for some i, $2 \leq i \leq n$. To show $x \notin (\mathbb{R}^n)^H$, we merely need to find some $h \in H$ so that $hx \neq x$. In fact, many h's will do the trick. Probably the simplest choice is $h = \text{diag}(1, \ldots, 1, 2, 1, \ldots, 1)$ with the 2 appearing in the ith position.

As a final example, consider the natural action of S_n on $X = \{1, \ldots, n\}$. The orbit through i is $\mathcal{O}_i = \{\sigma(i) \mid \sigma \in S_n\}$. Since any $i \in X$ can be moved to any other $j \in X$ using, say, the transposition (i, j), it follows that $\mathcal{O}_i = X$. In particular, X consists of a single orbit and the action is transitive.

As a special note, if we fix $\sigma \in S_n$ and look only at the restricted action of $\langle\sigma\rangle$ on X, then we have already extensively studied the orbit structure of $\langle\sigma\rangle$ on X in §7.2 and §7.3. Roughly speaking, we found that the orbits give rise to the cyclic decomposition of σ and also to its associated partition classifying its conjugacy class.

Returning to the action of S_n on X, the stabilizer of n is $(S_n)_n = \{\sigma \in S_n \mid \sigma(n) = n\}$. Since this means σ is really just a permutation of $\{1, \ldots, n-1\}$, it is easy to see that $(S_n)_n$ can be identified with S_{n-1}. In fact, $(S_n)_i \cong S_{n-1}$ for any $1 \leq i \leq n$ (Exercise 2.232).

Turning to fixed point sets, we calculate the order of X^σ for any $\sigma \in S_n$. Since $X^\sigma = \{i \mid \sigma(i) = i\}$, we already know from the proof of the Cyclic Decomposition Theorem that $|X^\sigma|$ is the number of 1's appearing in the associated partition of σ.

THEOREM 2.64. *Let G be a group with an action on a set X. Let $g \in G$ and $x, y \in X$. Define a relation on X by $x \sim y$ if there exists $g \in G$ so $y = g \cdot x$, i.e., if $y \in \mathcal{O}_x$.*

(1) *The relation $x \sim y$ is an equivalence relation on X whose equivalence classes are the orbits.*

(2) *G_x is a subgroup of G.*

(3) *$G_{g\cdot x} = gG_xg^{-1}$ so that the stabilizers of each point in an orbit are isomorphic.*

PROOF. For part (1), first observe that $e \cdot x = x$ and that $y = g \cdot x$ if and only if $g^{-1} \cdot y = x$. Finally, if $y = g_1 \cdot x$ and $z = g_2 \cdot y$ for $z \in X$ with $g_1, g_2 \in G$, then $z = (g_2 g_1) \cdot x$. Thus part (1) follows.

For part (2), first note that $e \in G_x$. Next, if $g_1, g_2 \in G_x$, then

$$(g_1 g_2) \cdot x = g_1 \cdot (g_2 \cdot x) = g_1 \cdot x = x$$

so that $g_1 g_2 \in G_x$. Finally, observe that $g_1 \cdot x = x$ if and only if $x = g_1^{-1} \cdot x$ so that $g_1^{-1} \in G_x$ and part (2) is done.

For part (3), recall that the notation gG_xg^{-1} means $\{ghg^{-1} \mid h \in G_x\}$. Now $k \in G_{g\cdot x}$ if and only if $k \cdot (g \cdot x) = g \cdot x$ if and only if $(g^{-1}kg) \cdot x = x$ if and only if $g^{-1}kg \in G_x$ if and only if $k \in gG_xg^{-1}$ as desired. □

A simple variant of the First Isomorphism Theorem is the following.

THEOREM 2.65 (Orbit Stabilizer Theorem). *Let G be a group with an action on a set X. Let $x \in X$.*

(1) *There is a bijection from G/G_x to $\mathcal{O}_x$ induced by mapping $g \to g \cdot x$.*
(2) *If G is finite, then $|\mathcal{O}_x| = |G| / |G_x|$.*

PROOF. Consider the surjective map $\varphi : G \to \mathcal{O}_x$ given by $\varphi(g) = g \cdot x$. If $g_x \in G_x$, then $(gg_x) \cdot x = g \cdot x$. Therefore the map $\overline{\varphi} : G/G_x \to \mathcal{O}_x$ given by $\overline{\varphi}(gG_x) = \varphi(g)$ is well defined and still surjective. It only remains to see $\overline{\varphi}$ is injective. For this, observe $\overline{\varphi}(g_1G_x) = \overline{\varphi}(g_2G_x)$ if and only if $g_1 \cdot x = g_2 \cdot x$ if and only if $g_2^{-1}g_1 \in G_x$ if and only if $g_1G_x = g_2G_x$ as desired. □

The Orbit Stabilizer Theorem can often be used to efficiently count things. The following theorem is an example.

COROLLARY 2.66. *Let p be a positive prime. Then*

$$|GL(2, \mathbb{Z}_p)| = (p^2 - 1)(p^2 - p).$$

PROOF. By standard matrix multiplication, $GL(2, \mathbb{Z}_p)$ acts on $\mathbb{Z}_p^2 = \mathbb{Z}_p \times \mathbb{Z}_p$. Since matrix multiplication is linear, this action actually descends to an action on nonzero lines. Precisely, a *line* in $\mathbb{Z}_p^2$ is a set of the form $l_{x,y} = \{(tx, ty) \mid t \in \mathbb{Z}_p\}$ for any fixed $x, y \in \mathbb{Z}_p$. It is *nontrivial* if $(x, y) \neq ([0], [0])$. Write $\mathcal{L}_p$ for the set of nontrivial lines in $\mathbb{Z}_p^2$. When x is nonzero, it follows that $l_{x,y} = l_{[1],x^{-1}y}$. From this it is easy to see there are exactly $p + 1$ nontrivial lines in $\mathbb{Z}_p^2$: $l_{[1],[0]}, l_{[1],[1]}, \ldots, l_{[1],[p-1]}$, and $l_{[0],[1]}$.

We claim the action of $GL(2, \mathbb{Z}_p)$ on $\mathcal{L}_p$ is transitive. For this, it suffices to see $GL(2, \mathbb{Z}_p) \cdot l_{[1],[0]} = \mathcal{L}_p$ (Exercise 2.234), which is easy:

$$\begin{pmatrix} [1] & [0] \\ [k] & [1] \end{pmatrix} \begin{pmatrix} [1] \\ [0] \end{pmatrix} = \begin{pmatrix} [1] \\ [k] \end{pmatrix},$$
$$\begin{pmatrix} [0] & [-1] \\ [1] & [0] \end{pmatrix} \begin{pmatrix} [1] \\ [0] \end{pmatrix} = \begin{pmatrix} [0] \\ [1] \end{pmatrix}.$$

As a result, the Orbit Stabilizer Theorem shows $|GL(2, \mathbb{Z}_p)| = \left|GL(2, \mathbb{Z}_p)_{l_{[1],[0]}}\right| |\mathcal{L}_p|$. But $g \in GL(2, \mathbb{Z}_p)_{l_{[1],[0]}}$ if and only if $g \cdot l_{[1],[0]} = l_{[1],[0]}$ if and only if $g \cdot ([1], [0]) = ([t], [0])$ for nonzero $[t] \in \mathbb{Z}_p$. It is easy to see that the only elements of $GL(2, \mathbb{Z}_p)$ satisfying this property are of the form

$$\begin{pmatrix} a & b \\ [0] & d \end{pmatrix}$$

for $a, d \in U_p$ and $b \in \mathbb{Z}_p$. Thus $\left|GL(2, \mathbb{Z}_p)_{l_{[1],[0]}}\right| = p(p-1)^2$ and $|GL(2, \mathbb{Z}_p)| = p(p-1)^2(p+1)$ as desired. □

This result is easily extended to show

$$|GL(n, \mathbb{Z}_p)| = (p^n - 1)(p^n - p)(p^n - p^2) \cdots (p^n - p^{n-1})$$

(Exercise 2.242). An alternate proof can also be obtained by using a linear algebra fact that a matrix is invertible if and only if the columns are linearly independent and the fact that an m-dimensional subspace of $\mathbb{Z}_p^n$ has p^m points.

We end with another counting theorem that will have a number of important applications, especially in Section 9.

THEOREM 2.67 (Orbit Decomposition and Burnside's Counting Theorem). *Let G be a finite group with an action on a finite set X. Let $\mathcal{O}_{x_1}, \ldots, \mathcal{O}_{x_m}$ be the distinct orbits indexed so that the orbits $\mathcal{O}_{x_1}, \ldots, \mathcal{O}_{x_n}$ consist of more than a single point and the orbits $\mathcal{O}_{x_{n+1}}, \ldots, \mathcal{O}_{x_m}$ are singleton orbits.*

(1)

$$\left|X\right| = \left|X^G\right| + \sum_{i=1}^{n} \left|G\right| / \left|G_{x_i}\right|.$$

(2)

$$m = \frac{1}{|G|} \sum_{g \in G} \left|X^g\right|.$$

PROOF. For (1), use three facts: the orbits are the equivalence classes of an equivalence relation so that the sum of their orders is $|X|$, the union of the singleton orbits is X^G, and the order of $\mathcal{O}_{x_i}$ is $|G| / |G_{x_i}|$. For part (2), the trick is to let $\delta : G \times X \to \mathbb{R}$ be given by $\delta(g, x)$ is 1 if $g \cdot x = x$ and 0 otherwise. Note that

$$|X^g| = |\{x \in X \mid g \cdot x = x\}| = \sum_{x \in X} \delta(g, x)$$

and that

$$|G_x| = |\{g \in G \mid g \cdot x = x\}| = \sum_{g \in G} \delta(g, x).$$

Also recall that $G_{g \cdot x} = g G_x g^{-1}$ so that the order of a stabilizer is constant along an orbit. In particular if $y \in \mathcal{O}_{x_i}$, then $|G_y| = |G_{x_i}|$ and $\sum_{y \in \mathcal{O}_{x_i}} |G_y| = |\mathcal{O}_{x_i}| \, |G_{x_i}|$. Thus

$$\begin{aligned}
\sum_{g \in G} |X^g| &= \sum_{g \in G} \sum_{x \in X} \delta(g, x) = \sum_{x \in X} \sum_{g \in G} \delta(g, x) = \sum_{x \in X} |G_x| \\
&= \sum_{x \in \left(\mathcal{O}_{x_1} \cup \cdots \cup \mathcal{O}_{x_m}\right)} |G_x| = \sum_{i=1}^{m} \sum_{y \in \mathcal{O}_{x_i}} |G_y| = \sum_{i=1}^{m} |\mathcal{O}_{x_i}| \, |G_{x_i}| \\
&= \sum_{i=1}^{m} |G| = m\,|G|
\end{aligned}$$

as desired. □

If G is a group with $g_1, g_2 \in G$, recall that g_1 and g_2 are conjugate if there exists an $h \in G$ so that $g_2 = c_h g_1$ and recall that the set of all elements in G conjugate to some fixed $g \in G$ is called a conjugacy class.

COROLLARY 2.68 (The Class Equation). *Let G be a finite group with $C_1, \ldots, C_n$ the distinct nonsingleton conjugacy classes and let $g_i \in C_i$. Then*

$$|G| = |Z(G)| + \sum_{i=1}^{n} |G| / |Z(g_i)|.$$

PROOF. Use the conjugation action of G on itself, $g \cdot h = g h g^{-1}$ for $g, h \in G$, and apply the Orbit Decomposition Theorem. □

Notice that if $n \geq 1$, then $|Z(g_i)| < |G|$ or else g_i would be central. In that case it follows that $|G| / |Z(g_i)| \geq 2$ so that $|G| \geq |Z(G)| + 2n$.

8.1. Exercises 2.230–2.248.

EXERCISE 2.230. For each of the following, (i) show an action is defined, (ii) describe each orbit, (iii) decide whether the action is transitive, (iv) calculate the stabilizer of each point, (v) calculate the kernel and decide whether the action is faithful, and (vi) calculate the fixed point set of the whole group:

(a) $r \cdot z = rz$ for $r \in \mathbb{R}^+$ and $z \in \mathbb{C}$.
(b) $x \cdot z = e^{ix} z$ for $x \in \mathbb{R}$ and $z \in \mathbb{C}$.
(c) $r \cdot s = s^r$ for $r, s \in \mathbb{R}^+$.
(d) $[a] \cdot x = (-1)^a x$ for $[a] \in \mathbb{Z}_2$ and $x \in \mathbb{R}$.
(e) $[0] \cdot (x, y) = (x, y)$ and $[1] \cdot (x, y) = (y, x)$ for $[0], [1] \in \mathbb{Z}_2$ and $(x, y) \in \mathbb{R}^2$.
(f) $\lambda \cdot v = \lambda v$ for $\lambda \in \mathbb{R}^+$ and $v \in \mathbb{R}^n$.
(g) $g \cdot v = gv$ for $g \in O(n, \mathbb{R})$ and $v \in \mathbb{R}^n$.

EXERCISE 2.231. Show directly that the formula

$$e^{2\pi i\theta} \cdot \begin{pmatrix} x \\ y \end{pmatrix} = \begin{pmatrix} \cos\theta & -\sin\theta \\ \sin\theta & \cos\theta \end{pmatrix} \begin{pmatrix} x \\ y \end{pmatrix} = \begin{pmatrix} x\cos\theta - y\sin\theta \\ x\sin\theta + y\cos\theta \end{pmatrix}$$

gives an action of S^1 on $\mathbb{R}^2$. *Hint:* Trigonometric identities.

EXERCISE 2.232. Let S_n act on $X = \{1, \ldots, n\}$ by $\sigma \cdot i = \sigma(i)$, $1 \leq i \leq n$. Show $(S_n)_i \cong S_{n-1}$.

EXERCISE 2.233. Suppose a group G has an action on a set X. Show the following are equivalent.

(a) The action is faithful.
(b) For every distinct $g, h \in G$, there is an $x \in X$ so that $g \cdot x \neq h \cdot x$.
(c) For any $g \in G$, $g \neq e$, there is an $x \in X$ so that $g \cdot x \neq x$.

EXERCISE 2.234. Suppose a group G has an action on a set X. Show the action is transitive if and only if there is an $x_0 \in X$ so $G \cdot x_0 = X$.

EXERCISE 2.235. Show a finite group G has $\frac{1}{|G|} \sum_{g \in G} |Z(g)|$ distinct conjugacy classes.

EXERCISE 2.236. If H is a subgroup of G, show H is normal if and only if it is a union of conjugacy classes.

EXERCISE 2.237 (Linear Fractional Transformations). Let $\widehat{\mathbb{C}} = \mathbb{C} \cup \{\infty\}$. If $g = \begin{pmatrix} a & b \\ c & d \end{pmatrix} \in GL(2, \mathbb{C})$, define

$$g \cdot z = \frac{az + b}{cz + d}$$

for $z \in \widehat{\mathbb{C}}$ where $\frac{a\infty + b}{c\infty + d} = \frac{a}{c}$ and $\frac{w}{0} = \infty$ for $w \neq 0$.

(a) Show this defines an action of $GL(2, \mathbb{C})$ on $\widehat{\mathbb{C}}$, called the *linear fractional action.*

(b) Show the kernel of the action is the center of $GL(2, \mathbb{C})$, $\{\begin{pmatrix} \lambda & 0 \\ 0 & \lambda \end{pmatrix} \mid \lambda \in \mathbb{C}^\times\}$.

(c) Find the stabilizer of 0 and find the stabilizer of ∞.
(d) Show the action is transitive (actually it is 3-transitive, Exercise 2.238).

EXERCISE 2.238 (n-transitive). An action of G on X is said to be *n-transitive* if for each distinct $x_1, \ldots, x_n \in X$ and distinct $y_1, \ldots, y_n \in X$ there is a $g \in G$ so $g \cdot x_i = y_i$, $1 \leq i \leq n$.

(a) Show the action of S_n on $\{1, \ldots, n\}$ is n-transitive.

(b) Show the linear fractional action of $GL(2, \mathbb{C})$ on $\widehat{\mathbb{C}}$ is 3-transitive (cf. Exercise 2.237). *Hint:* Given distinct $z_1, z_2, z_3 \in \widehat{\mathbb{C}}$, find a linear fractional transformation taking them to $1, 0, \infty$.

(c) Show the linear fractional action of $GL(2, \mathbb{C})$ on $\widehat{\mathbb{C}}$ is not 4-transitive.

EXERCISE 2.239. Let $\sigma \in S_n$ with associated partition $(n_1, n_2, \ldots, n_k)$ and let $m_j = |\{n_i \mid n_i = j,\ 1 \leq i \leq k\}|$. Write $\mathcal{C}_\sigma$ for the conjugacy class of σ in S_n.

(a) Using the conjugation action of S_n on itself, show

$$|(S_n)_\sigma| = 1^{m_1} 2^{m_2} \cdots n^{m_n}\, m_1! m_2! \cdots m_n!.$$

(b) Show

$$|\mathcal{C}_\sigma| = \frac{n!}{1^{m_1} 2^{m_2} \cdots n^{m_n}\, m_1! m_2! \cdots m_n!}.$$

EXERCISE 2.240. If $H \neq G$ is a subgroup of a finite group G, show $G \neq \bigcup_{g \in G} c_g H$. *Hint:* Reduce to the case of H not normal so that $|G/N(H)| \geq 2$, show the number of distinct sets of the form $c_g H$ is $|G| \,/\, |N(H)|$ (cf. Exercise 2.72), and show $\left|\bigcup_{g \in G} c_g H\right| \leq 1 + (|H - 1|)\,(|G| \,/\, |N(H)|)$.

EXERCISE 2.241 (Actions and $\mathrm{Bij}(X)$). Let X be a set and let $\mathrm{Bij}(X)$ be the set of bijections of X onto X. $\mathrm{Bij}(X)$ is a group under composition of functions.

(a) If $|X| = n < \infty$, show $\mathrm{Bij}(X) \cong S_n$.

(b) Let G be a group acting on X. For $g \in G$, define $\varphi_g : X \to X$ by $\varphi_g(x) = g \cdot x$, $x \in X$. Show $\varphi_g \in \mathrm{Bij}(X)$.

(c) Show the map $\varphi : G \to \mathrm{Bij}(X)$ given by $\varphi(g) = \varphi_g$ is a homomorphism.

(d) Show the kernel of the action is the same as $\ker \varphi$. Conclude that φ is injective if and only if the action is faithful.

(e) In any case, show (i) $G/\ker\varphi$ acts faithfully on X by $(g \ker \varphi) \cdot x = g \cdot x$, (ii) $G/\ker\varphi$ has the same orbits as G, and (iii) the corresponding homomorphism $G/\ker\varphi \to \mathrm{Bij}(X)$ is injective.

(f) Conversely, suppose $\psi : G \to \mathrm{Bij}(X)$ is a homomorphism. Show $g \cdot x = (\psi(g))\,(x)$ yields an action of G on X.

(g) Show there is a one-to-one correspondence between the set of actions of G on X and the set of homomorphisms $G \to \mathrm{Bij}(X)$.

EXERCISE 2.242 (Order of $GL(n, \mathbb{Z}_p)$ and Friends). **(a)** Show the number of nontrivial lines in $\mathbb{Z}_p^n$ is $p^{n-1} + p^{n-2} + \cdots + 1 = \frac{p^n - 1}{p-1}$ and that $GL(n, \mathbb{Z}_p)$ acts on this set by matrix multiplication.

(b) Let l be the line $\{(t, 0, \ldots, 0) \mid t \in \mathbb{Z}_p\}$. Show

$$|GL(n, \mathbb{Z}_p)_l| = (p-1)p^{n-1}\, |GL(n-1, \mathbb{Z}_p)|\,.$$

(c) Conclude that $|GL(n, \mathbb{Z}_p)| = (p^n - 1)p^{n-1}\, |GL(n-1, \mathbb{Z}_p)|$.

(d) Show

$$|GL(n, \mathbb{Z}_p)| = (p^n - 1)(p^n - p)(p^n - p^2) \cdots (p^n - p^{n-1}).$$

(e) Show $GL(n, \mathbb{Z}_p)/SL(n, \mathbb{Z}_p) \cong U_p$ and conclude

$$|SL(n, \mathbb{Z}_p)| = \frac{(p^n - 1)(p^n - p)(p^n - p^2) \cdots (p^n - p^{n-1})}{p - 1}.$$

(f) Recall $PGL(n, \mathbb{Z}_p) = GL(n, \mathbb{Z}_p)/Z(n, \mathbb{Z}_p)$ (cf. Exercise 2.169). Show

$$|PGL(n, \mathbb{Z}_p)| = \frac{(p^n - 1)(p^n - p)(p^n - p^2) \cdots (p^n - p^{n-1})}{p - 1}.$$

(g) Recall $PSL(n, \mathbb{Z}_p) = SL(n, \mathbb{Z}_p)/\left(Z(n, \mathbb{Z}_p) \cap SL(n, \mathbb{Z}_p)\right)$. Show

$$|PSL(n, \mathbb{Z}_p)| = \frac{(p^n - 1)(p^n - p)(p^n - p^2) \cdots (p^n - p^{n-1})}{(p - 1)\ (n, p - 1)}.$$

Hint: To calculate $|\{\zeta \in U_p \mid \zeta^n = [1]\}|$, use Exercise 2.179.

(h) Show

$$|PSL(2, \mathbb{Z}_p)| = \begin{cases} 6, & p = 2, \\ \frac{p^3 - p}{2}, & p \neq 2. \end{cases}$$

EXERCISE 2.243 (Symmetric Polynomials). Let $\mathbb{R}[x_1, \ldots, x_n]$ be the set of polynomials in the variables $x_1, \ldots, x_n$ with real coefficients. Define an action of S_n on $\mathbb{R}[x_1, \ldots, x_n]$ by

$$\sigma \cdot \left(\sum_{i=1}^{N} c_i \, x_1^{k_{1,i}} \cdots x_n^{k_{n,i}} \right) = \sum_{i=1}^{N} c_i \, x_{\sigma(1)}^{k_{1,i}} \cdots x_{\sigma(n)}^{k_{n,i}}$$

where $N \in \mathbb{N}$, $c_i \in \mathbb{R}$, and $k_{j,i} \in \mathbb{Z}_{\geq 0}$.

(a) Show this definition really defines an action of S_n on $\mathbb{R}[x_1, \ldots, x_n]$.

(b) A polynomial $p \in \mathbb{R}[x_1, \ldots, x_n]$ is said to be *symmetric* if $p \in \mathbb{R}[x_1, \ldots, x_n]^{S_n}$. For $0 \leq k \leq n$, let $q_k = x_1^k + \cdots + x_n^k$, called the *power sum symmetric polynomial* of degree k. Show q_k is symmetric.

(c) Define $s_0, s_1, \ldots, s_n \in \mathbb{R}[x_1, \ldots, x_n]$, called the *elementary symmetric polynomials*, so that

$$\prod_{j=1}^{n} (t - x_j) = \sum_{k=0}^{n} (-1)^k s_k \, t^{n-k}.$$

Show s_k is a symmetric polynomial of degree k satisfying

$$s_k = \sum_{1 \leq j_1 < j_2 < \cdots < j_k \leq n} x_{j_1} x_{j_2} \cdots x_{j_k}.$$

(d) Explicitly write out all the elementary symmetric polynomials for $n = 2, 3, 4$ (cf. Exercise 4.132).

EXERCISE 2.244. Suppose H and K are subgroups of G.

(a) Show $(h, k) \cdot g = hgk^{-1}$ yields an action of $H \times K$ on G. The orbits are called *double cosets*.

(b) Show $(H \times K)_g \cong H \cap c_g K \cong c_{g^{-1}} H \cap K$.

(c) Prove *Frobenius's Theorem*: If G is finite and $Hg_1K, \ldots, Hg_nK$ are the double cosets, then

$$|G| = \sum_{i=1}^{n} \frac{|H|\,|K|}{|H \cap c_{g_i} K|}.$$

EXERCISE 2.245 (Action on Function Spaces). Suppose G acts on X. Let $\mathcal{F}(X) = \{f : X \to \mathbb{C}\}$. For $g \in G$ and $f \in \mathcal{F}(X)$, define $g \cdot f \in \mathcal{F}(X)$ by

$$(g \cdot f)(x) = f(g^{-1} \cdot x)$$

for $x \in X$. Show this yields an action of G on $\mathcal{F}(X)$.

EXERCISE 2.246 (Colorings). Suppose G is a finite group acting on a finite set X. Suppose we have n "colors", $\{1, \ldots, n\}$. A *coloring* of X is then a map $f : X \to \{1, \ldots, n\}$ and we write $\mathcal{C}(X)$ for the set of colorings. Two colorings f and f' are called *indistinguishable* if there is a $g \in G$ so that $f(x) = f'(g \cdot x)$. They are called *distinct* otherwise.

(a) Show $|\mathcal{C}(X)| = n^{|X|}$.

(b) Show the following defines an action of G on $\mathcal{C}(X)$: $(g \cdot f)(x) = f(g^{-1} \cdot x)$ (cf. Exercise 2.245).

(c) Show the total number of distinct colorings is the number of orbits of G on $\mathcal{C}(X)$. Conclude that the number of distinct colorings is

$$\frac{1}{|G|} \sum_{g \in G} |\mathcal{C}(X)^g| .$$

(d) Suppose you are making bracelets with 4 beads. If you have n colors to choose from, show the total number of distinct bracelets you can make is $\frac{1}{8}(n^4 + 2n^3 + 3n^2 + 2n)$. *Hint:* What is the symmetry group of the bracelet?

EXERCISE 2.247 (G-isomorphisms). Suppose a group G acts on X and on Y. Say $\varphi : X \to Y$ is a *G-morphism* if

$$\varphi(g \cdot x) = g \cdot \varphi(x)$$

for $x \in X$. If φ is also a bijection, it is called a *G-isomorphism* and we write $X \cong Y$ as G-sets.

(a) If H is a subgroup of G, define $g_1 \cdot (g_2 H) = g_1 g_2 H$. Show this yields an action of G on G/H.

(b) Show G acts transitively on X if and only if $X \cong G/H$ for some subgroup H of G.

(c) Up to G-isomorphism, show that the most general space on which G acts is the disjoint union of sets of the form G/H.

(d) If K is a subgroup of G, show $G/H \cong G/K$ if and only if $K = c_{g_0} H$ for some $g_0 \in G$. In that case, show a G-isomorphism $\varphi : G/H \to G/K$ can be obtained using $\varphi(gH) = g g_0^{-1} K$.

(e) If $\varphi : X \to Y$ is a G-morphism, let $x_1 \sim x_2$ when $\varphi(x_1) = \varphi(x_2)$. Show this defines an equivalence relation on X. Let X/φ be the set of equivalence classes of ~ and show $g \cdot [x] = [g \cdot x]$ defines an action of G on X/φ. If φ is surjective, show $X/\varphi \cong Y$.

EXERCISE 2.248. With similar conventions to those in Exercise 2.237, let $\widehat{\mathbb{Z}}_p = \mathbb{Z}_p \cup \{\infty\}$ and if $g = \begin{pmatrix} [a] & [b] \\ [c] & [d] \end{pmatrix} \in GL(2, \mathbb{Z}_p)$, define

$$g \cdot z = \frac{[a]\, z + [b]}{[c]\, z + [d]}.$$

Show this gives an action of $GL(2, \mathbb{Z}_p)$ that is isomorphic to the action of $GL(2, \mathbb{Z}_p)$ on the set of nontrivial lines in $\mathbb{Z}_p^2$ (cf. Exercise 2.247).

9. Sylow Theorems

The converse to Lagrange's Theorem is false—for instance, $|A_4| = 12$ and $6 \mid 12$, but A_4 has no subgroup of order 6 (Exercise 2.211). However, there is a very powerful partial converse that we explore in this section.

It turns out that if p is a positive prime with p^k dividing the order of a finite group G, then G contains a subgroup of order p^k (cf. Lemma 2.39 for a special case). The proof of this theorem involves a more sophisticated use of induction than we have employed so far. It also uses the Class Equation from Section 8 in an essential way.

THEOREM 2.69. *Let G be a finite group and let p be a positive prime. If $p^k \mid |G|$ for some $k \in \mathbb{N}$, then G has a subgroup of order p^k.*

PROOF. We first prove the theorem in the special case where G is abelian. In that setting, the Fundamental Theorem of Abelian Groups shows that G has a subgroup isomorphic to $\mathbb{Z}_{p^{n_1}} \times \cdots \times \mathbb{Z}_{p^{n_m}}$ for $m \in \mathbb{N}$ with $k \leq n_1 + \cdots + n_m$. Write $k = n_1 + \cdots + n_{j-1} + k'$ with $0 < k' \leq n_j$. Observe that $\left\langle [p^{n_j - k'}] \right\rangle$ in $\mathbb{Z}_{p^{n_j}}$ has order $p^{k'}$. Thus the subgroup of $\mathbb{Z}_{p^{n_1}} \times \cdots \times \mathbb{Z}_{p^{n_m}}$ generated by the natural embeddings of $\mathbb{Z}_{p^{n_1}}, \ldots, \mathbb{Z}_{p^{n_{j-1}}}$ and $\left\langle [p^{n_j - k'}] \right\rangle$ generate a group of order p^k.

We turn now to the general case. Essentially we want to induct on the order of G. To make our meaning precise, fix a positive prime p and consider pairs (p^k, G) where $k \in \mathbb{N}$, G is a group, and $p^k \mid |G|$. Write $\mathcal{P}$ for the collection of all such pairs. We wish to show that if $(p^k, G) \in \mathcal{P}$, then G has a subgroup of order p^k.

At the very least, notice that $(p^k, G) \in \mathcal{P}$ implies that $|G| \geq p$. Of course if $|G| = p$, then $G \cong \mathbb{Z}_p$ and there is nothing really to do since the result is then obviously true. For $(p^k, G) \in \mathcal{P}$ with $|G| > p$, we use induction on $|G|$. Our inductive hypothesis is that any $(p^j, H) \in \mathcal{P}$ with $|H| < |G|$ satisfies the property that H has a subgroup of order p^j. Given that hypothesis, we must show that G has a subgroup of order p^k. There are two cases to consider: either $p \mid |Z(G)|$ or $p \nmid |Z(G)|$.

For the first case, suppose that $p \mid |Z(G)|$. Since $Z(G)$ is abelian, we have just seen there exists $z \in Z(G)$ so that the order of z is p. If $k = 1$, we are done since a subgroup of $Z(G)$ is also a subgroup of G. Otherwise, let $H = G / \langle z \rangle$ so that $|H| < |G|$ and $p^{k-1} \mid |H|$. By the inductive hypothesis, there is a subgroup K of H of order p^{k-1}. Writing $\pi : G \to G/\langle z \rangle$ for the surjective homomorphism $\pi(g) = g\langle z \rangle$, the subgroup $\pi^{-1}(K)$ of G has order p^k by the First Isomorphism Theorem and we are done.

For the second case, suppose that $p \nmid |Z(G)|$. Recall that the Class Equation shows $|G| = |Z(G)| + \sum_{i=1}^{n} |G| / |Z(g_i)|$ where $g_1, \ldots, g_n$ are representatives for the distinct nonsingleton conjugacy classes of G. If $p \mid (|G| / |Z(g_i)|)$ for all $1 \leq i \leq n$, the fact that $p \mid |G|$ and the Class Equation would lead to a contradiction of the assumption that $p \nmid |Z(G)|$. Thus there is an i so $p \nmid (|G| / |Z(g_i)|)$ which implies $p^k \mid |Z(g_i)|$. Since $|Z(g_i)| < |G|$ (g_i represents a nonsingleton conjugacy class), the inductive hypothesis shows that $Z(g_i)$, and therefore G, has a subgroup of order p^k. □

As an immediate consequence, we get the following (see Exercise 2.258 for a direct proof).

COROLLARY 2.70 (Cauchy's Theorem). *Let G be a finite group and let p be a positive prime. If $p \mid |G|$, then G has an element of order p.*

In light of Theorem 2.69, it makes sense to study subgroups of prime order. The results of that study will be quite illuminating.

DEFINITION 2.71. Let p be a positive prime and let G be a group. If each element of G has order a power of p, we say G is a *p-group*. If G is also a subgroup of K, then G is alternately called a *p-subgroup* of K.

For example, clearly $\mathbb{Z}_{p^n}$ is a p-group. For a nonabelian example, D_4 is a 2-group by Lagrange's Theorem and the fact that $|D_4| = 8$.

THEOREM 2.72. *Let p be a positive prime and let G be a finite group. Then G is a p-group if and only if $|G| = p^k$ for $k \in \mathbb{Z}_{\geq 0}$.*

PROOF. If G is a p-group and q is a positive prime dividing $|G|$, then Cauchy's Theorem says G has an element of order q. Since G is a p-group, q must be a power of p and so $q = p$. In particular, $|G| = p^k$ for $k \in \mathbb{Z}_{\geq 0}$. The converse follows from Lagrange's Theorem. □

DEFINITION 2.73. Let G be a finite group and let p be a positive prime dividing $|G|$. Let p^n be the largest power of p dividing $|G|$ (so that $|G| = p^n m$ with $n, m \in \mathbb{N}$ and $p \nmid m$). A *p-Sylow subgroup* is a subgroup of order p^n.

Just as the proof of Theorem 2.69 made essential use of the Class Equation, the proof of the following Sylow Theorems relies on counting tricks based on the more general Orbit Decomposition Theorem.

THEOREM 2.74 (The Three Sylow Theorems). *Let G be a finite group and let p be a positive prime dividing $|G|$. Write $|G| = p^n m$ with $p \nmid m$.*

(1) *A p-Sylow subgroup exists.*

(2) *All p-Sylow subgroups are conjugate; i.e., if P and P' are p-Sylow subgroups, there exists $g \in G$ so $c_g P' = P$. More generally, if H is a p-subgroup of G, there exists $g \in G$ so $c_g H \subseteq P$.*

(3) *If s is the number of p-Sylow subgroups, then*

$$s \mid m$$

and

$$s \equiv 1 \bmod(p).$$

Note that $s = 1$ if and only if the p-Sylow subgroup is normal.

PROOF. As part (1) is a special case of Theorem 2.69, turn to part (2). It suffices to prove the second half of the statement: if we find g so that $c_g P' \subseteq P$, we actually get $c_g P' = P$ since $|c_g P'| = |P'| = |P|$.

For the second half of (2), the trick is to look at the action of H on G/P given by left multiplication, $h \cdot gP = hgP$ for $h \in H$ and $g \in G$. If $(G/P)^H \neq \varnothing$, then there would be a $g^{-1} \in G$ so that $h \cdot g^{-1}P = g^{-1}P$ for all $h \in H$. As this is the same as saying $c_g h \in P$ for all $h \in H$, we would be done. To verify $(G/P)^H \neq \varnothing$, first use the Orbit Decomposition Theorem to write $|G/P| = \left|(G/P)^H\right| + \sum_{i=1}^N |H| / |H_{g_i P}|$ where $g_i P$, $1 \leq i \leq N$, are representatives of the distinct nonsingleton orbits of H on G/P. Since H is a p-group and since $\mathcal{O}_{g_i P}$ is a nonsingleton orbit, $p \mid (|H| / |H_{g_i P}|)$.

However, $p \nmid m = |G/P|$. Therefore $p \nmid \left|(G/P)^H\right|$ and, in particular, $\left|(G/P)^H\right| \neq 0$ as desired.

For part (3), to show $s \mid m$, consider the set of p-Sylow subgroups, $Q = \{P_1, \ldots, P_s\}$. Define an action of G on Q by $g \cdot P_i = c_g P_i$ for $g \in G$. By part (2), the action is transitive. Writing $N(P_i) = \{g \in G \mid c_g P_i = P_i\}$ for the *normalizer* of P_i (cf. Exercise 2.72), it follows that $|G| / |N(P_1)| = s$. Since P_1 is clearly a subgroup of $N(P_1)$, $|N(P_1)| = p^n m'$ for $m' \in \mathbb{N}$. Thus $s = \frac{m}{m'}$ so that $s \mid m$ as desired.

To show $s \equiv 1 \bmod(p)$, restrict the action of G on Q to just an action of P_1 on Q. By the Orbit Decomposition Theorem again, $s = \left|Q^{P_1}\right| + \sum_{j=1}^{M} |P_1| / \left|(P_1)_{P_{i_j}}\right|$ where P_{i_j}, $1 \leq j \leq M$, are representatives of the distinct nonsingleton orbits of P_1 on Q. Since P_1 is a p-group and since $\mathcal{O}_{P_{i_j}}$ is a nonsingleton orbit, $p \mid \left(|P_1| / \left|(P_1)_{P_{i_j}}\right|\right)$. Therefore it suffices to see $\left|Q^{P_1}\right| = 1$. For this, observe that $P_i \in Q^{P_1}$ if and only if $P_1 \subseteq N(P_i)$. In this case, by looking at the order of the groups, P_1 and P_i are both clearly p-Sylow subgroups of $N(P_i)$ and are therefore conjugate in $N(P_i)$ by part (2). However since P_i is a normal subgroup of $N(P_i)$ by construction, this happens if and only if $P_i = P_1$ as desired.

Finally, we address the remark that $s = 1$ if and only if the p-Sylow subgroup is normal. The point is that $S = \{c_g(P) \mid g \in G\}$ constitutes the set of all p-Sylow subgroups. Since P is normal if and only if $c_g(P) = P$ for all $g \in G$, it follows that P is normal if and only if $s = |S| = 1$. $\square$

As an example, let us apply Sylow's Theorems to groups of order 15. Suppose G is a group with $|G| = 15$. Since $15 = 3 \cdot 5$, there is a 3-Sylow subgroup Sy_3 with $|Sy_3| = 3$ and a 5-Sylow subgroup Sy_5 with $|Sy_5| = 5$. The only solution to the equations $s \mid 5$ and $s \equiv 1 \bmod(3)$ is $s = 1$ so that Sy_3 is normal. Similarly, the only solution to the equations $s \mid 3$ and $s \equiv 1 \bmod(5)$ is $s = 1$ so that Sy_5 is normal. As 3 and 5 are relatively prime, clearly $Sy_3 \cap Sy_5 = \{e\}$. Thus Theorem 2.33 shows that $G \cong Sy_3 \times Sy_5 \cong \mathbb{Z}_3 \times \mathbb{Z}_5 \cong \mathbb{Z}_{15}$. Up to isomorphism, there is therefore only one group of order 15.

For a more complicated example, consider a group G with $|G| = 30$. Here $30 = 2 \cdot 3 \cdot 5$ so there is a 2-Sylow subgroup Sy_2 with $|Sy_2| = 2$, a 3-Sylow subgroup Sy_3 with $|Sy_3| = 3$, and a 5-Sylow subgroup Sy_5 with $|Sy_5| = 5$. Regarding the number of such Sylow subgroups, all we can initially say is that the number of 2-Sylow subgroups is 1, 3, 5, or 15 since these are the only solutions to $s \mid 15$ and $s \equiv 1 \bmod(2)$. Similarly, the number of 3-Sylow subgroups is 1 or 10 since these are the only solutions to $s \mid 10$ and $s \equiv 1 \bmod(3)$. Finally, the number of 5-Sylow subgroups is 1 or 6 since these are the only solutions to $s \mid 6$ and $s \equiv 1 \bmod(5)$.

In fact, we can pull a bit more out of this information. For instance since $|Sy_2| = 2$ is prime, notice that any two 2-Sylow subgroups are either identical or the intersection is $\{e\}$. Moreover, every element of order 2 generates a 2-Sylow subgroup. As there are 1, 3, 5, or 15 distinct 2-Sylow subgroups, we see there are exactly 1, 3, 5, or 15 elements of order 2. Similarly, there are $1(3-1) = 2$ or $10(3-1) = 20$ elements of order 3 and $1(5-1) = 4$ or $6(5-1) = 24$ elements of order 5. Of course there is exactly 1 element of order 1. At the moment, we do not know how many elements (if any) have order 6, 10, 15, or 30.

In any case, since $|G| = 30$, we cannot have 20 elements of order 3 and 24 elements of order 5. Therefore either Sy_3 or Sy_5 is normal and so either $G_{10} = G/Sy_3$ or $G_6 = G/Sy_5$ is a group. Using Sylow's Theorems on a group of order 10 or on a group of order 6, it immediately follows that either G_{10} has a normal subgroup of order 5 or that G_6 has a normal subgroup of order 3. By the Correspondence Theorem, we therefore see that in either case G has a normal subgroup of order 15. Write H for this subgroup and recall that we have shown that $H \cong \mathbb{Z}_{15}$.

Since $|Sy_2| = 2$, it follows that $H \cap Sy_2 = \{e\}$ and therefore that $G = H\, Sy_2$ (Exercise 2.140). However, this is not enough to conclude that G is isomorphic to $\mathbb{Z}_{15} \times \mathbb{Z}_2 \cong \mathbb{Z}_{30}$ since Sy_2 may not be normal (Exercise 2.271). Instead, we need a slightly more general construction to take the place of the direct product used in Theorem 2.33.

DEFINITION 2.75. Let N and H be subgroups of G with N normal. Define the (internal) *semidirect product* of N and H as

$$N \rtimes H = \{(n, h) \mid n \in N,\ h \in H\}$$

with group law given by

$$(n_1, h_1)(n_2, h_2) = (n_1 c_{h_1}(n_2),\ h_1 h_2)$$

for $n_i \in N$ and $h_i \in H$ (cf. Exercise 2.150).

THEOREM 2.76. *Let N and H be subgroups of G with N normal.*

(1) *$N \rtimes H$ is a group.*

(2) *If N and H commute, then $N \rtimes H = N \times H$.*

(3) *If $NH = G$ and $N \cap H = \{e\}$, then $G \cong N \rtimes H$.*

PROOF. For part (1), notice the binary operation on $N \rtimes H$ is well defined since N is normal. To verify it produces a group, observe that the identity element is (e, e) and that the inverse to (n, h) is $\left(c_{h^{-1}}\left(n^{-1}\right), h^{-1}\right)$. To check associativity, calculate

$$\begin{aligned}(n_1, h_1)\left[(n_2, h_2)(n_3, h_3)\right] &= (n_1, h_1)(n_2 c_{h_2}(n_3), h_2 h_3) = \left(n_1 c_{h_1}\left[n_2 c_{h_2}(n_3)\right], h_1 h_2 h_3\right) \\ &= (n_1 c_{h_1}(n_2) c_{h_1 h_2}(n_3), h_1 h_2 h_3) \\ &= (n_1 c_{h_1}(n_2), h_1 h_2)(n_3, h_3) = \left[(n_1, h_1)(n_2, h_2)\right](n_3, h_3).\end{aligned}$$

Part (2) follows immediately from the definitions since then $c_h(n) = n$. For part (3), consider the map $\varphi : N \rtimes H \to G$ by $\varphi(n, h) = nh$. It is surjective by hypothesis. If φ is a homomorphism, then (n, h) is in $\ker \varphi$ if and only if $nh = e$. This forces $n = h^{-1}$ so that $n, h \in N \cap H = \{e\}$ so that φ is injective. It only remains to see that φ is a homomorphism:

$$\begin{aligned}\varphi(n_1, h_1)\varphi(n_2, h_2) &= n_1 h_1 n_2 h_2 = n_1 h_1 n_2 h_1^{-1} h_1 h_2 = n_1 c_{h_1}(n_2) h_1 h_2 \\ &= \varphi\left((n_1, h_1)(n_2, h_2)\right).\end{aligned}$$

□

As an example, let $n \in \mathbb{Z}$ with $n \geq 3$. Allowing a bit of ambiguity, write $\mathbb{Z}_n$ for the rotation subgroup $\mathcal{R}_n = \{R_i \mid i \in \mathbb{Z}\} \cong \mathbb{Z}_n$ of D_n and write $\mathbb{Z}_2$ for the subgroup $\{\mathrm{Id}, W_0\} \cong \mathbb{Z}_2$ of D_n (though any fixed W_i will work). Using Theorem 2.4, $\mathbb{Z}_n$ is a normal subgroup since

$$W_0 R_i W_0^{-1} = R_{-i}. \tag{2.77}$$

Moreover, clearly $\mathbb{Z}_n \cap \mathbb{Z}_2 = \{\mathrm{Id}\}$ and also $D_n = \mathbb{Z}_n\mathbb{Z}_2$ since $D_n = \{R_i, R_iW_0 \mid i \in \mathbb{Z}\}$. Thus Theorem 2.76 shows

$$D_n \cong \mathbb{Z}_n \rtimes \mathbb{Z}_2$$

where the nontrivial element of $\mathbb{Z}_2$ conjugates elements of $\mathbb{Z}_n$ according to (2.77).

THEOREM 2.78. *Let p be a positive prime with $p \geq 3$. Up to isomorphism, the only groups of order $2p$ are $\mathbb{Z}_{2p}$ and D_p.*

PROOF. Let G be a group with $|G| = 2p$. Write $Sy_2 \cong \mathbb{Z}_2$ and $Sy_p \cong \mathbb{Z}_p$ for a 2-Sylow and p-Sylow subgroup of G, respectively. The only solution to $s \mid 2$ and $s \equiv 1 \bmod(p)$ is $s = 1$. Therefore Sy_p is a normal subgroup. Moreover, $Sy_2 \cap Sy_p = \{e\}$ since the $(2,p) = 1$. Bearing this fact in mind, the equation $a_1b_1 = a_2b_2$ can be solved with $a_i \in Sy_2$ and $b_i \in Sy_p$ if and only if $a_2^{-1}a_1 = b_2b_1^{-1} \in Sy_2 \cap Sy_p = \{e\}$ if and only if $a_1 = a_2$ and $b_1 = b_2$. It follows that $|Sy_2Sy_p| = 2p$ so that $Sy_2Sy_p = G$. Theorem 2.76 therefore shows that $G \cong Sy_p \rtimes Sy_2$. If Sy_2 and Sy_p commute, then $G \cong Sy_p \times Sy_2 \cong \mathbb{Z}_p \times \mathbb{Z}_2 \cong \mathbb{Z}_{2p}$.

Otherwise, pick generators $w \in Sy_2$ and $r \in Sy_p$. Normality shows $c_w r = r^k$ for some $k \in \mathbb{Z}$. Since $w^2 = e$,

$$r = c_{w^2}r = c_w\left(c_w r\right) = c_w\left(r^k\right) = \left(c_w r\right)^k = \left(r^k\right)^k = r^{k^2}.$$

Thus $k^2 \equiv 1 \bmod(p)$ so that $(k-1)(k+1) \equiv 0 \bmod(p)$ and $k \equiv \pm 1 \bmod(p)$. Since r and w do not commute, $c_w r \neq r$ and we may therefore assume $k = -1$. Thus $wrw^{-1} = r^{-1}$. As this is essentially the same relation we used in (2.77), it follows that the map given by $r^k \to R_k$ and $w \to W_0$ induces an isomorphism $G \cong Sy_p \rtimes Sy_2 \cong \mathbb{Z}_p \rtimes \mathbb{Z}_2 \cong D_p$. □

We defined the semidirect product of groups N and H when they both sit as subgroups in a larger group G (in which N is normal) by the formula

$$(n_1, h_1)(n_2, h_2) = (n_1 c_{h_1}(n_2),\ h_1h_2)$$

for $n_i \in N$ and $h_i \in H$. More generally, it is often possible to define a notion of semidirect product without reference to a larger group G. All that is really needed is something to take the place of $c_{h_1}(n_2)$ in the above formula. Notice that the map $h \to c_h$ can be thought of as a homomorphism from H to $\mathrm{Aut}(N)$ (cf. Exercise 2.115). In fact, if the proof of Theorem 2.76 is examined, this is the only property of $c_{h_1}(n_2)$ that we use.

Therefore suppose N and H are groups with φ a homomorphism $\varphi : H \to \mathrm{Aut}(N)$. For $n \in N$ and $h \in H$, write $\varphi_h(n)$ for $(\varphi(h))(n)$. Define the (external) *semidirect product* of N and H as

$$N \rtimes H = \{(n, h) \mid n \in N,\ h \in H\}$$

with group law given by

$$(n_1, h_1)(n_2, h_2) = (n_1\varphi_{h_1}(n_2),\ h_1h_2).$$

It is straightforward to verify that this construction makes $N \rtimes H$ into a group and that Definition 2.75 is a special case (Exercise 2.150).

9.1. Exercises 2.249–2.281.

9.1.1. *p-groups.*

EXERCISE 2.249. Let p be a positive prime and let $U_n(p)$ be the subgroup of upper triangular matrices in $GL(n, \mathbb{Z}_p)$ with [1]'s on the main diagonal. Show $U_n(p)$ is a p-group.

EXERCISE 2.250 (Center of p-Groups). Let G be a finite p-group. Show $|Z(G)| = p^n$ for some $n \in \mathbb{N}$. In particular, $Z(G) \neq \{e\}$. *Hint:* Use Lagrange's Theorem to get $n \in \mathbb{Z}_{\geq 0}$ and the Class Equation to get $n \neq 0$.

EXERCISE 2.251. If H is a normal subgroup of a finite p-group G with $H \neq \{e\}$, show $H \cap Z(G) \neq \{e\}$. *Hint:* Let G act on H by conjugation and use the Orbit Decomposition Theorem similar to Exercise 2.250.

EXERCISE 2.252 (Groups of Order p^2). Let G be a group of order p^2.

(a) If $|G| = p^2$, show G is abelian. *Hint:* Exercises 2.152 and 2.250.

(b) Show G is isomorphic to $\mathbb{Z}_{p^2}$ or $\mathbb{Z}_p \times \mathbb{Z}_p$.

EXERCISE 2.253 (Groups of Order 8). If G is a group of order 8, this exercise shows G is isomorphic to $\mathbb{Z}_8$, $\mathbb{Z}_2 \times \mathbb{Z}_4$, $\mathbb{Z}_2 \times \mathbb{Z}_2 \times \mathbb{Z}_2$, D_4, or $Q \cong Q_4$ (cf. Exercises 2.13 and 2.60).

(a) Verify that all abelian groups of order 8 are listed above. Assume for the remainder that G is not abelian.

(b) Show there exists $i \in G$ so $|i| = 4$. *Hint:* Exercise 2.45.

(c) Write $H = \langle i \rangle$ and let $g \in G\backslash H$. Show $igi = g$. *Hint:* Look at $c_g(i)$ and use Exercise 2.141. Show $i^2 \in H$ so that $c_g^2(i) = i$.

(d) If there is a $g \in G\backslash H$ with $|g| = 2$, show $G \cong D_4$.

(e) If there is no $g \in G\backslash H$ with $|g| = 2$, show $|g| = 4$ for all $g \in G\backslash H$. Conclude $g^2 = i^2$ and show $G \cong Q$.

EXERCISE 2.254 (Structure of p-Groups). Let G be a group of order p^k. Show there are normal subgroups H_i of G, $0 \leq i \leq k$, so that $\{e\} = H_0 \subseteq H_1 \subseteq \cdots \subseteq H_k = G$ and so $H_i/H_{i-1} \cong \mathbb{Z}_p$. *Hint:* Induct on k and use the Correspondence Theorem and Exercise 2.250.

EXERCISE 2.255. Let G be a group of order p^k and let H be a normal subgroup of order p^m. Show the subgroups H_i in Exercise 2.254 can be modified so that $H_m = H$. *Hint:* Exercise 2.251.

EXERCISE 2.256. If $H \neq G$ is a subgroup of a finite p-group G, show $H \subsetneq N(H)$ (cf. Exercise 2.72). *Hint*: Induct on $|G|$ and look at whether $Z(G)$ is contained in H or not.

EXERCISE 2.257. If N is a normal subgroup of a finite p-group G with $|N| = p$, show $N \subseteq Z(G)$. *Hint:* Use Exercise 2.250 and induct on $|G|$ or use Exercise 2.251.

EXERCISE 2.258 (Direct Proof of Cauchy's Theorem). Let G be a finite group and let p be a positive prime with $p \mid |G|$.

(a) Write $n = |G|$ and let $X = \{(g_1, g_2, \ldots, g_p) \in \overbrace{G \times \cdots \times G}^{p \text{ copies}} \mid g_1 g_2 \cdots g_p = e\}$. Show $|X| = n^{p-1}$.

(b) For $[1] \in \mathbb{Z}_n$ and $x = (g_1, g_2, \ldots, g_p) \in X$, show that the rule $[1] \cdot x = (g_2, \ldots, g_p, g_1)$ extends to an action of $\mathbb{Z}_p$ on X.

(c) Show that every orbit of $\mathbb{Z}_p$ on X is of order 1 or p.

(d) Write s for the number of singleton orbits of $\mathbb{Z}_p$ on X and write t for the number of orbits with p elements. Show $n^{p-1} = s + tp$ and conclude that $p \mid s$.

(e) Show $s > 0$ and conclude that G has an element of order p.

9.1.2. *p-Sylow Subgroups.*

EXERCISE 2.259. Find an explicit 2-Sylow and 3-Sylow subgroup for the following groups and determine how many such subgroups there are:

(a) $\mathbb{Z}_6$,
(b) $\mathbb{Z}_{12}$,
(c) S_3,
(d) D_6,
(e) $GL(2, \mathbb{Z}_3)$ (cf. Exercise 2.242).

EXERCISE 2.260. Let G be a finite group, let p be a positive prime with $p \mid |G|$, and let H be a subgroup of G. If P is a p-Sylow subgroup of G satisfying $P \subseteq H$, show P is a p-Sylow subgroup of H.

EXERCISE 2.261. Let N be a normal subgroup of a finite group G. If P_N is a normal p-Sylow subgroup of N, show P_N is normal in G (cf. Exercise 2.138). *Hint:* Look at $c_g(P_N)$. How many subgroups of order $|P_N|$ does N have?

EXERCISE 2.262. Let N be a nontrivial normal subgroup of a finite group G and let P be a p-Sylow subgroup of G. Show $P \cap N$ is a p-Sylow subgroup of N. *Hint:* Start with a p-Sylow subgroup P_N of N and show there is a $g \in G$ so $c_g(P_N) \subseteq P$. Conclude $c_g(P_N) = P \cap N$ by showing $|P \cap N| \leq |P_N|$.

EXERCISE 2.263. Let P be a normal p-Sylow subgroup of a finite group G and let $\varphi : G \to G$ be a homomorphism. Show $\varphi(P) \subseteq P$.

EXERCISE 2.264. Let G be a finite group and let Q be the intersection of all p-Sylow subgroups of G. Show Q is the unique maximal normal p-subgroup of G.

EXERCISE 2.265. Let G be a finite abelian group with $n = |G|$.

(a) Show G is isomorphic to the direct product of all its Sylow subgroups (cf. Lemma 2.38).

(b) For any $d \in \mathbb{N}$ with $d \mid n$, show G has a subgroup of order d. *Hint:* See the proof of Theorem 2.69.

(c) Find an example to show that G may not have an element of order d.

9.1.3. *Classifications.*

EXERCISE 2.266. Choose a nontrivial element $\sigma \in S_n$ so that $\sigma^2 = \mathrm{Id}$ and write $\mathbb{Z}_2$ for $\langle \sigma \rangle$. Show $S_n \cong A_n \rtimes \mathbb{Z}_2$.

EXERCISE 2.267. Show $Q_6 \cong \mathbb{Z}_3 \rtimes \mathbb{Z}_4$ where the homomorphism $\varphi : \mathbb{Z}_4 \to \mathrm{Aut}(\mathbb{Z}_3)$ is given by $\varphi_{[a]_4}([b]_3) = [(-1)^a b]_3$ for $[a]_4 \in \mathbb{Z}_4$ and $[b]_3 \in \mathbb{Z}_3$ (cf. Exercise 2.60).

EXERCISE 2.268. Let G be a finite group with $|G| = \prod_{i=1}^n p_i^{k_i}$ where p_i are distinct positive primes and $k_i \in \mathbb{N}$. Let P_i be a p_i-Sylow subgroup.

(a) If $G \cong \prod_{i=1}^n P_i$, show each P_i is normal in G.

(b) Conversely, if each P_i is normal in G, show $G \cong \prod_{i=1}^n P_i$.

(c) Show any group of order 665 is isomorphic to $\mathbb{Z}_{665}$.

EXERCISE 2.269 ($\mathbb{Z}_q \rtimes \mathbb{Z}_p$). Let $p < q$ be positive primes with $p \mid (q-1)$ and let

$$G = \{\begin{pmatrix} x & y \\ [0] & [1] \end{pmatrix} \in GL(2, \mathbb{Z}_q) \mid x^p = [1]\}.$$

(a) Show G is a nonabelian group of order pq. *Hint:* To calculate $|\{x \in U_q \mid x^p = [1]\}|$, use Exercise 2.7 and Cauchy's Theorem or use Exercise 2.179.

(b) Show there exists $m \in U_q$ with order p. Let $w = \begin{pmatrix} m & [0] \\ [0] & [1] \end{pmatrix}$ and $r = \begin{pmatrix} [1] & [1] \\ [0] & [1] \end{pmatrix}$. Show $\langle w \rangle \cong \mathbb{Z}_p$, $\langle r \rangle \cong \mathbb{Z}_q$, and $c_w(r) = r^m$.

(c) Allowing a bit of ambiguity, write $\mathbb{Z}_q$ for $\langle r \rangle$ and $\mathbb{Z}_p$ for $\langle w \rangle$. Show $G \cong \mathbb{Z}_q \rtimes \mathbb{Z}_p$.

(d) If $p = 2$, show $G \cong D_q$.

EXERCISE 2.270 (Groups of Order pq). Let $p < q$ be positive primes.

(a) If G is a group of order pq, show there is $g \in G$ so $|g| = q$ with $\langle g \rangle$ normal in G.

(b) Modify Theorem 2.78 to prove the following result: If $p \nmid (q-1)$, G is isomorphic to $\mathbb{Z}_{pq}$. If $p \mid (q-1)$, then G is isomorphic to either $\mathbb{Z}_{pq}$ or $\mathbb{Z}_q \rtimes \mathbb{Z}_p$ (cf. Exercise 2.269).

EXERCISE 2.271 (Groups of Order 30). Let G be a group of order 30. This exercise shows that G is isomorphic to $\mathbb{Z}_{30}$, $\mathbb{Z}_5 \times D_3$, $\mathbb{Z}_3 \times D_5$, or D_{15}.

(a) Show $G \cong \mathbb{Z}_{15} \rtimes \mathbb{Z}_2$ for some homomorphism from $\mathbb{Z}_2$ to $\mathrm{Aut}(\mathbb{Z}_{15})$. *Hint:* See the discussion in the text.

(b) Show $\mathrm{Aut}(\mathbb{Z}_{15}) \cong \mathbb{Z}_2 \times \mathbb{Z}_4$. *Hint:* Exercise 2.116.

(c) Show there are four homomorphisms from $\mathbb{Z}_2$ to $\mathrm{Aut}(\mathbb{Z}_{15})$.

(d) Show G is isomorphic to exactly one of $\mathbb{Z}_{30}$, $\mathbb{Z}_5 \times D_3$, $\mathbb{Z}_3 \times D_5$, or D_{15}.

EXERCISE 2.272 (Groups of Order $2p^2$). Let $p \geq 3$ be a prime. This exercise shows that the only groups of order $2p^2$ are $\mathbb{Z}_{2p^2}$, $\mathbb{Z}_p \times \mathbb{Z}_p \times \mathbb{Z}_2$, D_{p^2}, $\mathbb{Z}_p \times D_p$, or $(\mathbb{Z}_p \times \mathbb{Z}_p) \rtimes \mathbb{Z}_2$ where the homomorphism $\varphi : \mathbb{Z}_2 \to \mathrm{Aut}(\mathbb{Z}_p \times \mathbb{Z}_p)$ is given by $\varphi_{[1]_2}(g) = g^{-1}$ for $g \in \mathbb{Z}_p \times \mathbb{Z}_p$.

(a) Let G be a group with $|G| = 2p^2$ and let P be a p-Sylow subgroup. Show P is normal and isomorphic to either $\mathbb{Z}_{p^2}$ or $\mathbb{Z}_p \times \mathbb{Z}_p$ (cf. Exercise 2.252).

(b) Let $Sy_2 = \langle w \rangle \subseteq G$ be a 2-Sylow subgroup. Show $G \cong P \rtimes Sy_2$.

(c) If $P \cong \mathbb{Z}_{p^2}$, show G is isomorphic to $\mathbb{Z}_{2p^2}$ or D_{p^2}.

(d) If $P \cong \mathbb{Z}_p \times \mathbb{Z}_p$, show that $c_w = \mathrm{Id}$ or that $c_w(p) = p^{-1}$ for $p \in P$ or that there exist $p_1, p_{-1} \in P$ so that $P \cong \langle p_1 \rangle \times \langle p_{-1} \rangle$ with $\langle p_i \rangle \cong \mathbb{Z}_p$ and $c_w(p_i) = p_i^i$. Conclude that G is isomorphic to $\mathbb{Z}_p \times \mathbb{Z}_p \times \mathbb{Z}_2$, $(\mathbb{Z}_p \times \mathbb{Z}_p) \rtimes \mathbb{Z}_2$, or $\mathbb{Z}_p \times D_p$.

EXERCISE 2.273 (Groups of Order 12). This exercise shows that the only groups G of order 12 are $\mathbb{Z}_{12}$, $\mathbb{Z}_2 \times \mathbb{Z}_2 \times \mathbb{Z}_3$, D_6, A_4, or Q_6 (cf. Exercise 2.60).

(a) Verify that all order 12 abelian groups are on the list. Assume for the remainder that G is nonabelian.

(b) Write P_3 for a 3-Sylow subgroup of G. Show there is a homomorphism $\varphi : G \to S_4$ with $\ker \varphi \subseteq P_3$. If φ is injective, show $G \cong A_4$. *Hint:* Exercises 2.215 and 2.207.

(c) For the remainder, show we may assume that P_3 is normal. Write P_2 for a 2-Sylow subgroup of G. Show there is a homomorphism $\psi : G \to S_3$ with $\ker\psi \subseteq P_2$. Show $|\ker\psi| = 2$ and conclude $H = P_3 \ker\psi \cong \mathbb{Z}_6$.

(d) Let a be a generator for H and let $b \in G\backslash H$. Show that $aba = b$ and that $b^2 \in H$.

(e) Show that either $b^2 = 1$ or $b^2 = a^3$. Conclude G is isomorphic to either D_6 or Q_6.

EXERCISE 2.274 (Groups of Order p^2q, $p < q < 2p$). Suppose p and q are positive primes satisfying $p < q < 2p$ and that G is a group of order $p^2q \neq 12$ (cf. Exercise 2.273). This exercise shows G is isomorphic to either $\mathbb{Z}_{p^2q}$ or $\mathbb{Z}_p \times \mathbb{Z}_{pq}$ and, in particular, is abelian.

(a) Let P be a p-Sylow subgroup. Show P is normal and conclude that $G \cong P \rtimes \mathbb{Z}_q$. *Hint:* Use Sylow's Theorems and the conditions on p and q to show P is normal.

(b) Show that the only homomorphism of $\mathbb{Z}_q$ into U_{p^2} or $GL(2, \mathbb{Z}_p)$ is trivial. *Hint:* Look at the order of the groups; cf. Exercise 2.242.

(c) Conclude G is abelian. *Hint:* Use Exercise 2.252 and then Exercise 2.116 or 2.117.

EXERCISE 2.275. If G is a group of order n, show all possibilities are listed below up to isomorphism.

(a) $n = 18:$ $\mathbb{Z}_{18}$, $\mathbb{Z}_3 \times \mathbb{Z}_6$, D_9, $D_3 \times \mathbb{Z}_3$, $(\mathbb{Z}_3 \times \mathbb{Z}_3) \rtimes \mathbb{Z}_2$.

(b) $n = 28:$ $\mathbb{Z}_{28}$, $\mathbb{Z}_2 \times \mathbb{Z}_{14}$, D_{14}, $D_7 \times \mathbb{Z}_2$.

EXERCISE 2.276. Classify groups of the following order:

(a) 361,
(b) 323,
(c) 301,
(d) 134,
(e) 1547,
(f) 539,
(g) 1058.

EXERCISE 2.277 (Groups of Order pqr). Let G be a group of order pqr where $p < q < r$ are positive primes. This exercise shows that (i) the r-Sylow subgroup is normal and (ii) G has a normal subgroup of order qr.

(a) If an r-Sylow subgroup P_r is not normal, show G contains $pq(r-1)$ elements of order r. If a q-Sylow subgroup P_q is not normal, show G contains at least $r(q-1)$ elements of order q. Conclude that either P_r or P_q is normal.

(b) If P_r or P_q is normal, show G has a normal subgroup N of order qr. *Hint:* Exercise 2.270, part (a), and the Correspondence Theorem.

(c) Show N contains a normal subgroup of order r. Conclude that P_r is normal. *Hint:* Exercise 2.261.

EXERCISE 2.278. Suppose p and q are positive primes satisfying $p < q < 2p$ and that G is a group of order $p^n q$ with $n \in \mathbb{Z}_{\geq 0}$. This exercise shows G must have (i) a normal subgroup of order either p^n or $p^{n-1}q$ and (ii) a normal subgroup of order either p^n or p^{n-1}.

(a) Verify that the exercise holds when $n = 0, 1$. For the remainder of this exercise, proceed by induction on n.

(b) Let P be a p-Sylow subgroup. Show $|G| \nmid q!$ and conclude P has a subgroup $N \neq \{e\}$ that is normal in G. *Hint:* Exercise 2.217.

(c) If $N \neq P$, apply the induction hypothesis to G/N and use this to finish the exercise. *Hint:* Use the Correspondence Theorem and Exercise 2.137.

EXERCISE 2.279. Let G be a group of order p^2q^2 for positive primes p, q. This exercise shows G must have a proper normal subgroup.

(a) If $p = q$, show G has a proper normal subgroup. *Hint:* Exercise 2.250.

(b) If $p < q$, let Q be a q-Sylow subgroup and suppose it is not normal. If so, show $|G| = 36$. *Hint:* Sylow's Theorems.

(c) If $|G| = 36$, show there is a noninjective homomorphism from G to S_4. Conclude G has a proper normal subgroup. *Hint:* Exercise 2.215.

EXERCISE 2.280 (Schur). Let A be an abelian normal subgroup of a finite group G. If $|A|$ and $|G/A|$ are relatively prime, this exercise shows G has a subgroup H of order $|G/A|$ so that $G \cong A \rtimes H$. To that end, let $n = |A|$ and $m = |G/A|$ so $|G| = nm$ with $(n, m) = 1$.

(a) Choose representatives $g_1, \ldots, g_m$ for the left cosets in G/A. If $(g_iA)(g_jA) = g_{h_{i,j}}A$, $1 \leq h_{i,j} \leq m$, show there is a unique $a_{i,j} \in A$ so that $g_ig_j = g_{h_{i,j}}a_{i,j}$. If it is possible to find $a_i \in A$, $1 \leq i \leq m$, so that $a_{h_{i,j}} = a_{i,j}c_{g_j^{-1}}(a_i)\,a_j$, show $H = \{g_ia_i\}$ is a subgroup of G isomorphic to G/A.

(b) Expand $g_i(g_jg_k) = (g_ig_j)\,g_k$ to show $a_{i,h_{j,k}}a_{j,k} = a_{h_{i,j},k}c_{g_k^{-1}}(a_{i,j})$.

(c) If j is fixed and i varies over $1, \ldots, m$, show $h_{i,j}$ also varies over $1, \ldots, m$. Let $b_j = (\prod_{i=1}^{m} b_{i,j})$ and show $b_{h_{j,k}}a_{j,k}^m = b_kc_{g_k^{-1}}(b_j)$.

(d) If $a \in A$, show a has a unique m^{th} root in A. Write $a^{\frac{1}{m}}$ for this element. Let $a_j = b_j^{-\frac{1}{m}}$ and show $a_{h_{j,k}}a_{j,k}^{-1} = a_kc_{g_k^{-1}}(a_j)$.

(e) Conclude $H = \{g_ia_i\}$ is a subgroup and that $G \cong A \rtimes H$.

EXERCISE 2.281 (Schur–Zassenhaus Theorem). Let N be a normal subgroup of a finite group G. If $|N|$ and $|G/N|$ are relatively prime, this exercise shows G has a subgroup H of order $|G/N|$ so that $G \cong N \rtimes H$. To that end, let $n = |N|$ and $m = |G/N|$ so $|G| = nm$ with $(n, m) = 1$. Proceed by induction on $|G|$.

(a) If N_0 is a normal subgroup of G properly contained in N with $n_0 = |N_0|$, use the inductive hypothesis on $\overline{N} = N/N_0$ in $\overline{G} = G/N_0$ to show $\overline{G}$ has a subgroup of order m. *Hint:* Exercise 2.168

(b) If N_0 is a normal subgroup of G properly contained in N with $n_0 = |N_0|$, conclude that G has a subgroup of order mn_0 containing N_0 and use the inductive hypothesis on this subgroup to conclude that it (and therefore G) has a subgroup of order m. *Hint:* Correspondence Theorem.

(c) If P is a p-Sylow subgroup of N, show it is also a p-Sylow subgroup of G.

(d) Writing $N_P = \{g \in G \mid c_gP = P\}$ for the normalizer of P, conclude that $|G/N_P| = |N/(N \cap N_P)|$, $|G| = |NN_P|$, and $G/N \cong N/(N \cap N_P)$ so that $m = |N/(N \cap N_P)|$.

(e) If $N_P \neq G$, use the inductive hypothesis on N in N_P to show N_P (and therefore G) has a subgroup of order m.

(f) For the remaining case, show we may assume P is normal in G and that N contains no proper subgroups that are normal in G. Conclude $N = P$.

(g) Writing Z_P for the center of P, show Z_P is a nontrivial normal subgroup of G and conclude $N = Z_P$. *Hint:* Exercises 2.250 and 2.138.

(h) Use Exercise 2.280 to show G has a subgroup of order m.
(i) In any case, write H for the subgroup of G of order m and show $G \cong N \rtimes H$.

10. Simple Groups and Composition Series

10.1. Simple Groups. In Section 6 we classified finite abelian groups and found that each was a direct product of finite cyclic groups. In other words, groups of the form $\mathbb{Z}_{p^n}$, p a positive prime and $n \in \mathbb{N}$, constitute a complete collection of building blocks for the construction of finite abelian groups. This means, more or less, that if we completely understand $\mathbb{Z}_{p^n}$, then we can essentially understand all finite abelian groups.

Ideally, we would like a similar classification theorem for any finite group. Unfortunately, the world of noncommutative groups is ridiculously more complicated. Nevertheless, much can be said. One approach is to set up some sort of inductive classification on the order of the group. With this strategy, if a finite group G happens to have a proper normal subgroup N, then N and G/N are both groups of smaller order and can therefore be understood inductively. While knowledge of N and G/N does not always enable us to perfectly describe G (e.g., Exercise 2.285), information on N and G/N nevertheless tells us much of what is going on in G. The only time this inductive procedure completely fails is when G has no proper normal subgroups. In that case, such a group should be viewed as an elemental building block of finite groups and gets a special name.

DEFINITION 2.79. A nontrivial group is *simple* if it has no proper normal subgroups.

The only easy example of a simple group we know of so far is $\mathbb{Z}_p$, p a positive prime. Finding all other simple finite groups is a Herculean task. The effort took many thousands of pages by hundreds of authors over many decades. It turns out that besides $\mathbb{Z}_p$, there are four other families of simple groups. The first family is A_n, $n \neq 2, 4$, which we will prove is simple in Theorem 2.80 below. The second family consists of so-called classical groups. We will give an example of these types of groups in Theorem 2.81 below where we prove that $PSL(2, \mathbb{Z}_p)$ is simple for $p \neq 2, 3$. The third family consists of groups of exceptional Lie type. The origin of such groups is motivated by the study of gadgets called Lie groups, which are groups on which we can do multivariable calculus. In total, groups from these first three categories comprise 18 infinite families of simple groups. Members of the final family of simple groups are called *sporadic groups* and consist of 26 miracles of nature that do not fit into any of the 18 aforementioned respectable families. The largest of these is called the *Monster Group* and consists of roughly $8 \cdot 10^{53}$ elements!

Turning back to more tame groups, we show A_n is simple for $n \geq 5$. Besides making A_n one of the basic building blocks of all finite groups, this fact has rather deep and remarkable connections to our ability to solve polynomial equations. Roughly speaking, the following theorem plays an instrumental role in showing it is impossible to find a formula that solves equations of degree five or higher (cf. §5.2 of Chapter 4). For the sake of completeness, note that A_2 is trivial and so not simple by technicality, $A_3 \cong \mathbb{Z}_3$ and is thus simple, and A_4 is genuinely not simple by Exercise 2.210.

THEOREM 2.80. *For $n \geq 5$, A_n is simple.*

PROOF. Suppose that N is a normal subgroup of A_n with $N \neq \{\mathrm{Id}\}$. To finish the proof, we must show that $N = A_n$. We will see below that N contains some 3-cycle. Temporarily assuming this fact, by Corollary 2.58 it only remains to verify that N contains all 3-cycles. To see this, suppose $(i,j,k) \in N$ and (i',j',k') is any other 3-cycle. Let σ be the unique element of S_n mapping $i \to j$, $j \to j'$, $k \to k'$ and fixing everything else. Equation (2.51) shows $\sigma(i,j,k)\sigma^{-1} = (i',j',k')$. If σ is even, then $\sigma \in A_n$ and the normality of N finishes the proof. If σ is not even, choose $l, m \notin \{i,j,k\}$ in $\{1,2,\ldots,n\}$ (here we need $n \geq 5$) and modify σ by the transposition (l,m). The new and improved σ will still map $i \to j$, $j \to j'$, and $k \to k'$, but it will now be even as needed.

To finish the proof, it remains to see that N contains a 3-cycle. Fix $\sigma \in N$, $\sigma \neq \mathrm{Id}$. Write σ as a product of disjoint cycles of length at least two, $\sigma = c_1 \cdots c_r$. There are 4 cases to consider.

Case I: Suppose one of the cycles in $\{c_i\}$ has length at least four. After relabeling if necessary, write $c_1 = (i,j,k,l,\ldots,m)$. Then $c_{(i,j,k)}(\sigma)\,\sigma^{-1} \in N$ by normality. Since

$$(i,j,k)\sigma(i,j,k)^{-1}\sigma^{-1} = (j,k,i,l,\ldots,m)(m,\ldots,l,k,j,i) = (i,j,l),$$

Case I is complete. We may therefore assume all the cycles in $\{c_i\}$ have length at most three.

Case II: Suppose at least two of the cycles have length three and write $c_1 = (i,j,k)$ and $c_2 = (i',j',k')$. Then N contains $\sigma^{-1}c_{(i,j,k')}(\sigma)$ and since

$$\begin{aligned}\sigma^{-1}(i,j,k')\sigma(i,j,k')^{-1} &= (k',j',i')(k,j,i)\,(i,j,k')(i',j',k') \\ &= (k',j',i')(j,k',k)(k',j',i')^{-1} = (j,j',k),\end{aligned}$$

Case II is complete.

Case II: Suppose exactly one of the cycles has length three and write $c_1 = (i,j,k)$. Note that this forces the rest of the cycles to square to the identity. Then $c_{(j,i,k)}(\sigma)\sigma \in N$ and since

$$c_{(j,i,k)}(\sigma)\sigma = c_{(j,i,k)}((i,j,k))(i,j,k) = (k,i,j)(i,j,k) = (i,k,j),$$

Case III is done.

Case IV: Suppose all the cycles are transpositions. By parity, there must be an even number of them—and in particular, at least two. Write $c_1 = (i,j)$ and $c_2 = (k,l)$. Then $c_{(i,j,k)}(\sigma)\sigma \in N$ and so

$$c_{(i,j,k)}(\sigma)\sigma = (j,k)(i,l)\,(i,j)(k,l) = (i,k)(j,l) \in N.$$

Again using $n \geq 5$, there is $(i,k,m) \in A_n$. Then $c_{(i,k,m)}\left((i,k)(j,l)\right)(i,k)(j,l) \in N$ and since

$$c_{(i,k,m)}\left((i,k)(j,l)\right)(i,k)(j,l) = (k,m)(j,l)\,(i,k)(j,l) = (i,m,k),$$

Case IV is complete. □

Our next example of a family of simple groups comes from the quotient

$$PSL(2,\mathbb{Z}_p) = SL(2,\mathbb{Z}_p)/\{\pm I_2\}$$

which has order $\frac{p^3-p}{2}$ for primes $p \geq 3$ (Exercises 2.169 and 2.242).

THEOREM 2.81. *$PSL(2,\mathbb{Z}_p)$ is simple for positive primes $p \neq 2,3$.*

PROOF. For $x \in \mathbb{Z}_p$ and $\zeta \in U_p$, consider the following elements of $SL(2, \mathbb{Z}_p)$:

$$R_x = \begin{pmatrix} [1] & x \\ [0] & [1] \end{pmatrix}, \ L_x = \begin{pmatrix} [1] & [0] \\ x & [1] \end{pmatrix}, \ T = \begin{pmatrix} [1] & [1] \\ [-1] & [0] \end{pmatrix}, \ D_\zeta = \begin{pmatrix} \zeta & [0] \\ [0] & \zeta^{-1} \end{pmatrix}.$$

As a first step, we show that $SL(2, \mathbb{Z}_p)$ is generated by $\{L_{[a]}, R_{[a]} \mid [a] \in \mathbb{Z}_p\}$. To see this, first multiply matrices to verify

$$T = L_{[-1]} R_{[1]} L_{[-1]} \quad \text{and} \quad D_\zeta = L_{-\zeta^{-1}} R_{\zeta - [1]} L_{[1]} R_{\zeta^{-1} - [1]}.$$

Then if $g = \begin{pmatrix} [a] & [b] \\ [c] & [d] \end{pmatrix} \in SL(2, \mathbb{Z}_p)$, check that

$$g = \begin{cases} L_{[c][a]^{-1}} D_{[a]} R_{[b][a]^{-1}}, & [a] \neq [0], \\ T D_{[b]^{-1}} R_{[-bd]}, & [a] = [0]. \end{cases}$$

For the second step, we show that when N is a normal subgroup of $SL(2, \mathbb{Z}_p)$ containing an element of the form R_ζ, $\zeta \neq [0]$, then $N = SL(2, \mathbb{Z}_p)$. To see this, observe that $\langle R_\zeta \rangle = \{R_{[a]} \mid [a] \in \mathbb{Z}_p\}$ since p is prime and $\{R_{[a]} \mid [a] \in \mathbb{Z}_p\} \cong \mathbb{Z}_p$. The fact that $c_T(R_\zeta) = L_{-\zeta}$ and step one finish the argument.

For step three, we show $PSL(2, \mathbb{Z}_p)$ is simple if we can prove that any normal subgroup $N \nsubseteq \{\pm I_2\}$ of $SL(2, \mathbb{Z}_p)$ contains an element of the form R_ζ, $\zeta \neq [0]$. To see this, suppose $\overline{N}$ is a nontrivial normal subgroup of $PSL(2, \mathbb{Z}_p)$. We need to see $\overline{N} = PSL(2, \mathbb{Z}_p)$. For this, consider the natural homomorphism $\varphi : SL(2, \mathbb{Z}_p) \to PSL(2, \mathbb{Z}_p)$ mapping g to $g\{\pm I_2\}$. Using the Correspondence Theorem, let $N = \varphi^{-1}(\overline{N})$, a normal subgroup of $SL(2, \mathbb{Z}_p)$ containing $\{\pm I_2\}$. Since $\overline{N}$ is nontrivial, $N \nsubseteq \{\pm I_2\}$. Thus by hypothesis here, N contains some R_ζ and so step two shows $N = SL(2, \mathbb{Z}_p)$ so that $\overline{N} = PSL(2, \mathbb{Z}_p)$ as desired.

For the final step, it remains to show that any normal subgroup $N \nsubseteq \{\pm I_2\}$ of $SL(2, \mathbb{Z}_p)$ contains an element of the form R_ζ. This is accomplished by matrix multiplication. Fix $g = \begin{pmatrix} [a] & [b] \\ [c] & [d] \end{pmatrix} \in N$ with $g \neq \pm I_2$.

In the case of $[c] = 0$, $c_{R_{[1]}}(g) g^{-1} = R_{[1-a^2]}$, which finishes this case unless $[a] = \pm[1]$. However, if $[a] = \pm[1]$, then $[b] \neq [0]$ by the assumption on g. In this situation, $g^2 = R_{[\pm 2b]}$, which takes care of the case of $[c] = [0]$.

Turn now to the case of $[c] \neq [0]$. Let $h = c_{R_{[d][c]^{-1}}}(g)$. For $p > 5$, there exists $\zeta \in U_p$ so $\zeta^2 \neq \pm[1]$ since $\left|U_{p-1}^2\right| = \frac{p-1}{2}$. For such a ζ,

$$c_{D_\zeta}(h) = \begin{pmatrix} \zeta^2 & [a+d]\,[c]^{-1}(1 - \zeta^2) \\ [0] & \zeta^{-2} \end{pmatrix},$$

which is now handled by the first case. If $p = 5$, then $\left(c_{D_{[2]}}(h)\right)^2 = R_{[c+d][c]^{-1}}$, which finishes matters unless $[c+d] = [0]$. If $[c+d] = [0]$, then

$$c_{L_{[2c]}}(h) = \begin{pmatrix} [2] & -[c]^{-1} \\ [0] & [3] \end{pmatrix},$$

which is again handled by the first case. □

For the sake of completeness, $PSL(2, \mathbb{Z}_2) \cong S_3$ and $PSL(2, \mathbb{Z}_3) \cong A_4$ and so they are not simple (Exercise 2.289). However, it turns out that $PSL(n, \mathbb{Z}_p)$ is always simple when $n \geq 3$ (Exercise 2.291). It is also worth noting that it is a theorem (which we do not prove here) that $PSL(2, \mathbb{Z}_p)$ is the only simple group of order $\frac{p^3 - p^2}{2}$, $p \geq 5$, up to isomorphism. For instance, this gives rise to the rather interesting identity $A_5 \cong PSL(2,5)$, which can be quite painful to write down explicitly. It should also be noted that $\mathbb{Z}_p$ can be generalized in all of these theorems to what is known as a finite field (cf. Section 3 in Chapter 4).

10.2. Composition Series. Now that we at least have a long list of simple groups, we start examining how they fit together in a general group.

DEFINITION 2.82. A *composition series* of a group G is a series of subgroups G_i of G so that

$$\{e\} = G_0 \subseteq G_1 \subseteq \cdots \subseteq G_n = G$$

with G_i normal in G_{i+1} and G_{i+1}/G_i simple, $0 \leq i \leq n-1$. The quotient groups G_{i+1}/G_i are called the *composition factors* and n is called the *length* of G.

When a composition series exists, it turns out that the length and the set of composition factors are unique even though the series of subgroups G_i is not. As an example, consider $\mathbb{Z}_{12}$. One composition series is

$$\{[0]\} \subseteq \langle [6] \rangle \subseteq \langle [2] \rangle \subseteq \mathbb{Z}_{12}$$

which has length 3 and composition factors $\langle [6] \rangle / \{[0]\} \cong \mathbb{Z}_2$, $\langle [2] \rangle / \langle [6] \rangle \cong \mathbb{Z}_3$, and $\mathbb{Z}_{12} / \langle [2] \rangle \cong \mathbb{Z}_2$. On the other hand,

$$\{[0]\} \subseteq \langle [6] \rangle \subseteq \langle [3] \rangle \subseteq \mathbb{Z}_{12}$$

is a different composition series but also has length 2 and the same composition factors $\langle [6] \rangle / \{[0]\} \cong \mathbb{Z}_2$, $\langle [3] \rangle / \langle [6] \rangle \cong \mathbb{Z}_2$, and $\mathbb{Z}_{12} / \langle [3] \rangle \cong \mathbb{Z}_3$—though in a different order.

For a nonabelian case, S_n has a composition series

$$\{\mathrm{Id}\} \subseteq A_n \subseteq S_n$$

with length 2 and composition factors A_n and $\mathbb{Z}_2$ $(n \geq 5)$. Of course, not all groups have composition series. For instance, suppose $\mathbb{Z}$ had a composition series

$$\{0\} = G_0 \subseteq G_1 \subseteq \cdots \subseteq G_n = \mathbb{Z}.$$

Then since G_1 is of the form $\langle k \rangle$ for some $k \in \mathbb{N}$ (Exercise 2.63), $G_1/G_0 \cong \mathbb{Z}$, which is not simple since it has tons of (automatically normal) proper subgroups. This yields a contradiction. In general, an abelian group has a composition series if and only if it is finite (Exercise 2.296). In fact, finite groups always have composition series.

THEOREM 2.83 (Jordan-Hölder). *Let G be a finite nontrivial group. Then:*

(1) *G has a composition series.*

(2) *Any two composition series have the same length and the same composition factors up to reordering.*

PROOF. Begin with part (1). If G is simple, a composition series of length 1 will do and so there is nothing to prove. Otherwise, suppose G is not simple and induct on $|G|$. Since G is finite and nonsimple, there exists a proper maximal normal subgroup N. Then N has a composition series

$$\{e\} = G_0 \subseteq G_1 \subseteq \cdots \subseteq G_{n-1} = N$$

by induction. Since N is maximal, the Correspondence Theorem shows that G/N must be simple. Thus

$$\{e\} = G_0 \subseteq G_1 \subseteq \cdots \subseteq G_{n-1} \subseteq G_n = G$$

is a composition series for G.

Turn now to part (2). Clearly G has a composition series of length 1 if and only if it is simple. In that case, there is only one composition series and so there is nothing to do. Otherwise, suppose G is not simple and induct on the minimal length of one of its composition series. Suppose G has two composition series, the first with minimal length $n+1$:

$$\{e\} = G_0 \subseteq G_1 \subseteq \cdots \subseteq G_n \subseteq G_{n+1} = G,$$
$$\{e\} = G'_0 \subseteq G'_1 \subseteq \cdots \subseteq G'_m \subseteq G'_{m+1} = G.$$

If $G_n = G'_m$, then

$$\{e\} = G_0 \subseteq G_1 \subseteq \cdots \subseteq G_n,$$
$$\{e\} = G'_0 \subseteq G'_1 \subseteq \cdots \subseteq G'_m$$

are both composition series for $G_n = G'_m$, the first with minimal length n. Induction shows $n = m$ and that, up to reordering, they have the same composition factors. This finishes the case of $G_n = G'_m$.

Now suppose $G_n \neq G'_m$. Since G_n and G'_m are both maximal normal subgroups, neither is contained in the other. Thus $G_n G'_m$ is a normal subgroup properly containing both and so $G_n G'_m = G$ (e.g., Exercise 2.140).

Next let $H = G_n \cap G'_m$ and choose a composition series for H,

$$\{e\} = H_0 \subseteq H_1 \subseteq \cdots \subseteq H_{k-1} = H.$$

By the Second Isomorphism Theorem (Exercise 2.167),

$$G_n/H = G_n/\left(G_n \cap G'_m\right) \cong \left(G_n G'_m\right)/G'_m = G'_{m+1}/G'_m$$

and so it is simple. Thus

$$\{e\} = H_0 \subseteq H_1 \subseteq \cdots \subseteq H_{k-1} = H \subseteq G_n,$$
$$\{e\} = G_0 \subseteq G_1 \subseteq \cdots \subseteq G_n$$

are both composition series for G_n, the bottom with minimal length n. Therefore induction shows $n = k$ and that the composition factors of the two composition series agree up to reordering. In gory detail, this latter fact means there is an i_0, $1 \leq i_0 \leq n$, so $G_{i_0}/G_{i_0-1} \cong G_n/H \cong G'_{m+1}/G'_m$ and that the sets (counted with multiplicity) $\{H_{i+1}/H_i \mid 1 \leq i \leq n-1\}$ and $\{G_{i+1}/G_i \mid 1 \leq i \leq n,\ i \neq i_0\}$ agree.

Similarly, $G'_m/H \cong G_{n+1}/G_n$ is simple so that

$$\{e\} = H_0 \subseteq H_1 \subseteq \cdots \subseteq H_{k-1} = H \subseteq G'_m,$$
$$\{e\} = G'_0 \subseteq G'_1 \subseteq \cdots \subseteq G'_m$$

are both composition series for G'_m, the top with minimal length k. Again induction shows $m = k$ and that the composition factors of the two composition series agree up to reordering. Thus there is a j_0, $1 \leq j_0 \leq m$, so $G'_{j_0}/G'_{j_0-1} \cong G'_m/H \cong G_{m+1}/G_m$ and so that the sets (counted with multiplicity) $\{H_{i+1}/H_i \mid 1 \leq i \leq m-1\}$ and $\{G'_{j+1}/G'_j \mid 1 \leq j \leq m,\ j \neq j_0\}$ agree.

As $n = k$, we get $n = m$ as desired. This proof is finished by observing

$$\begin{aligned}
&\{G_{i+1}/G_i \mid 1 \leq i \leq n+1\} \\
&\quad = \{G_{i+1}/G_i \mid 1 \leq i \leq n,\ i \neq i_0\} \cup \{G_{i_0}/G_{i_0-1}\} \cup \{G_{n+1}/G_n\} \\
&\quad = \{H_{i+1}/H_i \mid 1 \leq i \leq n-1\} \cup \{G'_{m+1}/G'_m\} \cup \{G'_{j_0}/G'_{j_0-1}\} \\
&\quad = \{H_{i+1}/H_i \mid 1 \leq i \leq m-1\} \cup \{G'_{j_0}/G'_{j_0-1}\} \cup \{G'_{m+1}/G'_m\} \\
&\quad = \{G'_{j+1}/G'_j \mid 1 \leq j \leq m,\ j \neq j_0\} \cup \{G'_{j_0}/G'_{j_0-1}\} \cup \{G'_{m+1}/G'_m\} \\
&\quad = \{G'_{j+1}/G'_j \mid 1 \leq j \leq m+1\}.
\end{aligned}$$

□

Since the list of all simple finite groups is known, the Jordan-Hölder Theorem gives a procedure for attempting to understand any finite group in terms of its simple composition terms. As has already been mentioned, knowing the composition factors of a group G does not equate to a complete understanding of G. In fact, this too is an extraordinarily hard problem in general. However if one is given a single explicit group, the problem is usually manageable.

10.3. Exercises 2.282–2.301.

EXERCISE 2.282. Show that any group of the following order is not simple:
(a) 77,
(b) 104,
(c) 30.

EXERCISE 2.283. Show that any simple abelian group is finite and isomorphic to $\mathbb{Z}_p$ for p a positive prime. *Hint:* Choose nontrivial $g \in G$, look at the possibilities for $\langle g \rangle$, then show $\langle g \rangle = G$.

EXERCISE 2.284. Let G be a group of order $2n$ with n odd. Show G has a normal subgroup of order n and so is not simple. *Hint:* Exercises 2.213, 2.206, and 2.141 and Cauchy's Theorem with $p = 2$.

EXERCISE 2.285. Let p be a positive prime. Find two nonisomorphic groups G_i containing normal subgroups N_i isomorphic to $\mathbb{Z}_p$ so that G_i/N_i is also isomorphic to $\mathbb{Z}_p$. Equivalently, find two nonisomorphic groups of length 2 with composition factors $\mathbb{Z}_p$ and $\mathbb{Z}_p$.

EXERCISE 2.286. Show $G = \langle \{\sigma^2 \mid \sigma \in S_n\} \rangle$ is a subgroup of A_n and is normal in S_n. Conclude $G = A_n$ for $n \geq 5$.

EXERCISE 2.287 (Normal Subgroups of S_n). For $n \geq 5$, show A_n is the only proper normal subgroup of S_n. *Hint:* Proof of Theorem 2.80 or Exercise 2.206.

EXERCISE 2.288. Let G be a simple group and let H be a subgroup with $n = |G/H| < \infty$. From Exercise 2.215, let $\sigma : G \to S_n$ be the permutation representation homomorphism.
(a) If $H \neq G$, show σ is injective. *Hint:* $\ker \sigma \subseteq H$.
(b) If $H \neq G$ and $\sigma^{-1}(A_n) = \{e\}$, show $G \cong \mathbb{Z}_2$. *Hint:* Exercises 2.45 and 2.283 or try Exercise 2.206.
(c) If $H \neq G$ and $G \ncong \mathbb{Z}_2$, show $\sigma : G \hookrightarrow A_n$.

EXERCISE 2.289. Show the following:
(a) $PSL(2, \mathbb{Z}_2) \cong S_3$. *Hint:* Exercise 2.215.
(b) $PSL(2, \mathbb{Z}_3) \cong A_4$. *Hint:* Exercises 2.215 and 2.207.

EXERCISE 2.290. Let p be a positive prime. A matrix of the form $T_{i,j}(a) = I_n + aE_{i,j}$ for $a \in \mathbb{Z}_p$ and $1 \leq i \neq j \leq n$ is called a *transvection* (here $E_{i,j}$ is the $n \times n$ matrix with all $[0]$ entries except for a $[1]$ in the ij^{th} position).

(a) Show $T \in SL(n, \mathbb{Z}_p)$.

(b) Show that multiplying a matrix on the left by a transvection produces an elementary row operation (cf. Exercise 2.184).

(c) Show that it is possible to multiply any element of $SL(n, \mathbb{Z}_p)$ on the left by a series of transvections in order to get a diagonal matrix (cf. Exercise 2.185).

(d) Conclude that $SL(n, \mathbb{Z}_p)$ is generated by transvections.

(e) Show $Z(SL(n, \mathbb{Z}_p)) = \{\zeta I_n \mid \zeta \in U_p, \zeta^n = [1]\}$.

EXERCISE 2.291 ($PSL(n, \mathbb{Z}_p)$ Is Simple, $n \geq 3$). Let p be a positive prime. Recall $PSL(n, \mathbb{Z}_p) = SL(n, \mathbb{Z}_p)/Z(SL(n, \mathbb{Z}_p))$ (cf. Exercises 2.169 and 2.242). This exercise shows $PSL(n, \mathbb{Z}_p)$ is simple for $n \geq 3$.

(a) Let $N \not\subseteq Z(SL(n, \mathbb{Z}_p))$ be a normal subgroup of $SL(n, \mathbb{Z}_p)$. In order to show $PSL(n, \mathbb{Z}_p)$ is simple, show it suffices to show N contains a transvection $T_{i,j}(a)$, $a \neq [0]$. *Hint:* Show $T_{i,j}(a) \in N$ implies all transvections are in N and then use Exercise 2.290.

(b) Conjugating by transvections, show there exists $g \in N \backslash Z(SL(n, \mathbb{Z}_p))$ so $g_{n,1} \neq [0]$.

(c) Show

$$h = c_{T_{1,2}([1])}(g)g^{-1} = \begin{pmatrix} [1] & [1] & & & & \\ & [1] & & & & \\ & & [1] & & & \\ & & & \ddots & & \\ & & & & [1] & \\ h_{n,1} & h_{n,2} & h_{n,3} & \cdots & h_{n,n-1} & [1] \end{pmatrix}$$

where empty slots are $[0]$. (N.B.: $n \geq 3$ is used here.)

(d) Finish the proof by showing $c_{T_{n,1}([1])}(h)h^{-1} = T_{n,2}([1])$.

EXERCISE 2.292. Show that the only simple groups of order less than 60 are abelian. N.B.: $|A_5| = 60$.

EXERCISE 2.293. Find a composition series and determine the composition length and factors of the following groups:

(a) $\mathbb{Z}_{3 \cdot 5 \cdot 7}$,

(b) $\mathbb{Z}_{8 \cdot 9}$,

(c) D_{14},

(d) D_8,

(e) A_4 (cf. Exercise 2.210),

(f) Q (cf. Exercise 2.13).

EXERCISE 2.294. Find all composition series for $\mathbb{Z}_{24}$.

EXERCISE 2.295. If G is an abelian group of order $\prod_{i=1}^{k} p_i^{n_i}$ for distinct positive primes p_i and $n_i \in \mathbb{N}$, show the length of G is $\sum_{i=1}^{k} n_i$ and find the composition factors.

EXERCISE 2.296. Show that an abelian group has a composition series if and only if it is finite. *Hint:* Exercise 2.283 and the Fundamental Theorem of Finite Abelian Groups.

EXERCISE 2.297. Let N be a normal subgroup of G.

(a) Show G has a composition series if and only if N and G/N have composition series.

(b) In this case, show the length of G is the length of N plus the length of G/N and that the composition factors of G are the composition factors of N and G/N.

EXERCISE 2.298. Let G_i, $1 \le i \le n$, be groups and let $G = \prod_{i=1}^n G_i$.

(a) Show G has a composition series if and only if each G_i has a composition series.

(b) In this case, show the length of G is the sum of the lengths of G_i.

EXERCISE 2.299. **(a)** Let N be a simple normal subgroup of G so that G/N is simple. Show that either N is the unique proper normal subgroup or that $G \cong N \times (G/N)$. *Hint:* If $H \ne N$ is another proper normal subgroup, look at $H \cap N$ and the image of H in G/N.

(b) Show that groups G of length 2 come in two varieties: (i) G has a unique proper normal subgroup N and both N and G/N are simple or (ii) G is the direct product of two simple groups.

EXERCISE 2.300. Let G_i, $1 \le i \le n$, be subgroups of a finite group G with $G_i \subseteq G_{i+1}$ and G_i normal in G_{i+1}. Show there is a composition series of G containing the groups G_i (called a *refinement*).

EXERCISE 2.301 (Solvable). A group G is *solvable* if there exists a series of subgroups G_i so that

$$\{e\} = G_0 \subseteq G_1 \subseteq \cdots \subseteq G_n = G$$

with G_i normal in G_{i+1} and G_{i+1}/G_i abelian, $0 \le i \le n-1$.

(a) Show abelian groups are solvable.

(b) Show finite p-groups are solvable. *Hint:* Exercise 2.254.

(c) Show D_n is solvable.

(d) Show A_4 is solvable (cf. Exercise 2.210).

CHAPTER 3

Rings

Now that we have gained some mastery over working with a single binary operation, it is time to look at objects with two binary operations. Of course our prototypes are $\mathbb{Z}$ and $\mathbb{Z}_n$, which are equipped with both addition and multiplication operations. Many generalizations of this are found in mathematics and they are called *rings*. This chapter begins their study.

1. Examples and Basic Properties

1.1. Definition. It might be instructive to compare the following definition to Axiom 1 and Theorem 1.22.

DEFINITION 3.1. A *ring* is a nonempty set R equipped with two binary operations $+$ and $\cdot$ so that the following are true:

(1) $(R, +)$ is an abelian group.

(2) (*Associativity*) For $r_1, r_2, r_3 \in R$,

$$r_1 \cdot (r_2 \cdot r_3) = (r_1 \cdot r_2) \cdot r_3.$$

(3) (*Distributivity*) For $r_1, r_2, r_3 \in R$,

$$r_1 \cdot (r_2 + r_3) = (r_1 \cdot r_2) + (r_1 \cdot r_3),$$
$$(r_1 + r_2) \cdot r_3 = (r_1 \cdot r_3) + (r_2 \cdot r_3).$$

DEFINITION 3.2. **(1)** A ring R is called a *ring with identity* (also called a *ring with unity*) if there exists an element $1_R \in R$ so that

$$1_R \cdot r = r \cdot 1_R = r$$

for all $r \in R$.

(2) A ring R is called *commutative* if

$$r_1 \cdot r_2 = r_2 \cdot r_1$$

for all $r_i \in R$.

The identity element for the additive structure of a ring R is usually written as 0_R or simply as 0 if there is little possibility of confusion. Of course additive group notation is used. Similarly, the identity element in a ring R with identity is usually written as 1 instead of 1_R and multiplicative notation is employed. Most of the rings we study will have a multiplicative identity and it is worth noting that the identity element is unique (Exercise 3.1). Just as in the case with multiplicative group notation, it is common to suppress the explicit writing of the binary operation $\cdot$ in rings. Thus $r \cdot s$ is usually written as rs.

1.2. Examples. An easy example of a ring is to take any abelian group R, with group law written additively, and define a second binary operation by $r \cdot s = 0$ for all $r, s \in R$. Such rings are called *trivial rings*. Of course trivial rings are really just abelian groups with a fancy name and so are already well studied. As a special case, there is also the boring *zero ring*, $R = \{0\}$, denoted simply $R = 0$, where $0 + 0 = 0$ and $0 \cdot 0 = 0$. It is a mildly interesting fact that $R = 0$ is the only ring with identity for which $1 = 0$ (Exercise 3.27).

Turning to more honest examples, the most familiar rings are $\mathbb{Z}$, $\mathbb{Z}_n$, $\mathbb{Q}$, $\mathbb{R}$, and $\mathbb{C}$. The additive and multiplicative structures give the two binary operations. Each of these examples are rings since they are abelian groups with respect to addition and satisfy multiplicative associativity and the distributive law. In fact, these are all examples of commutative rings with identity.

Of course there are many variations of arithmetical examples. For instance,

$$\mathbb{Q}(\sqrt{2}) = \{a + b\sqrt{2} \mid a, b \in \mathbb{Q}\}$$

is easily seen to be a ring under addition and multiplication (cf. §1.4 below).

As an example of a ring without identity, consider the set of even integers, $2\mathbb{Z}$. It is easily seen to be a ring since addition and multiplication preserve evenness, but clearly there is no multiplicative identity since $2\mathbb{Z}$ is missing the number 1. In contrast, the set of odd integers does not even qualify to be a ring since it is not closed under addition.

To get noncommutative rings, the most natural place to looks is in the set of matrices. For example, it is well known from linear algebra that $M_{n,n}(\mathbb{F})$ is a ring with identity under matrix addition and multiplication. Though $M_{n,n}(\mathbb{F})$ has an identity, it is noncommutative (for $n \geq 2$). More generally, if R is any commutative ring, then $M_{n,n}(R)$, the set of $n \times n$ matrices with entries in R, is a ring (Exercise 3.18). Gadgets such as $GL(n, \mathbb{R})$ are not rings since they are not closed under addition.

Another type of surprisingly useful ring is a polynomial ring. For instance, $\mathbb{R}[x]$, the ring of polynomials in the variable x with coefficients in $\mathbb{R}$, is a ring under the usual rules of adding and multiplying polynomials. More generally, if R is any ring, $R[x]$, the ring of polynomials in the variable x with coefficients in R, is a ring (cf. Section 3). These examples can be generalized further by looking at many indeterminates or by constructions such as formal power series.

Even more generally, function spaces are often rings. For instance, $\mathcal{F}(\mathbb{R})$, the set of functions from $\mathbb{R}$ to $\mathbb{R}$ is a ring under pointwise addition and multiplication of functions. There are endless variations on this theme involving different domains, codomains, or adjectives such as continuous, differentiable, and so on.

1.3. Basic Properties. Just as was the case for $\mathbb{Z}_n$ (Definition 1.23), it is useful to study zero divisors and units.

DEFINITION 3.3. Let R be a ring with $r, s \in R$.

(1) An element r is called a *zero divisor* if there is a nonzero s so that $rs = 0$ or $sr = 0$ (in the first case r is called a *left zero divisor* and in the second case it is called a *right zero divisor*).

(2) If $R \neq 0$, R is called a *domain* if R has no zero divisors. It is called an *integral domain* if it is also commutative and has an identity.

(3) If R is a ring with identity, an element r is called a *unit* (or is said to be *invertible*) if there is an s so that $rs = sr = 1$. In that case, s is called the *inverse* of r and is denoted by r^{-1}. The set of units is denoted by $R^\times$ and is called the *group of units.*

(4) If $R \neq 0$ has an identity, R is called a *division ring* (or *skew field*) if every nonzero r is a unit, i.e., if $R^\times = R\backslash\{0\}$. If R is also commutative, it is called a *field.*

As the reader no doubt expects, zero divisors and units are mutually exclusive definitions, though they are not always the only possibilities (Exercise 3.3). In particular, a field is always an integral domain. It is also worth noting that inverses, when they exist, are unique and that the group of units really is a group under multiplication (Exercise 3.1).

Turning to some examples, begin with $\mathbb{Z}$ and $\mathbb{Z}_n$. For instance, $\mathbb{Z}$ has no zero divisors and so is a domain. Since $\mathbb{Z}$ is also commutative and has an identity, $\mathbb{Z}$ is in fact an integral domain. However, its group of units is fairly small, namely $\mathbb{Z}^\times = \{\pm 1\}$, and so $\mathbb{Z}$ is not a field.

On the other hand, $\mathbb{Z}_n$ has a richer variety of possibilities. Here we have seen (Theorem 1.24) that $[k] \in \mathbb{Z}_n$ is zero if $(k, n) = n$, it is a zero divisor if $1 < (k, n) < n$, and it is a unit if $(k, n) = 1$. In particular, $\mathbb{Z}_n^\times = U_n$ and $\mathbb{Z}_n$ has zero divisors if and only if n is not prime. Of course, $\mathbb{Z}_n$ is always commutative and has an identity so that $\mathbb{Z}_n$ is an integral domain exactly when n is prime. In fact when $n = p$ is prime, $\mathbb{Z}_p$ is actually a field since all nonzero elements are invertible, i.e., $\mathbb{Z}_p^\times = \mathbb{Z}_p\backslash\{[0]\}$. Finite fields such as $\mathbb{Z}_p$ are very important objects in mathematics (cf. Section 3).

In order to show no favoritism, infinite fields are also extremely important and are very well studied. Some of the most important examples include $\mathbb{Q}$, $\mathbb{R}$, and $\mathbb{C}$.

To find an example of a division ring that is not a field, we turn to a construction that is often first seen in a multivariable calculus or physics course while working with cross products. Namely, the *quaternions* are the ring

$$\mathbb{H} = \{a\mathbf{1} + b\mathbf{i} + c\mathbf{j} + d\mathbf{k} \mid a, b, c, d \in \mathbb{R}\}$$

where $\mathbf{1}, \mathbf{i}, \mathbf{j}, \mathbf{k} \in GL(2, \mathbb{C})$ are given by

$$\mathbf{1} = \begin{pmatrix} 1 & 0 \\ 0 & 1 \end{pmatrix}, \quad \mathbf{i} = \begin{pmatrix} i & 0 \\ 0 & -i \end{pmatrix}, \quad \mathbf{j} = \begin{pmatrix} 0 & 1 \\ -1 & 0 \end{pmatrix}, \quad \mathbf{k} = \begin{pmatrix} 0 & i \\ i & 0 \end{pmatrix}$$

(cf. Exercise 2.13). The quaternions are a noncommutative ring with identity (e.g., $\mathbf{ij} = -\mathbf{ji}$). In fact, every nonzero element is a unit (Exercise 3.14), which makes $\mathbb{H}$ into a division ring but not a field.

Turning to more matrix examples, $M_{n,n}(R)$, $n \geq 2$ and R nontrivial, is never commutative and has lots of (left or right) zero divisors. For instance in $M_{2,2}(\mathbb{R})$ we have

$$\begin{pmatrix} 0 & 1 \\ 0 & 0 \end{pmatrix}\begin{pmatrix} 1 & 0 \\ 0 & 0 \end{pmatrix} = \begin{pmatrix} 0 & 0 \\ 0 & 0 \end{pmatrix} \quad \text{but} \quad \begin{pmatrix} 1 & 0 \\ 0 & 0 \end{pmatrix}\begin{pmatrix} 0 & 1 \\ 0 & 0 \end{pmatrix} = \begin{pmatrix} 0 & 1 \\ 0 & 0 \end{pmatrix}.$$

Of course the group of units here is precisely $M_{2,2}(\mathbb{R})^\times = GL(2, \mathbb{R})$ by definition.

We end this subsection with a few remarks on multiplicative and exponential notation in a ring. As already noted, additive notation is used for the abelian group structure on a ring R. For instance for $m \in \mathbb{N}$ and $r \in R$, we therefore have

$$mr = \overbrace{r + r + \cdots + r}^{m \text{ copies}},$$

$(-m)r = m(-r)$, and $(0)r = 0_R$. Notice that this agrees with $0_R \cdot r = 0_R$ and, if there is a multiplicative identity, $1_R \cdot r = (1)r$ so that leaving off the subscripts R will not cause undue confusion.

As with the case of multiplicative group notation, we employ similar conventions for exponentials with respect to the multiplicative ring structure. Namely, for $m \in \mathbb{N}$ and $r \in R$, we define

$$r^m = \overbrace{r \cdot r \cdot \cdots \cdot r}^{m \text{ copies}}.$$

If R has an identity and if r is a unit, we also write $r^0 = 1$ and $r^{-m} = (r^{-1})^m$.

1.4. Subrings. As with groups and subgroups, there is an analogous notion of subrings for rings. Besides being flat out useful, this idea will also allow us to easily construct many new rings.

DEFINITION 3.4. Let R be a ring and let $S \subseteq R$, $S \neq \varnothing$. We say S is a *subring* if S is a ring with respect to the binary operations of R restricted to S.

As an example, $2\mathbb{Z} \subseteq \mathbb{Z} \subseteq \mathbb{Q} \subseteq \mathbb{R} \subseteq \mathbb{C} \subseteq \mathbb{H} \subseteq M_{2,2}(\mathbb{C})$ is a chain of subrings. Just as with subgroups, there is an easy criterion for determining when a subset is a subring. The proof of the following theorem is virtually identical to that of Theorem 2.13 and so the proof is left to Exercise 3.34.

THEOREM 3.5. *Let R be a ring and let $S \subseteq R$.*

(1) *S is a subring if and only if:*
  (a) *S is closed under both binary operations of R, i.e., if $s_1, s_2 \in S$, then $s_1 + s_2 \in S$ and $s_1 s_2 \in S$,*
  (b) *$0 \in S$, and*
  (c) *when $s \in S$, then $-s \in S$.*

(2) *Alternately, S is a subring if and only if $S \neq \varnothing$ and whenever $s_1, s_2 \in S$, then $s_1 - s_2 \in S$ and $s_1 s_2 \in S$.*

As an example of Theorem 3.5, recall $\mathbb{Q}(\sqrt{2}) = \{a + b\sqrt{2} \mid a, b \in \mathbb{Q}\} \subseteq \mathbb{R}$. To see $\mathbb{Q}(\sqrt{2})$ is a subring, first note that it is manifestly nonempty. Next, suppose $a + b\sqrt{2}$, $a' + b'\sqrt{2} \in \mathbb{Q}(\sqrt{2})$. Then

$$\left(a + b\sqrt{2}\right) - \left(a' + b'\sqrt{2}\right) = (a - a') + (b - b')\sqrt{2} \in \mathbb{Q}(\sqrt{2}),$$

$$\left(a + b\sqrt{2}\right)\left(a' + b'\sqrt{2}\right) = (aa' + 2bb') + (ab' + ba')\sqrt{2} \in \mathbb{Q}(\sqrt{2})$$

so that $\mathbb{Q}(\sqrt{2})$ is a subring of $\mathbb{R}$ and, in particular, it is a ring in its own right.

As with groups, the analogous notions of a centralizer and generators are very useful.

DEFINITION 3.6. Let R be a ring with $T \subseteq R$, $T \neq \varnothing$.

(1) The *centralizer* of T in R is

$$Z(T) = Z_R(T) = \{r \in R \mid rt = tr \text{ for all } t \in T\}.$$

In the special case of $T = R$,

$$Z(R) = \{r \in R \mid rt = tr \text{ for all } t \in R\}$$

is called the *center* of R.

(2) Let $\mathcal{T} = \{\text{subrings } S \subseteq R \mid T \subseteq S\}$. The *subring generated by* T is

$$\langle T \rangle = \bigcap_{S \in \mathcal{T}} S.$$

The elements of T are called *generators* of $\langle T \rangle$.

As we saw in the case of groups, these constructions yield subrings. Moreover, there is a more concrete method of working with $\langle T \rangle$. Since the proof of the following theorem is virtually identical to that of Theorem 2.15, we leave it to Exercise 3.35.

THEOREM 3.7. *Let R be a ring with $T \subseteq R$, $T \neq \varnothing$.*

(1) *$Z(T)$ and $\langle T \rangle$ are subrings of R.*

(2) *$\langle T \rangle$ is the smallest subring of R containing T and can alternately be written as*

$$\langle T \rangle = \{\textit{finite sums of products of the form } t_1 t_2 \cdots t_k \mid k \in \mathbb{N} \textit{ with } t_i \in T \cup \{-T\}\}.$$

Note that some of this notation can be ambiguous when an object is sometimes viewed as a group and sometimes viewed as a ring. For instance, suppose R is a ring and $T \subseteq R$, $T \neq \varnothing$. If the multiplication structure is temporarily forgotten and R is simply viewed as an abelian group under $+$, then the notation $\langle T \rangle$ stands for the set of all *finite sums* of elements from T or $-T$, Theorem 2.15. However, if R is viewed as a ring under both its addition and multiplication structures, then $\langle T \rangle$ stands for the set of all *finite sums of products* of elements from T or $-T$, Theorem 3.7. For example, if $\mathbb{Z}[x]$ is viewed simply as an abelian group under polynomial addition, then $\langle x \rangle = \{mx \mid m \in \mathbb{Z}\}$. However, if $\mathbb{Z}[x]$ is viewed as a ring under polynomial addition and multiplication, then $\langle x \rangle = \{m_1 x + m_2 x^2 + \cdots + m_N x^N \mid N \in \mathbb{N}, m_i \in \mathbb{Z}\}$, i.e., the set of all polynomials with constant term 0. Of course, sometimes the two notations happily coincide. This is the case with $\mathbb{Z}$ where, for $k \in \mathbb{Z}_{\geq 0}$, $\langle k \rangle = k\mathbb{Z}$ whether $\mathbb{Z}$ is viewed as a group or as a ring (Exercise 3.16).

1.5. Direct Products. Analogous to the situation of a direct product of groups, there is the notion of a direct product of rings. As with groups, the following construction allows us to take rings and to glue them together in an infinite number of new ways as well as allowing us to recognize when a complicated ring is really just pieced together from simpler rings.

DEFINITION 3.8. Let R_i, $1 \leq i \leq N$, be a ring. Define binary operations on the *direct product* $\prod_{i=1}^N R_i = R_1 \times R_2 \times \cdots \times R_N$ by

$$(r_1, r_2, \ldots, r_N) + (r_1', r_2', \ldots, r_N') = (r_1 + r_1', r_2 + r_2', \ldots, r_N + r_N'),$$

$$(r_1, r_2, \ldots, r_N)(r_1', r_2', \ldots, r_N') = (r_1 r_1', r_2 r_2', \ldots, r_N r_N')$$

for $r_i, r_i' \in R_i$.

The proof of the following theorem is straightforward and left to Exercise 3.36.

THEOREM 3.9. *With the above structure, $\prod_{i=1}^N R_i$ is a ring.*

1.6. Exercises 3.1–3.39.

EXERCISE 3.1. Let R be a ring with identity.

(a) Show the multiplicative identity is unique, i.e., if $u \in R$ satisfies $ur = ru = r$ for all $r \in R$, then $u = 1$. *Hint:* Try $r = 1$.

(b) Show inverses are unique, i.e., if $r \in R$ is a unit with $rs = sr = 1$ and $rs' = s'r = 1$ for some $s, s' \in R$, then $s = s'$. *Hint:* Look at $s'rs$.

(c) Suppose $r, s \in R^\times$. Show $rs \in R^\times$ with $(rs)^{-1} = s^{-1}r^{-1}$. Conclude that $R^\times$ is a group with respect to multiplication.

* EXERCISE 3.2. Let R be a ring with $r, s \in R$. Show the following:

(a) $(-r)s = r(-s) = -(rs)$.

(b) $-r = (-1_R)r$.

EXERCISE 3.3. Let R be a ring with identity.

(a) Show $\{\text{zero divisors}\} \cap \{\text{units}\} = \varnothing$.

(b) Conclude that fields are integral domains.

(c) Show $x \in \mathbb{R}[x]$ is neither a zero divisor nor a unit.

EXERCISE 3.4 (Cancellation in Domains). **(a)** Let R be a ring with $r \in R$ nonzero and not a zero divisor. If $ra = rb$ for $a, b \in R$, show $a = b$. *Hint:* $r(a - b)$.

(b) If D is a domain and $ra = rb$ with $r, a, b \in D$, show that either $r = 0$ or $a = b$.

(c) If D is commutative and $r, s \in D$ satisfy $r^2 = s^2$, show $r = s$ or $r = -s$.

(d) Show by example that parts (b) and (c) can be false if D is not a domain.

EXERCISE 3.5. **(a)** Show $\mathbb{R}E_{1,1}$ is a subring of $M_{2,2}(\mathbb{R})$.

(b) Show $\mathbb{R}E_{1,1}$ has an identity but that $1_{\mathbb{R}E_{1,1}} \neq 1_{M_{2,2}(\mathbb{R})}$.

EXERCISE 3.6. Let $R = \{\begin{pmatrix} a & 0 \\ 0 & 0 \end{pmatrix} \mid a \in \mathbb{R}\}$ and $S = \{\begin{pmatrix} a & b \\ 0 & 0 \end{pmatrix} \mid a, b \in \mathbb{R}\}$. Show that R is a subring of S and that R has an identity but that S does not have an identity.

EXERCISE 3.7. Show that every finite integral domain is a field. *Hint:* If nonzero $r \in R$, use use Exercise 3.4 to show $|rR| = |R|$.

EXERCISE 3.8. Show that the set of upper triangular matrices is a subring of $M_{n,n}(\mathbb{R})$.

EXERCISE 3.9. **(a)** Let $n, m \in \mathbb{Z}$. Show $\begin{pmatrix} \mathbb{Z} & n\mathbb{Z} \\ m\mathbb{Z} & \mathbb{Z} \end{pmatrix} = \{\begin{pmatrix} a & b \\ c & d \end{pmatrix} \mid a, d \in \mathbb{Z}, b \in n\mathbb{Z}, c \in m\mathbb{Z}\}$ is a subring of $M_{2,2}(\mathbb{Z})$.

* **(b)** Show $\{\begin{pmatrix} a & b \\ -b & a \end{pmatrix} \mid a, b \in \mathbb{R}\}$ is a subring of $M_{2,2}(\mathbb{R})$.

EXERCISE 3.10. Let S_α and S_n be subrings of R, $\alpha \in \mathcal{A}$ and $n \in \mathbb{N}$.

(a) Show $\bigcap_{\alpha \in \mathcal{A}} S_\alpha$ is a subring.

(b) If $S_n \subseteq S_{n+1}$, show $\bigcup_{n \in \mathbb{N}} S_n$ is a subring.

(c) If $S_1 \not\subseteq S_2$ and $S_2 \not\subseteq S_1$, show $S_1 \cup S_2$ is not a subring (cf. Exercise 2.62).

EXERCISE 3.11 ($\mathbb{Z}[i]$). The *Gaussian integers* are defined as $\mathbb{Z}[i] = \{a + ib \mid a, b \in \mathbb{Z}\}$.

(a) Show $\mathbb{Z}[i]$ is the subring of $\mathbb{C}$ generated by 1 and i. In particular, conclude $\mathbb{Z}[i]$ is an integral domain.

(b) Show $\mathbb{Z}[i]^\times = \pm\{1, i\}$. *Hint:* If $z \in \mathbb{Z}[i]$, show $|z|^2 \in \mathbb{Z}$ and then that $|z| = 1$ if z is a unit.

EXERCISE 3.12. For $n \in \mathbb{N}$, let $\mathbb{Z}_n[i] = \{a + ib \mid a, b \in \mathbb{Z}_n\}$ with addition and multiplication given by

$$(a + ib) + (a' + ib') = (a + a') + i(b + b'),$$
$$(a + ib)(a' + ib') = (aa' - bb') + i(ab' + ba').$$

(a) Show $\mathbb{Z}_n[i]$ is a commutative ring with identity.

(b) Show $\mathbb{Z}_2[i]$ and $\mathbb{Z}_5[i]$ have zero divisors and so are not integral domains.

(c) Show $\mathbb{Z}_3[i]$ is a field with 9 elements.

EXERCISE 3.13. Let $m \in \mathbb{Z}$ and define $\mathbb{Z}(\sqrt{m}) = \{a + b\sqrt{m} \mid a, b \in \mathbb{Z}\}$.

(a) Show $\mathbb{Z}(\sqrt{m})$ is the subring of $\mathbb{R}$ generated by 1 and $\sqrt{m}$. In particular, conclude $\mathbb{Z}(\sqrt{m})$ is a ring.

(b) Show $\mathbb{Q}(\sqrt{m}) = \{a + b\sqrt{m} \mid a, b \in \mathbb{Q}\}$ is a field. *Hint:* Multiply by $\overline{a + b\sqrt{m}} = a - b\sqrt{m}$.

EXERCISE 3.14. **(a)** Show $\mathbb{H}$ is the subring of $M_{2,2}(\mathbb{R})$ generated by $\mathbb{R}\mathbf{1}$, $\mathbf{i}$, and $\mathbf{j}$ and therefore is a ring in its own right (cf. Exercise 2.13).

(b) Define the *conjugate* of $z = a\mathbf{1} + b\mathbf{i} + c\mathbf{j} + d\mathbf{k}$ as $\overline{z} = a\mathbf{1} - b\mathbf{i} - c\mathbf{j} - d\mathbf{k}$ and the *norm* as $N(z) = a^2 + b^2 + c^2 + d^2$. Verify that $z\overline{z} = \overline{z}z = N(z)$. Conclude that $\mathbb{H}$ is a division ring but not a field.

(c) Show $\overline{z_1 z_2} = \overline{z_2}\,\overline{z_1}$ and conclude that $N(z_1 z_2) = N(z_1)N(z_2)$ for $z_i \in \mathbb{H}$. *Hint:* Work first with $z_1 \in \{\mathbf{1}, \mathbf{i}, \mathbf{j}, \mathbf{k}\}$.

(d) Let $\mathbb{H}_\mathbb{Z} = \{a\mathbf{1} + b\mathbf{i} + c\mathbf{j} + d\mathbf{k} \mid a, b, c, d \in \mathbb{Z}\}$. Show $\mathbb{H}_\mathbb{Z}$ is a subring of $\mathbb{H}$ and that $\mathbb{H}_\mathbb{Z}^\times = \{z \in \mathbb{H}_\mathbb{Z} \mid N(z) = 1\}$. Conclude that $\mathbb{H}_\mathbb{Z}^\times \cong Q$ (cf. Exercise 2.13).

EXERCISE 3.15 ($\mathcal{F}(S, R)$). Let S be a nonempty set and let R be a ring. Define the function space $\mathcal{F}(S, R) = \{f : S \to R\}$. For $f_i \in \mathcal{F}(S, R)$, let $f_1 + f_2 \in \mathcal{F}(S, R)$ by given by $(f_1 + f_2)(s) = f_1(s) + f_2(s)$, $s \in S$, and let $f_1 f_2 \in \mathcal{F}(S, R)$ be given by $(f_1 f_2)(s) = f_1(s) f_2(s)$

(a) Show $\mathcal{F}(S, R)$ is a ring.

(b) Show $\mathcal{F}(S, R)$ has a multiplicative identity if and only if R does. *Hint:* Examine the constant functions $f_r(s) = r$ for $r \in R$ and $s \in S$.

(c) Show $\mathcal{F}(S, R)$ is commutative if and only if R is.

EXERCISE 3.16. **(a)** Let $k \in \mathbb{Z}$. Show $\langle k \rangle = k\mathbb{Z}$ whether $\mathbb{Z}$ is viewed as an abelian group under $+$ or viewed as a ring under $+$ and $\cdot$.

(b) Let $n \in \mathbb{N}$ and $[k] \in \mathbb{Z}_n$. Show $\langle [k] \rangle = k\mathbb{Z}_n$ whether $\mathbb{Z}_n$ is viewed as an abelian group under $+$ or viewed as a ring under $+$ and $\cdot$.

(c) Explicitly calculate $\langle x^2 \rangle$ in $\mathbb{Z}[x]$ first viewed as an abelian group under polynomial addition and then viewed as a ring under polynomial addition and multiplication. Show the two answers are different.

EXERCISE 3.17 (Subrings of $\mathbb{Z}$ and $\mathbb{Z}_n$). **(a)** Show that the set of subrings of $\mathbb{Z}$ is $\{\langle k \rangle \mid k \in \mathbb{Z}_{\geq 0}\}$ (cf. Exercises 2.63 and 3.16).

(b) Show that the set of subrings of $\mathbb{Z}_n$ is $\{\langle [k] \rangle \mid k \in \mathbb{N} \text{ and } k \mid n\}$.

EXERCISE 3.18 ($M_{n,n}(R)$). If R is a commutative ring, let $M_{n,n}(R) = \{(r_{i,j}) \mid 1 \le i,j \le n,\ r_{i,j} \in R\}$ be the set of $n \times n$ matrices with entries in R equipped with the operations of matrix addition and multiplication,

$$(r_{i,j}) + (r'_{i,j}) = (r_{i,j} + r'_{i,j}),$$

$$(r_{i,j})(r'_{i,j}) = (s_{i,j}) \text{ where } s_{i,j} = \sum_{k=1}^{n} r_{i,k} r'_{k,j}.$$

(a) Show $M_{n,n}(R)$ is a ring.

(b) Show that $M_{n,n}(R)$ is not commutative and that it has zero divisors when $n \ge 2$ and R is nontrivial

(c) Show $M_{n,n}(R)$ has an identity if and only if R does.

(d) Show $M_{2,2}(R)^{\times} = GL(2,R)$ where $GL(2,R) = \{\begin{pmatrix} a & b \\ c & d \end{pmatrix} \in M_{2,2}(R) \mid \det \begin{pmatrix} a & b \\ c & d \end{pmatrix} = ad - bc \in R^{\times}\}$. N.B.: The analogous result holds for $M_{n,n}(R)^{\times}$.

EXERCISE 3.19 (Characteristic). Let R be a ring with identity. Let $C = \{k \in \mathbb{Z} \mid kr = 0, \text{ all } r \in R\}$ (cf. Definition 4.30).

(a) Show C is a subgroup of $\mathbb{Z}$ and conclude that $C = \langle n \rangle$, some unique $n \in \mathbb{Z}_{\ge 0}$. *Hint:* Exercise 2.63.

The integer n is called the *characteristic* of R and is denoted $\operatorname{char} R = n$.

(b) Calculate the characteristic of $\mathbb{Z}$ and $\mathbb{Z}_n$ for $n \in \mathbb{N}$.

(c) Calculate the characteristic of $M_{m,m}(\mathbb{Z})$ and $M_{m,m}(\mathbb{Z}_n)$ for $m \in \mathbb{N}$.

(d) Show that $k \in C$ if and only if $k1_R = 0$. *Hint:* $r = 1_R \cdot r$.

EXERCISE 3.20 (Binomial Theorem and Nilpotent). Let R be a ring with $r, s \in R$ so that $rs = sr$.

(a) For $n \in \mathbb{N}$, show by induction (cf. Exercise 1.2) that

$$(r+s)^n = \sum_{k=0}^{n} \binom{n}{k} r^{n-k} s^k.$$

(b) If R is commutative, let $N(R) = \{r \in R \mid r^k = 0 \text{ for some } k \in \mathbb{N}\}$, the set of all *nilpotent* elements of R. Show N is a subring of R.

(c) If $r \in N(R)$, show $1 - r$ is a unit. *Hint:* $1 + r + r^2 + \cdots$.

(d) Show by example that $N(R)$ need not be a subring if R is not commutative.

EXERCISE 3.21 (Idempotents). Let R be a ring. An element $e \in R$ is called an *idempotent* if $e^2 = e$. Roughly speaking, idempotents behave like projections in linear algebra.

(a) Show $[0]$, $[1]$, $[3]$, and $[4]$ are idempotents in $\mathbb{Z}_6$.

(b) Show $\begin{pmatrix} 1 & 0 \\ 0 & 0 \end{pmatrix}$ and $\begin{pmatrix} 0 & 1 \\ 0 & 1 \end{pmatrix}$ are idempotents in $M_{2,2}(\mathbb{R})$.

(c) If e is an idempotent, show $1 - e$ is also one.

(d) If e is an idempotent, show $eRe = \{ere \mid r \in R\}$ is a subring of R.

(e) Show e is a multiplicative identity for eRe and that $eRe = \{r \in R \mid er = re = r\}$. If R has a unit, conclude $1 \in eRe$ if and only if $e = 1$.

EXERCISE 3.22 ($\operatorname{End}(G)$). Let G be an abelian group (written additively) and let $\operatorname{End}(G) = \{\varphi : G \to G \mid \varphi \text{ is a homomorphism}\}$. For $\varphi_i \in \operatorname{End}(G)$, define

$\varphi_1+\varphi_2 \in \text{End}(G)$ by $(\varphi_1+\varphi_2)(g) = \varphi_1(g)+\varphi_2(g)$, $g \in G$, and define $\varphi_1\varphi_2 \in \text{End}(G)$ by $\varphi_1\varphi_2 = \varphi_1 \circ \varphi_2$.

(a) Show $\text{End}(G)$ is a ring with additive identity given by the trivial map.

(b) Show $\text{End}(G)$ has a multiplicative identity given by the identity map.

(c) Show $\text{End}(G)^\times = \text{Aut}(G)$ (cf. Exercise 2.115).

(d) Show $\text{End}(G)$ may not be commutative. *Hint:* Try $G = \mathbb{Z}_2 \times \mathbb{Z}_2$ (cf. Exercise 2.117).

EXERCISE 3.23. Show that $Z(M_{n,n}(\mathbb{R})) = \{\lambda I_n \mid \lambda \in \mathbb{R}^\times\}$.

EXERCISE 3.24. **(a)** Let $x, y \in \mathbb{R}$ with $x < y$. Show $\mathbb{R} = \langle (x, y) \rangle$.

(b) As a ring, show $\mathbb{Q} = \left\langle \{\frac{1}{p} \mid p \text{ is prime}\} \right\rangle$.

EXERCISE 3.25. Let R_i, $1 \leq i \leq N$, be rings with identity. Show $\left(\prod_{i=1}^N R_i\right)^\times = \prod_{i=1}^N R_i^\times$.

EXERCISE 3.26. Let R be a ring with identity and let S be a subring containing 1_R.

(a) If $u \in S$ is a unit in S, show u is a unit in R.

(b) Show by example that it is possible to have $u \in S$ be a unit in R but not in S.

EXERCISE 3.27. Show $R = 0$ is the only ring with identity in which $1 = 0$.

EXERCISE 3.28. Show that the abelian hypothesis on the additive structure in the definition of a ring with identity is actually forced by the distributive law. *Hint:* Expand $(1 + 1)(r + s)$ in two different ways.

EXERCISE 3.29. Let R be a ring with $r \in R$. A *right inverse* to r is an element $y \in R$ so that $ry = 1$ and a *left inverse* is an element $z \in R$ so that $zr = 1$.

(a) Let G be the group $\prod_{i=1}^\infty \mathbb{R}$ (cf. Exercise 2.91) and consider the ring $\text{End}(G)$ (cf. Exercise 3.22). Define $\varphi, \psi \in \text{End}(G)$ by $\varphi(r_1, r_2, \ldots) = (0, r_1, r_2, \ldots)$ and $\psi(r_1, r_2, r_3, \ldots) = (r_2, r_3, \ldots)$. Show $\psi\varphi = \text{Id}$ but that $\varphi\psi \neq \text{Id}$.

(b) Show φ and ψ are not units in $\text{End}(G)$. Conclude that it is possible for elements of rings to have left or right inverses without being units.

(c) Show r is a unit if and only if there exists a right inverse and a left inverse. In that case, show they are the same.

(d) If R is finite, show that r is a unit if and only if it has a right inverse if and only if it has a left inverse. *Hint:* Count rR and Rr.

EXERCISE 3.30. Show every finite domain is a division ring (cf. Exercises 3.7 and 3.29). N.B.: *Wedderburn's Theorem* says that every finite division ring is a field (Exercise 4.178).

EXERCISE 3.31 (Boolean Rings). A ring $R \neq 0$ is called *Boolean* if $r^2 = r$ for all $r \in R$.

(a) Let $T \neq \varnothing$ be a set and consider the power set $\mathcal{P}(T)$ equipped with the binary operations $U + V = (U \backslash V) \cup (V \backslash U)$ and $U \cdot V = U \cap V$ for $U, V \in \mathcal{P}(T)$. Show that these operations make $\mathcal{P}(T)$ into a Boolean ring (cf. Exercise 2.27).

(b) Show that Boolean rings are commutative and that they have characteristic 2 (cf. Exercises 2.45 and 3.19).

EXERCISE 3.32. Define the *Lie bracket* on $M_{n,n}(\mathbb{R})$ by $[A, B] = AB - BA$ for $A, B \in M_{n,n}(\mathbb{R})$. Prove the *Jacobi identity* for $A, B, C \in M_{n,n}(\mathbb{R})$:

$$[A, [B, C]] = [[A, B], C] + [B, [A, C]].$$

Conclude that the Lie bracket is not associative ($n \geq 2$).

EXERCISE 3.33. Let R be a ring with binary operations $+$ and $\cdot$. Let R^{opp} be the same set as R equipped with the same addition, $+$, but with a new multiplication, $\#$, defined by $r \# s = s \cdot r$. Show R^{opp} is a ring (cf. Exercise 2.52).

EXERCISE 3.34. Prove Theorem 3.5.

EXERCISE 3.35. Prove Theorem 3.7.

EXERCISE 3.36. Prove Theorem 3.9.

EXERCISE 3.37 (Group Ring, $R[G]$). Let G be a group and let R be a ring with identity. Define the *group ring* of G with coefficients in R as the set of formal sums

$$R[G] = \{\sum_{g \in G} r_g g \mid r_g \in R \text{ and } r_g = 0 \text{ for all but finitely many } g\}$$

with binary operations defined by

$$\sum_{g \in G} r_g g + \sum_{g \in G} r'_g g = \sum_{g \in G} \left(r_g + r'_g\right) g,$$

$$\left(\sum_{g \in G} r_g g\right) \left(\sum_{g \in G} r'_g g\right) = \sum_{g \in G} \left(\sum_{h \in G} r_{gh^{-1}} r'_h\right) g.$$

Note that the multiplication definition is generated by setting $(r_g g)(r_h h) = (r_g r_h) gh$ and then extending with the distributive law.

(a) For $R = \mathbb{R}$ and $G = S_4$, compute $[9(4,3,2,1) + 7(3,4,1)]\,[(2,3,1) - 6(3,1)]$ in $\mathbb{R}[S_4]$.

(b) For $R = M_{2,2}(\mathbb{R})$ and $G = \mathbb{Z}_4$, compute

$$\left[\begin{pmatrix} 1 & 2 \\ 2 & 3 \end{pmatrix} [3] + \begin{pmatrix} 4 & 5 \\ 6 & 7 \end{pmatrix} [1]\right] \left[\begin{pmatrix} 0 & -1 \\ -2 & -3 \end{pmatrix} [2] + \begin{pmatrix} 1 & 1 \\ 1 & 1 \end{pmatrix} [0]\right].$$

(c) If G is a finite group, let $c \in \mathbb{Q}[G]$ be given by $c = \frac{1}{|G|} \sum_{g \in G} g$. Show $c^2 = c$.

(d) In general, show $R[G]$ has a multiplicative identity given by $(1_R)e_G$.

(e) In general, show $R[G]$ is a ring with additive identity given by $\sum_{g \in G} (0_R)\, g$.

EXERCISE 3.38 (Discrete Valuation Rings). Let F be a field and let $\nu : F^{\times} \to \mathbb{Z}$ satisfy (i) ν is a homomorphism, i.e., $\nu(xy) = \nu(x) + \nu(y)$ for $x, y \in F^{\times}$, (ii) ν is surjective, and (iii) $\nu(x + y) \geq \min\{\nu(x), \nu(y)\}$ when $x + y \neq 0$. Such ν are called *discrete valuations* on F.

(a) Let p be a positive prime and define $\nu_p : \mathbb{Q} \to \mathbb{Z}$ by $\nu_p(\frac{a}{b}) = n$ where $\frac{a}{b} = p^n \frac{c}{d}$ with $p \nmid c$ and $p \nmid d$ for $n, a, b, c, d \in \mathbb{Z}$ with $bd \neq 0$. Show ν_p is a discrete valuation of $\mathbb{Q}$.

A *discrete valuation ring* is one of the form $R = \{x \in F^{\times} \mid \nu(x) \geq 0\} \cup \{0_F\}$.

(b) Show R is a subring of F.

(c) Show R contains 1 and so is an integral domain.

(d) For $x \in F^{\times}$, show x or x^{-1} is in R.

(e) Show that $R^\times = \{x \in F^\times \mid \nu(x) = 0\}$.

(f) Fix $t \in R$ so $\nu(t) = 1$. Show that any $r \in R$ can be written as $r = ut^n$ for $u \in R^\times$ and $n \in \mathbb{Z}_{\geq 0}$. The element t is called a *uniformizing* or *local parameter*.

(g) For the valuation ν_p of $\mathbb{Q}$ from part (a), show $R = \mathbb{Z}_{(p)} \equiv \{\frac{a}{b} \in \mathbb{Q} \mid (b,p) = 1\}$.

EXERCISE 3.39 (Infinite Products). If R_a, $\alpha \in \mathcal{A}$, is an infinite collection of rings, there are two different products. In this setting,

$$\prod_{\alpha \in \mathcal{A}} R_\alpha = \{(r_\alpha) \mid r_\alpha \in R_\alpha\}$$

is called the *direct product* and

$$\sum_{\alpha \in \mathcal{A}} R_\alpha = \{(r_\alpha) \mid r_\alpha \in R_\alpha \text{ and } r_a = 0_{R_\alpha} \text{ for all but finitely many } \alpha\}$$

is called the *direct sum* (cf. Exercise 2.91). For many applications, the infinite direct sum is more useful since it turns out that the infinite direct product is often too big and pathological.

(a) Show $\prod_{\alpha \in \mathcal{A}} R_\alpha$ and $\sum_{\alpha \in \mathcal{A}} R_\alpha$ are rings.

(b) Show $\sum_{\alpha \in \mathcal{A}} R_\alpha$ is a subring of $\prod_{\alpha \in \mathcal{A}} R_\alpha$ with equality if and only if $|\mathcal{A}| < \infty$.

2. Morphisms and Quotients

2.1. Morphisms. As with groups, we need a way of saying two rings are the same up to relabeling. The only real difference with rings is that we now have two operations to preserve instead of just one.

DEFINITION 3.10. Let R and S be rings and let $\varphi : R \to S$ be a function.

(1) If

$$\varphi(r_1 + r_2) = \varphi(r_1) + \varphi(r_2)$$

and

$$\varphi(r_1 r_2) = \varphi(r_1)\varphi(r_2)$$

for all $r_1, r_2 \in R$, then φ is called a *homomorphism*.

(2) A bijective homomorphism is called an *isomorphism*. In that case R and S are said to be *isomorphic* and we write $R \cong S$.

(3) If $R = S$, an isomorphism is called an *automorphism*.

For example, consider the subring $R = \{\begin{pmatrix} a & b \\ -b & a \end{pmatrix} \mid a, b \in \mathbb{R}\}$ of $M_{2,2}(\mathbb{R})$ (cf. Exercise 3.9). At first blush, R and $\mathbb{C}$ appear to be quite different sorts of objects. Nevertheless, consider the map $\varphi : R \to \mathbb{C}$ given by $\varphi\begin{pmatrix} a & b \\ -b & a \end{pmatrix} = a + ib$. Then φ is a homomorphism since

$$\begin{aligned}
\varphi\left(\begin{pmatrix} a & b \\ -b & a \end{pmatrix} + \begin{pmatrix} a' & b' \\ -b' & a' \end{pmatrix}\right) &= \varphi\begin{pmatrix} a + a' & b + b' \\ -b - b' & a + a' \end{pmatrix} \\
&= (a + a') + i(b + b') = (a + ib) + (a' + ib') \\
&= \varphi\begin{pmatrix} a & b \\ -b & a \end{pmatrix} + \varphi\begin{pmatrix} a' & b' \\ -b' & a' \end{pmatrix}
\end{aligned}$$

and

$$\varphi\left(\begin{pmatrix} a & b \\ -b & a \end{pmatrix}\begin{pmatrix} a' & b' \\ -b' & a' \end{pmatrix}\right) = \varphi\begin{pmatrix} aa'-bb' & ab'+ba' \\ -ab'-ba' & aa'-bb' \end{pmatrix} = (aa'-bb') + i(ab'+ba')$$
$$= (a+ib)(a'+ib') = \varphi\begin{pmatrix} a & b \\ -b & a \end{pmatrix}\varphi\begin{pmatrix} a' & b' \\ -b' & a' \end{pmatrix}.$$

Since φ is also clearly bijective, it follows that φ is an isomorphism, i.e., $R \cong \mathbb{C}$. Thus R and $\mathbb{C}$ are essentially the same object only differing in their labels. More poetically, "A rose by any other name would smell as sweet."

In general, it is much harder to be a ring homomorphism than it is to be a group homomorphism. For example, viewing $\mathbb{Z}$ as only an abelian group under addition, the most general group homomorphism from $\mathbb{Z}$ to $\mathbb{Z}$ is given by $\varphi_n(k) = nk$ for $n, k \in \mathbb{Z}$ (Exercise 2.99). In particular, there are a $\mathbb{Z}$'s worth of group homomorphisms. Viewing $\mathbb{Z}$ now as a ring under addition and multiplication, observe that a ring homomorphism from $\mathbb{Z}$ to $\mathbb{Z}$ is, at the very least, still a homomorphism of groups and therefore is of the form φ_n for some n. However, to be a ring homomorphism, φ_n must also satisfy the multiplicative property

$$nk_1k_2 = \varphi_n(k_1k_2) = \varphi_n(k_1)\varphi_n(k_2) = n^2k_1k_2$$

for all $k_i \in \mathbb{Z}$. Clearly this condition holds if and only if $n = n^2$ if and only if $n \in \{0, 1\}$. Thus there are only two ring homomorphisms: the trivial map and the identity map.

As with groups, ring homomorphisms give rise to two special subrings.

DEFINITION 3.11. Let $\varphi : R \to S$ be a homomorphism between rings.

(1) The *image* of φ is

$$\operatorname{Im}\varphi = \{\varphi(r) \mid r \in R\}.$$

(2) The *kernel* of φ is

$$\ker\varphi = \{r \in R \mid \varphi(r) = 0_S\}.$$

As with Theorem 2.25 in the group setting, ring homomorphisms also satisfy a number of nice properties.

THEOREM 3.12. *Let $\varphi : R \to S$ be a homomorphism between rings and let $r \in R$.*

(1) *The composition of homomorphisms is a homomorphism.*
(2) *If φ is an isomorphism, then so is φ^{-1}.*
(3) $\varphi(0_R) = 0_S$.
(4) $\varphi(-r) = -\varphi(r)$.
(5) $\varphi(r_1 + r_2 + \cdots + r_N) = \varphi(r_1) + \varphi(r_2) + \cdots + \varphi(r_N)$ *for $r_i \in R$.*
(6) $\varphi(r_1r_2\cdots r_N) = \varphi(r_1)\varphi(r_2)\cdots\varphi(r_N)$ *for $r_i \in R$.*
(7) $\operatorname{Im}\varphi$ *is a subring of S.*
(8) $\ker\varphi$ *is a subring of R.*
(9) *φ is one-to-one if and only if* $\ker\varphi = \{0_R\}$.
(10) *If R is a ring with identity, then* $\operatorname{Im}\varphi$ *is a ring with identity and* $\varphi(1_R) = 1_{\operatorname{Im}\varphi}$.

(11) *If R is a ring with identity with $r \in R^\times$, then $\varphi(r) \in (\operatorname{Im}\varphi)^\times$ and $\varphi(r^{-1}) = \varphi(r)^{-1}$.*

PROOF. The proof of this theorem is almost identical to that of Theorem 2.25 and so the bulk of the proof is left to Exercise 3.44. The only new idea that is needed here is found in the proof of part (10), which we give here. For this, simply observe that the general element of $\operatorname{Im}\varphi$ is $\varphi(r)$, $r \in R$, and that

$$\varphi(1_R)\varphi(r) = \varphi(1_R \cdot r) = \varphi(r)$$

and similarly for $\varphi(r)\varphi(1_R) = \varphi(r)$. □

Note that even if R and S both have identities, it can easily happen that $1_{\operatorname{Im}\varphi} \neq 1_S$ when φ is not surjective (cf. Exercise 3.42).

2.2. Ideals. Recall that the notion of a quotient group was a very important and useful construction in group theory. In that construction, a crucial technicality came into play. Namely, if G is a group and H is a subgroup, it is always possible to form the quotient G/H. However, G/H inherits a group structure from G if and only if H is normal in G. A similar story plays out for rings. If R is a ring and S is a subring, we will see that we can always form the quotient R/S, but an additional requirement is needed to ensure that R/S inherits a ring structure. The extra condition, it turns out, is given in the definition below.

DEFINITION 3.13. Let R be a ring and let $I \subseteq R$ be a subring. For $r \in R$, write $rI = \{ra \mid a \in I\}$ and $Ir = \{ar \mid a \in I\}$. We say I is an *ideal* (or a *two-sided ideal*) if $rI \subseteq I$ and $Ir \subseteq I$ for all $r \in R$. An ideal is called *proper* if $I \neq \{0\}$ and $I \neq R$.

For example, in $\mathbb{Z}$, the most general subring is $\langle k \rangle = k\mathbb{Z}$, $k \in \mathbb{Z}_{\geq 0}$ (Exercise 3.17). In this case, clearly $n\langle k \rangle = \langle k \rangle n \subseteq \langle k \rangle$, $n \in \mathbb{Z}$, so that $\langle k \rangle$ is an ideal. Ideals such as these get a special name (cf. Exercise 3.57).

DEFINITION 3.14. If I is an ideal of a commutative ring R with identity such that $I = aR$, $a \in I$, then I is called a *principal ideal* and a is called a *generator* for the ideal. In this case, we write $(a) = aR$.

For example, consider the polynomial ring $\mathbb{R}[x]$. Then the principal ideal generated by x^2+1 is written $\left(x^2+1\right)$ and is given by $\{(x^2+1)f \mid f \in \mathbb{R}[x]\}$.

Of course, in general, not all subrings are ideals. For instance, the set of upper triangular matrices in $M_{2,2}(\mathbb{R})$ is a subring (Exercise 3.8) that is not an ideal. In other words, the set of upper triangular matrices is not closed under left (or right) multiplication by arbitrary elements of $M_{2,2}(\mathbb{R})$. To see this, simply multiply:

$$\begin{pmatrix} 0 & 1 \\ 1 & 0 \end{pmatrix}\begin{pmatrix} a & b \\ 0 & c \end{pmatrix} = \begin{pmatrix} 0 & c \\ a & b \end{pmatrix}$$

which is no longer upper triangular when $a \neq 0$. In fact, it turns out that $M_{2,2}(\mathbb{R})$ has no proper ideals (Exercise 3.60). Just as with the notion of simple groups, rings containing no proper ideals are basic building blocks for all other rings.

DEFINITION 3.15. Let $R \neq 0$ be a ring. Then R is called *simple* if R has no proper ideals.

As already noted, ideals are designed to be used in quotient ring constructions. However, they also have many other applications. One such example is the following decomposition theorem.

THEOREM 3.16. *Suppose S and T are ideals of the ring R. Writing $S+T$ for $\{s+t \mid s \in S \text{ and } t \in T\}$, suppose $R = S+T$ and $S \cap T = \{0\}$. Then*

$$R \cong S \times T.$$

In particular, $ST = TS = \{0\}$.

PROOF. Consider the map $\varphi : S \times T \to R$ given by $\varphi(s,t) = s+t$. By Theorem 2.33 we already know $R \cong S \times T$ as abelian groups under $+$. Therefore it only remains to check the multiplicative part of the homomorphism condition. For this, observe that

$$\varphi((s,t)(s',t')) = \varphi(ss',tt') = ss' + tt'$$

and

$$\varphi(s,t)\varphi(s',t') = (s+t)(s'+t') = ss' + st' + ts' + tt',$$

$s, s' \in S$ and $t, t' \in T$. Thus it remains to verify that $st' + ts' = 0$. However, ST and TS are contained in $S \cap T = \{0\}$ since both are ideals and we are done. □

2.3. Quotients. Since rings are, at the very least, abelian groups, quotients of rings can be constructed as they were for groups. The only difference is that additive group notation is used.

DEFINITION 3.17. Let R be a ring and let S be a subring of R. Let $r \in R$.

(1) The *left coset* of r with respect to S is $r + S = \{r + s \mid s \in S\}$.

(2) If C is a left coset with respect to S and $C = r + S$, then r is called a *representative* of C.

(3) The set of left cosets is denoted by R/S and is called the *quotient* of R by S.

(4) The *index* of R in S is $|R/S|$ and is denoted $[R:S]$.

It is worth noting that the left cosets are also right cosets since a ring is an abelian group under $+$. Along that same vein, it follows that a subring S of R is, at the very least, a normal subgroup under $+$. Therefore we already know that R/S carries a natural group structure induced by $+$, namely

$$(r_1 + S) + (r_2 + S) = (r_1 + r_2) + S$$

for $r_i \in R$. It is important to find out when it also carries a ring structure induced by $\cdot$. The following result is analogous to Theorem 2.31.

THEOREM 3.18. *Let R be a ring and let S be a subring of R. The following are equivalent.*

(1) *S is an ideal.*

(2) *The equation*

$$(r_1 + S) \cdot (r_2 + S) = r_1 r_2 + S$$

gives a well defined binary operation on R/S where $r_1, r_2 \in R$.

PROOF. We have already seen (Lemma 2.27) that the general representative for the coset $r_i + S$ is $r_i + s_i$ where $s_i \in S$. Therefore the equation $(r_1 + S)(r_2 + S) = r_1 r_2 + S$ is well defined if and only if

$$r_1 r_2 + S = (r_1 + s_1)(r_2 + s_2) + S$$

if and only if

$$(r_1 + s_1)(r_2 + s_2) - r_1 r_2 \in S.$$

But since $(r_1 + s1)(r_2 + s_2) - r_1 r_2 = r_1 s_2 + r_2 s_1 + s_1 s_2$, it follows that the binary operation is well defined if and only if $r_1 s_2 + r_2 s_1 \in S$ for all $r_i \in R$ and $s_i \in S$. Hence, (1) clearly implies (2). To see (2) implies (1), work first with $s_1 = 0$ and then with $s_2 = 0$. □

With Theorem 3.18 in hand, it follows that R/S inherits the structure of a ring exactly when S is an ideal. This result is similar to Definition 2.32 and is left to Exercise 3.72. Of course if R has an identity, then so does R/S, namely $1_R + S$.

DEFINITION 3.19. Let R be a ring and let I be an ideal of R. Under the bilinear operations

$$\begin{aligned}(r_1 + I) + (r_2 + I) &= (r_1 + r_2) + I,\\ (r_1 + I) \cdot (r_2 + I) &= r_1 r_2 + I,\end{aligned}$$

$r_i \in R$, R/I is a ring and is called the *quotient ring* or *factor ring.*

Of course our first example is our old friend $\mathbb{Z}/(n\mathbb{Z})$, otherwise known as $\mathbb{Z}_n$, equipped with the usual addition and multiplication structures. However, we have studied $\mathbb{Z}_n$ to death and so it hardly counts as an example any more. To get a genuinely new example, consider the polynomial ring $\mathbb{R}[x]$, the principal ideal

$$I = \left(x^2 + 1\right) = \{(x^2 + 1)f \mid f \in \mathbb{R}[x]\},$$

and its quotient $\mathbb{R}[x]/I$.

One place to start the study of $\mathbb{R}[x]/I$ is to get a handle on the nature of the set of cosets. For this, recall the Division Algorithm from Theorem 1.3 developed for the study of $\mathbb{Z}$. As the reader no doubt recalls from elementary mathematics, the type of long division used for $\mathbb{Z}$ works equally well for polynomials. Therefore there is a type of division algorithm suitable for polynomials. We will develop this carefully in Section 3. For now, we simply assume the result and state the fact that any polynomial $p \in \mathbb{R}[x]$ can be uniquely written as $p = (x^2+1)q+(a+bx)$ for some $q \in \mathbb{R}[x]$. In particular, the set of cosets in $\mathbb{R}[x]/I$ is given by $\{(a+bx)+I \mid a, b \in \mathbb{R}\}$.

Since this collection bears a striking resemblance to $\mathbb{C} = \{a + ib \mid a, b \in \mathbb{R}\}$, consider the map $\varphi : \mathbb{C} \to \mathbb{R}[x]/I$ given by $\varphi(a + ib) = (a + bx) + I$, $a, b \in \mathbb{R}$. By construction, φ is a bijection. Hopefully φ is a homomorphism. To verify this, calculate

$$\begin{aligned}\varphi((a_1 + ib_1) + (a_2 + ib_2)) &= \varphi((a_1 + a_2) + i(b_1 + b_2))\\ &= [(a_1 + a_2) + (b_1 + b_2)x] + I\\ &= [(a_1 + b_1 x) + I] + [(a_2 + b_2 x) + I]\\ &= \varphi(a_1 + ib_1) + \varphi(a_2 + ib_2)\end{aligned}$$

and

$$\begin{aligned}\varphi((a_1 + ib_1)(a_2 + ib_2)) &= \varphi((a_1 a_2 - b_1 b_2) + i(a_1 b_2 + b_1 a_2))\\ &= [(a_1 a_2 - b_1 b_2) + (a_1 b_2 + b_1 a_2)x] + I\\ &= \left[a_1 a_2 + (a_1 b_2 + b_1 a_2)x + x^2 b_1 b_2 - b_1 b_2(x^2 + 1)\right] + I\\ &= \left[a_1 a_2 + (a_1 b_2 + b_1 a_2)x + x^2 b_1 b_2\right] + I\\ &= [(a_1 + b_1 x) + I]\,[(a_2 + b_2 x) + I]\\ &= \varphi(a_1 + ib_1)\varphi(a_2 + ib_2),\end{aligned}$$

$a_i, b_i \in \mathbb{R}$. Thus φ is an isomorphism and $\mathbb{R}[x]/I \cong \mathbb{C}$.

A less rigorous but more intuitive way to look at the isomorphism $\mathbb{R}[x]/\left(x^2+1\right) \cong \mathbb{C}$ is the following. Modding out by $\left(x^2+1\right)$ is tantamount to forcing $x^2+1=0$ which is equivalent to viewing $x^2=-1$. In other words, modding out by $\left(x^2+1\right)$ is basically the same as setting $x=i$. Once this is done, polynomials in x become polynomials in i. Reducing the various powers of i to $\pm\{1,i\}$, any polynomial in x turns into an expression of the form $a+ib$. Thus modding out by $\left(x^2+1\right)$ is a fancy way of constructing the set of complex numbers, $\mathbb{C}$.

Just as there was a Correspondence Theorem for group quotients, there is one for ring quotients. The proof is nearly identical and is left to Exercise 3.73.

THEOREM 3.20 (Correspondence Theorem). *Let I be an ideal of R. Let*

$$\mathcal{R}_{R,I} = \{\textit{subrings of } R \textit{ containing } I\},$$

$$\mathcal{R}_{R/I} = \{\textit{subrings of } R/I\},$$

and write $\pi : R \to R/I$ for the map given by $\pi(r) = r+I$, $r \in R$.

(1) *The map π is a surjective homomorphism.*

(2) *There is a bijection from $\mathcal{R}_{R,I}$ to $\mathcal{R}_{R/I}$ that maps $S \in \mathcal{R}_{R,I}$ to $\overline{S} \in \mathcal{R}_{R/I}$ where*

$$\overline{S} = \pi(S) = \{s+I \mid s \in S\}.$$

The inverse maps $\overline{S} \in \mathcal{R}_{R/I}$ to $S \in \mathcal{R}_{R,I}$ where

$$S = \pi^{-1}(\overline{S}) = \{r \in R \mid r+I \in \overline{S}\}.$$

(3) *The above bijection preserves ideals; i.e., $S \in \mathcal{R}_{R,I}$ is an ideal in R if and only if the corresponding $\overline{S} \in \mathcal{R}_{R/I}$ is an ideal in R/I.*

We now give conditions on ideals that guarantee that quotient spaces have properties such as being an integral domain or a field.

DEFINITION 3.21. Let P and M be ideals of R.

(1) If R is commutative and if it has an identity, P is called *prime* if $P \neq R$ and whenever $rr' \in P$ for $r, r' \in R$, then either $r \in P$ or $r' \in P$.

(2) M is called *maximal* if $M \neq R$ and the only ideals of R containing M are M itself and R.

THEOREM 3.22. *Let P and M be ideals of R.*

(1) *If R is commutative and if it has an identity, P is prime if and only if R/P is an integral domain.*

(2) *M is maximal if and only if R/M is simple. If R is commutative and if it has an identity, M is maximal if and only if R/M is a field.*

PROOF. Consider part (1) first. The equation $(r+P)(r'+P) = 0+P$, $r, r' \in R$, is equivalent to the condition $rr' \in P$. Therefore R/P is a domain if and only if the condition $rr' \in P$ implies either $r \in P$ or $r' \in P$. In other words, R/P is a domain if and only if P is prime. Finally, since R is commutative and since it has an identity, so does R/P.

Turn now to part (2). By the Correspondence Theorem, there is a one-to-one correspondence between ideals of R containing M and ideals of R/M. Therefore M is maximal if and only if R/M is simple and the first statement is complete.

For the second half, assume R is commutative and that it has an identity. We must show that R/M is simple if and only if it is a field. Switching notation slightly, we we will actually prove the following general result: A ring R is a simple commutative ring with identity if and only if R is a field.

If R is simple with $r \in R$, $r \neq 0$, then (r) is a nonzero ideal and therefore $(r) = R$. In particular, $1 \in R$ so that there exists $b \in R$ so that $rb = 1$. Thus R is a field. Conversely suppose R is a field with I a nonzero ideal. Pick $r \neq 0$ in I. Then $1 = r^{-1}r \in r^{-1}I \subseteq I$. Thus $R \cdot 1 \subseteq I$ so that $I = R$ and R is simple as desired. □

As an example, the fact that $\mathbb{R}[x]/\left(x^2+1\right) \cong \mathbb{C}$ is a field shows $\left(x^2+1\right)$ is a maximal ideal in $\mathbb{R}[x]$.

2.4. Isomorphism Theorem.

THEOREM 3.23 (First Isomorphism Theorem). *Let $\varphi : R \to S$ be a homomorphism of rings.*

(1) $\ker\varphi$ *is an ideal of* R.

(2) $R/\ker\varphi \cong \operatorname{Im}\varphi$. *In particular if φ is surjective, then* $R/\ker\varphi \cong S$.

PROOF. For part (1), we already know that $\ker\varphi$ is a subring. It remains to see that $r\ker\varphi \subseteq \ker\varphi$ and $(\ker\varphi)\, r \subseteq \ker\varphi$ for $r \in R$. For the first containment, let $k \in \ker\varphi$ and calculate

$$\varphi(rk) = \varphi(r)\varphi(k) = \varphi(r) \cdot 0 = 0$$

as desired. The second containment is similar.

Turning to part (2), we already know from Theorem 2.35 that $\overline{\varphi} : R/\ker\varphi \to \operatorname{Im}\varphi$ given by $\overline{\varphi}(r + \ker\varphi) = r$ is a well defined map establishing an isomorphism of the underlying group structures. It only remains to see $\overline{\varphi}$ is a homomorphism of rings. For this, calculate

$$\begin{aligned}\overline{\varphi}((r + \ker\varphi)(r' + \ker\varphi)) &= \overline{\varphi}(rr' + \ker\varphi) = \varphi(rr') \\ &= \varphi(r)\varphi(r') = \overline{\varphi}(r + \ker\varphi)\overline{\varphi}(r' + \ker\varphi),\end{aligned}$$

$r, r' \in R$. □

As an example, consider $\varphi : M_{n,n}(\mathbb{Z}) \to M_{n,n}(\mathbb{Z}_m)$, $m \in \mathbb{N}$, given by $\varphi(A_{i,j}) = ([A_{i,j}]_m)$, $(A_{i,j}) \in M_{n,n}(\mathbb{Z})$. The fact that φ is a homomorphism follows immediately from the fact that the map $\mathbb{Z} \to \mathbb{Z}_m$ given by $a \to [a]_m$, $a \in \mathbb{Z}$, is a homomorphism. Of course φ is also clearly surjective and $\ker\varphi = mM_{n,n}(\mathbb{Z})$. Thus $M_{n,n}(\mathbb{Z})/\left(mM_{n,n}(\mathbb{Z})\right) \cong M_{n,n}(\mathbb{Z}_m)$.

2.5. Exercises 3.40–3.86.

2.5.1. *Morphisms.*

EXERCISE 3.40. **(a)** If R and S are rings, show that the *trivial map* $\varphi : R \to S$ given by $\varphi(r) = 0_S$, $r \in R$, is a homomorphism.

(b) Show that the *identity map* $\operatorname{Id} : R \to R$ given by $\operatorname{Id}(r) = r$ is an automorphism.

(c) Show that $P : M_{2,2}(\mathbb{Q}) \to \mathbb{Q}$ given by $P(A_{i,j}) = A_{1,1}$ is not a homomorphism.

(d) Show that $C : M_{2,2}(\mathbb{C}) \to M_{2,2}(\mathbb{C})$ given by $C(Z_{i,j}) = \left(\overline{Z}_{i,j}\right)$ is a homomorphism.

(e) Show that $\tau : M_{2,2}(\mathbb{R}) \to M_{2,2}(\mathbb{R})$ given by $\tau(B) = B^T$ is not a homomorphism.

(f) Show that the *projection map* $\pi : R \times S \to R$ given by $\pi(r,s) = r$ is a surjective homomorphism.

(g) If $\psi : R \to S$ is a homomorphism, show that $\widetilde{\psi} : R[x] \to S[x]$ given by $\widetilde{\psi}(\sum_i a_i x^i) = \sum_i \psi(a_i)x^i$ is a homomorphism.

(h) If $\psi : R \to S$ is a homomorphism, show that $\Psi : M_{n,n}(R) \to M_{n,n}(S)$ given by $\Psi(A_{i,j}) = (\psi(A_{i,j}))$ is a homomorphism.

EXERCISE 3.41. **(a)** Show that $R = \{\begin{pmatrix} a & b \\ [0] & a \end{pmatrix} \mid a, b \in \mathbb{Z}_2\}$ is a subring of $M_{2,2}(\mathbb{Z}_2)$.

(b) Viewed as groups under addition (ignoring the multiplication structure), show $R \cong \mathbb{Z}_2 \times \mathbb{Z}_2$.

(c) As rings, show $R \not\cong \mathbb{Z}_2 \times \mathbb{Z}_2$. *Hint:* Find an element $r \in R$ satisfying $r^2 = 0$.

EXERCISE 3.42. Let $\varphi : \mathbb{R} \to M_{2,2}(\mathbb{R})$ be given by $\varphi(r) = rE_{1,1}$.

(a) Show φ is a homomorphism.

(b) Show that $\varphi(1_{\mathbb{R}}) = 1_{\mathbb{R}E_{1,1}}$ but that $\varphi(1_{\mathbb{R}}) \neq 1_{M_{2,2}(\mathbb{R})}$ (cf. Exercise 3.5).

EXERCISE 3.43 (Frobenius Automorphism). **(a)** Let p be a positive prime and let $k \in \mathbb{Z}$, $1 \leq k \leq p-1$. Show $p \mid \binom{p}{k}$. *Hint:* Write $\binom{p}{k} = \frac{p(p-1)\cdots(p-k+1)}{k(k-1)\cdots 1}$.

(b) If R is a ring with $\operatorname{char} R = p$ and if $r, s \in R$ commute, show $(r+s)^p = r^p + s^p$ (cf. Exercise 3.19).

(c) If F is a finite field with $\operatorname{char} F = p$, show that $\varphi : F \to F$ given by $\varphi(x) = x^p$, $x \in F$, is an automorphism of F. It is called the *Frobenius automorphism* (cf. the proof of Theorem 4.33). *Hint:* To see injectivity, show $\ker \varphi = 0$ by using the fact that F is a domain. Next, use the finiteness of F to get surjectivity.

EXERCISE 3.44. Prove Theorem 3.12.

2.5.2. *Ideals.*

EXERCISE 3.45. Show the following are ideals:

(a) (3) in $\mathbb{Z}$,

(b) (x^2+2) in $\mathbb{R}[x]$,

(c) $\begin{pmatrix} \mathbb{Z} & 5\mathbb{Z} \\ 3\mathbb{Z} & 15\mathbb{Z} \end{pmatrix}$ in $\begin{pmatrix} \mathbb{Z} & 5\mathbb{Z} \\ 3\mathbb{Z} & \mathbb{Z} \end{pmatrix}$,

(d) $\begin{pmatrix} 0 & \mathbb{Q} \\ 0 & \mathbb{Q} \end{pmatrix}$ in $\begin{pmatrix} \mathbb{Q} & \mathbb{Q} \\ 0 & \mathbb{Q} \end{pmatrix}$,

(e) $\{f \in \mathcal{F}(\mathbb{R},\mathbb{R}) \mid f(\pi) = 0\}$ in $\mathcal{F}(\mathbb{R},\mathbb{R})$ (cf. Exercise 3.15).

EXERCISE 3.46. Show the following are not ideals:

(a) $\mathbb{Z}$ in $\mathbb{Q}$,

(b) $\{(n,n) \in \mathbb{Z} \times \mathbb{Z}\}$ in $\mathbb{Z} \times \mathbb{Z}$,

(c) $\begin{pmatrix} \mathbb{R} & \mathbb{R} \\ 0 & \mathbb{R} \end{pmatrix}$ in $M_{2,2}(\mathbb{R})$.

EXERCISE 3.47 (Ideals of $\mathbb{Z}$ and $\mathbb{Z}_n$). **(a)** Show that the set of ideals of $\mathbb{Z}$ is $\{\langle k \rangle = (k) = k\mathbb{Z} \mid k \in \mathbb{Z}_{\geq 0}\}$ (cf. Exercise 3.17).

(b) Show that the set of ideals of $\mathbb{Z}_n$ is $\{\langle [k] \rangle = ([k]) = k\mathbb{Z}_n \mid k \in \mathbb{N} \text{ and } k \mid n\}$.

EXERCISE 3.48. **(a)** Let $m, n \in \mathbb{Z}$. In $\mathbb{Z}$, show $(m) \subseteq (n)$ if and only if $n \mid m$ (cf. Exercise 3.47).

(b) Let $n, k, j \in \mathbb{N}$ with $k, j \mid n$. In $\mathbb{Z}_n$, show $([k]) \subseteq ([j])$ if and only if $j \mid k$. *Hint:* Write $kk' = n$ and $jj' = n$. Show $([k]) \subseteq ([j])$ if and only if $k = jm + nq$ and that this implies $\frac{j'}{k'} = m + j'q \in \mathbb{Z}$.

EXERCISE 3.49. Let R be a ring with identity and let I be an ideal. Show $I = R$ if and only if I contains a unit.

* EXERCISE 3.50. **(a)** If I is an ideal and if S is a subring of R, show $I + S = \{a + s \mid a \in I,\ s \in S\}$ is a subring of R.

(b) If I is only a subring, show that part (a) can be false.

EXERCISE 3.51 (Operations on Ideals). Let I_α and I_n be ideals of R, $\alpha \in \mathcal{A}$ and $n \in \mathbb{N}$.

(a) Show $\bigcap_{\alpha \in \mathcal{A}} I_\alpha$ is an ideal.

(b) If $I_n \subseteq I_{n+1}$, show $\bigcup_{n \in \mathbb{N}} I_n$ is an ideal.

(c) If $I_1 \not\subseteq I_2$ and $I_2 \not\subseteq I_1$, show $I_1 \cup I_2$ is not an ideal (cf. Exercise 3.10).

(d) Write $\sum_{\alpha \in A} I_\alpha = \{a_1 + \cdots + a_N \mid N \in \mathbb{N},\ a_i \in \bigcup_{\alpha \in \mathcal{A}} I_\alpha\}$. Show $\sum_{\alpha \in A} I_\alpha$ is an ideal.

(e) For $N \in \mathbb{N}$, write $I_1 I_2 \cdots I_N = \{\sum_{i=1}^{M} a_{1,i} a_{2,i} \cdots a_{N,i} \mid M \in \mathbb{N},\ a_{j,i} \in I_j\}$. Show $I_1 I_2 \cdots I_N$ is an ideal.

EXERCISE 3.52. **(a)** Let $n, m \in \mathbb{N}$. Show $n\mathbb{Z} + m\mathbb{Z} = \mathbb{Z}$ if and only if $(n, m) = 1$ (cf. Exercise 3.51).

(b) Show $(n\mathbb{Z})(m\mathbb{Z}) = nm\mathbb{Z}$ and $(n\mathbb{Z}) \cap (m\mathbb{Z}) = [n, m]\mathbb{Z}$ (cf. Exercise 1.28).

(c) Conclude $(n\mathbb{Z})(m\mathbb{Z}) = (n\mathbb{Z}) \cap (m\mathbb{Z})$ if and only if $(n, m) = 1$.

EXERCISE 3.53. Let R be a commutative ring with identity with ideals I, J satisfying $I + J = R$ (cf. Exercise 3.51). This exercise shows $IJ = I \cap J$.

(a) Show $IJ \subseteq I \cap J$.

(b) Show $I \cap J = (I \cap J)\, R \subseteq (I \cap J)\, I + (I \cap J)\, J \subseteq JI + IJ = IJ$. Conclude $IJ = I \cap J$.

EXERCISE 3.54. Let I be an ideal in R and suppose I has an identity, 1_I. (N.B.: This does not make 1_I the identity for R; cf. Exercise 3.5.) This exercise shows 1_I is a central idempotent (cf. Exercise 3.21) of R satisfying $I = 1_I R = R 1_I = 1_I R 1_I$.

(a) Show $1_I^2 = 1_I$.

(b) For $r \in R$, let $a = 1_I r - 1_I r 1_I$. Show that $a \in I$ and that $a = a 1_I = 0$. Conclude that $1_I r = 1_I r 1_I$. Similarly show $r 1_I = 1_I r 1_I$ and conclude $1_I \in Z(R)$.

(c) Show $I \subseteq 1_I R \subseteq I$ and $I \subseteq R 1_I \subseteq I$ so that $I = 1_I R = R 1_I$. Using parts (a) and (b), show $I = 1_I R 1_I$.

EXERCISE 3.55. Let I, J be ideals of R so that $I + J = R$ and $I \cap J = \{0\}$.

(a) Show $R \cong I \times J$.

(b) If R has an identity, show I has an identity and conclude $1_I R = I$ with 1_I a central idempotent of R. *Hint:* Write $1_R = a + b$ with $a \in I$ and $b \in J$. Look at a and then use Exercise 3.54.

EXERCISE 3.56. **(a)** Let R be a ring and let $a \in Z(R)$. Show that the *annihilator* of a, $\operatorname{ann}(a) = \{r \in R \mid ra = 0\}$, is an ideal.

(b) If $a \notin Z(R)$, show part (a) can be false.

EXERCISE 3.57 (Principal Ideals). Let R be a ring with $a \in Z(R)$.

(a) Show $aR = Ra$ is an ideal, called the *principal ideal generated by a*.

(b) If R has an identity, show aR is the smallest ideal in R containing a. The ideal aR is written $(a) = aR$.

(c) If R does not have an identity, show aR might not contain a. *Hint:* Try $R = 2\mathbb{Z}$.

(d) Write $\mathbb{Z}a + aR = \{na + ar \mid n \in \mathbb{Z},\ r \in R\}$, called the *ideal generated by* a. Show $\mathbb{Z}a + aR$ is an ideal satisfying (i) $a \in \mathbb{Z}a + aR$, (ii) $aR \subseteq \mathbb{Z}a + aR$ with equality if and only if $a \in aR$, and (iii) if R has an identity, $aR = \mathbb{Z}a + aR$.

(e) If R does not have an identity, show $\mathbb{Z}a + aR$ is the smallest ideal in R containing a.

(f) Show rR may not be an ideal if $r \notin Z(R)$ (cf. Exercises 3.61 and 3.62).

EXERCISE 3.58. **(a)** Show that division rings are simple. (N.B.: There exist plenty of noncommutative simple rings that are not division rings; cf. Exercise 3.60.)

(b) Let R be a commutative ring with identity. Show R is simple if and only if it is a field (cf. Theorem 3.22).

EXERCISE 3.59 (Ideals of $\mathbb{F}[x]$). Let $\mathbb{F}$ be a field and let $I \neq 0$ be an ideal in $\mathbb{F}[x]$. This exercise shows that all ideals in $\mathbb{F}[x]$ are principal.

(a) Choose nonzero $f \in I$ of minimal degree n, $n \in \mathbb{Z}_{\geq 0}$, and write $f = \sum_{k=0}^{n} a_k x^k$, $a_k \in \mathbb{F}$ with $a_n \neq 0$. Show that $(f) = f\,\mathbb{F}[x] \subseteq I$.

(b) By way of contradiction, suppose $I \backslash (f) \neq \varnothing$. Choose $g \in I \backslash (f)$ of minimal degree m, $m \in \mathbb{Z}_{\geq 0}$, and write $g = \sum_{k=0}^{m} b_k x^k$, $b_k \in \mathbb{F}$. Show $m \geq n$ and let $h = g - b_m a_n^{-1} x^{m-n} f$. Conclude $h \in (f)$ and use this to obtain a contradiction. *Hint:* Show $h \in I$ and $\deg h < m$.

(c) Conclude that $I = (f)$ and that all ideals in $\mathbb{F}[x]$ are principal.

(d) If D is an integral domain but not a field, show by example that part (c) can fail for $D[x]$. *Hint:* Try $I = \{\sum_{k=0}^{n} a_k x^k \in \mathbb{Z}[x] \mid a_0 \in 2\mathbb{Z}\}$ and observer $2 \in I$ and $x \in I$. (Actually, part (c) always fails for such D.)

(e) Show that part (c) fails if $\mathbb{F}[x]$ is replaced by polynomials in two variables, $\mathbb{F}[x, y]$. *Hint:* Look at the set of polynomials with zero constant term.

EXERCISE 3.60 (Ideals and Simplicity of $M_{n,n}(R)$). Let R be a ring with identity. If S is a subring of R, view $M_{n,n}(S) \cong \{(A_{i,j}) \in M_{n,n}(R) \mid A_{i,j} \in S\}$ as a subring of $M_{n,n}(R)$. This exercise shows that the only ideals of $M_{n,n}(R)$ are of the form $M_{n,n}(I)$ where I is an ideal of R. In particular, when R is simple, then so is $M_{n,n}(R)$ (cf. Exercise 3.58).

(a) If $\mathcal{I}$ is an ideal of $M_{n,n}(R)$, let $I = \{a \in R \mid aE_{1,1} \in \mathcal{I}\}$. Show I is an ideal of R.

(b) If $(A_{i,j}) \in M_{n,n}(I)$, show that $(A_{i,j}) \in \mathcal{I}$. *Hint:* Check that $A_{i,j}E_{i,j} = E_{i,1}\,(A_{i,j}E_{1,1})\,E_{1,j}$.

(c) If $A = (A_{i,j}) \in \mathcal{I}$, show that $(A_{i,j}) \in M_{n,n}(I)$. Conclude $\mathcal{I} = M_{n,n}(I)$. *Hint:* Check that $A_{i,j}E_{1,1} = E_{1,i}AE_{j,1}$.

(d) Conclude $M_{n,n}(R)$ is simple if and only if R is simple.

(e) If D is a division ring, show $M_{n,n}(D)$ is simple (cf. Exercise 3.58). In particular if F is field, show $M_{n,n}(F)$ is simple.

(f) If R does not have an identity, show that part (d) can fail. *Hint:* Try $R = 2\mathbb{Z}$ and $\mathcal{I} = \{(A_{i,j}) \in M_{2,2}(R) \mid A_{1,2} \in 4\mathbb{Z}\}$.

EXERCISE 3.61 (Principal Ideals in General). Let R be a ring with $a \in R$. Write $RaR = \{r_1 a r_1' + \cdots + r_N a r_N' \mid N \in \mathbb{N},\ r_i, r_i' \in R\}$, $aR + Ra + RaR =$

$\{ar + r'a + b \mid r, r' \in R,\ b \in RaR\}$, called the *principal ideal generated by* a, and $\mathbb{Z}a + aR + Ra + RaR = \{na + c \mid n \in \mathbb{Z},\ c \in aR + Ra + RaR\}$, called the *ideal generated by* a (cf. Exercise 3.57 for the special case of $a \in Z(R)$).

(a) Show $RaR \subseteq aR + Ra + RaR \subseteq \mathbb{Z}a + aR + Ra + RaR$ are ideals satisfying (i) $RaR = aR + Ra + RaR$ if and only if $aR \subseteq RaR$ and $Ra \subseteq RaR$, (ii) $aR + Ra + RaR = \mathbb{Z}a + aR + Ra + RaR$ if and only if $a \in aR + Ra + RaR$, and (iii) if R has an identity, all three ideals are equal.

(b) If R has an identity, show that the smallest ideal in R containing a is RaR.

(c) If R does not have an identity, show that the smallest ideal in R containing a is $\mathbb{Z}a + aR + Ra + RaR$.

EXERCISE 3.62 (Left and Right Ideals). Let R be a ring and let $I \subseteq R$ be a subring. We say I is a *left ideal* if $rI \subseteq I$ and we say I is a *right ideal* if $Ir \subseteq I$ for all $r \in R$.

(a) Show $\{\begin{pmatrix} 0 & b \\ 0 & d \end{pmatrix} \mid b, d \in \mathbb{R}\}$ is a left ideal of $M_{2,2}(\mathbb{R})$ but not a two-sided ideal.

(b) Show $\{\begin{pmatrix} 0 & 0 \\ c & d \end{pmatrix} \mid c, d \in \mathbb{R}\}$ is a right ideal of $M_{2,2}(\mathbb{R})$ but not a two-sided ideal.

EXERCISE 3.63 (Ideals in Discrete Valuation Rings). Let F be a field, let $\nu : F^\times \to \mathbb{Z}$ be a discrete valuation on F, and let $R = \{x \in F^\times \mid \nu(x) \geq 0\} \cup \{0_F\}$ be the corresponding discrete valuation ring (cf. Exercise 3.38). For $n \in \mathbb{Z}_{\geq 0}$, let $I_n = \{r \in R \mid \nu(r) \geq n\} \cup \{0\}$.

(a) Show I_n is a principal ideal with $I_{n+1} \subseteq I_n$.

(b) Show that every nonzero ideal is of the form I_n, $n \in \mathbb{Z}_{\geq 0}$. *Hint:* Look at (t^n) where $t \in R$ with $\nu(t) = 1$.

2.5.3. *Quotient Rings.*

✻ EXERCISE 3.64. Let R be a ring. Show $R[x]/(x) \cong R$.

EXERCISE 3.65. **(a)** Show there is no solution to the equation $X^2 = 2$ in the field $\mathbb{Q}$. *Hint:* Exercise 1.25.

(b) Let I be the principal ideal $I = (x^2 - 2)$. Show $\mathbb{Q}[x]/I = \{a + bx + I \mid a, b \in \mathbb{Q}\}$. *Hint:* Use the fact that $p \in \mathbb{Q}[x]$ can be uniquely written as $p = (x^2 - 2)q + (a + bx)$ for some $q \in \mathbb{Q}[x]$.

(c) Show $\mathbb{Q}[x]/I$ is a field. *Hint:* For inverses, look at $a - bx$.

(d) Show there is a solution to the equation $X^2 = 2$ in $\mathbb{Q}[x]/I$. *Hint:* $x + I$.

(e) Show $\mathbb{Q}[x]/I \cong \mathbb{Q}(\sqrt{2})$ by explicitly constructing an isomorphism (cf. §1.2).

EXERCISE 3.66. **(a)** Let I be the principal ideal $I = (x^2 - 1)$. Show $\mathbb{Q}[x]/I = \{a + bx + I \mid a, b \in \mathbb{Q}\}$. *Hint:* Use the fact that $p \in \mathbb{Q}[x]$ can be uniquely written as $p = (x^2 - 1)q + (a + bx)$ for some $q \in \mathbb{Q}[x]$.

(b) Show $\mathbb{Q}[x]/I$ is not a field. *Hint:* Look for zero divisors.

(c) Show $\mathbb{Q}[x]/I \cong \mathbb{Q} \times \mathbb{Q}$ by explicitly constructing an isomorphism. *Hint:* Look at the map $c(x - 1) + d(x + 1) + I \to (-2c, 2d)$.

EXERCISE 3.67. Let F be a field, let $p = \sum_i a_i x^i \in F[x]$, and let $n \in \mathbb{Z}_{\geq 0}$. Let I be the principal ideal $I = (x^n)$ and consider the ring $R = F[x]/I$.

(a) If $a_0 = 0_F$, show $(p + I)^n = 0_R$. *Hint:* Write $p = xp'$, $p' \in F[x]$.

(b) If $a_0 \neq 0_F$, show $p+I$ is a unit. *Hint:* Reduce to the case of $a_0 = 1_F$, write $p = 1 + xp'$, and look at $\sum_i (-xp')^i$.

EXERCISE 3.68 ($|\mathbb{Z}[i]/(z)|$). Let $z = n + mi \in \mathbb{Z}[i]$, $z \neq 0$ (cf. Exercise 3.11), and let $I = (z)$. This exercise shows $|\mathbb{Z}[i]/I| = N(z)$ where $N(z) = n^2 + m^2$ is the norm of z.

(a) Let $d = (n, m)$. Show $\frac{N(z)}{d} \in I$. *Hint:* Use $\frac{\overline{z}}{d}$.

(b) Choose $x, y \in \mathbb{Z}$ so $nx + my = d$. Let $u = (y + ix)z$. Show $u \in I$ with $\operatorname{Im} u = d$.

(c) Show every coset of $\mathbb{Z}[x]/I$ has a representative of the form $a + ib$ with $a \in [0, \frac{N(z)}{d})$ and $b \in [0, d)$. *Hint:* Use part (b) and then (a).

(d) Suppose $(a + ib) + I = (a' + ib') + I$ with $a, a' \in [0, \frac{N(z)}{d})$ and $b, b' \in [0, d)$. Show $b = b'$. *Hint:* If $w \in I$, show $d \mid \operatorname{Im} w$.

(e) Show $I \cap \mathbb{R} = \frac{N(z)}{d}\mathbb{Z}$. Conclude that $a = a'$ so that each representative $a + ib$ above is unique and $|\mathbb{Z}[i]/I| = N(z)$.

* EXERCISE 3.69. **(a)** Show $\mathbb{Z}[i]/(2) \cong \mathbb{Z}_2 \times \mathbb{Z}_2$ (cf. Exercise 3.68).

(b) Show $\mathbb{Z}[i]/(3)$ is a field with 9 elements.

(c) Show $\mathbb{Z}[i]/(5) \cong \mathbb{Z}_5 \times \mathbb{Z}_5$.

(d) Show $\mathbb{Z}[i]/(2 + i) \cong \mathbb{Z}_5$ (cf. Exercise 3.82).

EXERCISE 3.70. **(a)** If R is a ring and if I is an ideal, show that the *projection map* $\pi : R \to R/I$ given by $\pi(r) = r + I$, $r \in R$, is a surjective homomorphism with $\ker \pi = I$.

(b) Show that a subset of a ring is an ideal if and only if it is the kernel of a homomorphism (cf. Exercise 2.165).

EXERCISE 3.71 (Rings of Order 4). **(a)** Show that $\mathbb{Z}_2[x]/(x^2 + x + 1)$ is a field of order 4. *Hint:* Use the fact that $\{0, 1, x, 1 + x\}$ gives a complete list of coset generators.

(b) Up to isomorphism, show there are exactly four nontrivial rings with identity of order 4: $\mathbb{Z}_4$, $\mathbb{Z}_2 \times \mathbb{Z}_2$, $\{\begin{pmatrix} a & b \\ [0] & a \end{pmatrix} \mid a, b \in \mathbb{Z}_2\}$ (cf. Exercise 3.41), and $\mathbb{Z}_2[x]/(x^2 + x + 1)$. *Hint:* Start with the abelian groups of order 4 and see what multiplicative structures are possible.

EXERCISE 3.72. Prove R/I is a ring in Definition 3.19.

EXERCISE 3.73. Prove Theorem 3.20.

2.5.4. *Prime and Maximal Ideals.*

EXERCISE 3.74. Let R be a commutative ring with identity. Show $\{0\}$ is prime if and only if R is an integral domain.

EXERCISE 3.75. Let R be a commutative ring with identity. Show maximal ideals are prime. *Hint:* Theorem 3.22.

EXERCISE 3.76 (Prime Ideals of $\mathbb{Z}$ and $\mathbb{Z}_n$). **(a)** Show that the set of prime ideals of $\mathbb{Z}$ is $\{(p) \mid p \text{ is a positive prime}\} \cup \{(0)\}$ (cf. Exercise 3.47). Except for (0), show that this coincides with the set of maximal ideals in $\mathbb{Z}$.

(b) Show that the set of prime ideals of $\mathbb{Z}_n$ is $\{([p]) \mid p \text{ is a positive prime with } p \mid n\}$. Except for $([n])$ when n is prime, show that this coincides with the set of maximal ideals in $\mathbb{Z}_n$.

EXERCISE 3.77. Let $C([0,1]) = \{f : [0,1] \to \mathbb{R} \mid f \text{ is continuous}\}$. This exercises establishes a bijection between $[0,1]$ and maximal ideals of $C([0,1])$.

(a) For $x \in [0,1]$, let $M_x = \{f \in C([0,1]) \mid f(x) = 0\}$. Show M_x is a maximal ideal. *Hint:* Show $C([0,1])/M_x \cong \mathbb{R}$.

(b) Show that every maximal ideal is of the form M_x for some x. *Hint:* If an ideal is contained in no M_x, use bump functions to construct a unit, i.e., a nonvanishing f.

EXERCISE 3.78 (Prime Ideals in General). Let R be a ring with identity and let P be an ideal, $P \neq R$. For possibly noncommutative R, P is called *prime* if whenever I, J are ideals in R with $IJ \subseteq P$, then $I \subseteq P$ or $J \subseteq P$ (cf. Exercise 3.51).

(a) Show P is prime if and only if whenever $rRs \subseteq P$, $r, s \in R$, then $r \in P$ or $s \in P$ (cf. Exercise 3.57). *Hint:* Show $rRs \subseteq P$ implies $(RrR)(RsR) \subseteq P$. For the other direction, suppose $IJ \subseteq P$ with $a \in I \backslash P$. Show $aRb \in P$, $b \in J$.

(b) If R is commutative, show P is prime if and only if whenever $rs \in P$, $r, s \in R$, then $r \in P$ or $s \in P$, i.e., the usual definition.

EXERCISE 3.79 (Local Rings). A nonzero commutative ring R with identity is called *local* if $J(R) = R \backslash R^{\times}$ is an ideal.

(a) Show R is local if and only if R has a unique maximal ideal, namely $J(R)$.

(b) Show R is local if and only if R has a maximal ideal M so that $1 + m \in R^{\times}$ for all $m \in M$.

(c) Show that fields are local rings.

(d) Show $\mathbb{Z}_{p^n}$ is local, p a positive prime.

(e) Show $F[x]/(x^n)$, F a field, is local (cf. Exercise 3.67).

(f) Let $C_0 = \{f : (a,b) \to \mathbb{R} \mid a, b \in \mathbb{R} \text{ with } a < 0 < b,\ f \text{ is continuous}\}$. Define an equivalence relation on C_0 by saying $f \sim g$, $f, g \in C_0$, when f and g agree on some open interval around 0. Write C_0^0 for the set of equivalence classes of C_0. Elements of C_0^0 are called *germs of continuous functions* at 0. Show C_0^0 has the structure of a local ring. *Hint:* Show $[f] \in C_0^0$ is a unit if and only if $f(0) \neq 0$.

(g) Show that discrete valuation rings are local (cf. Exercises 3.63 and 3.38).

(h) Suppose D is an integral domain, $\mathrm{QF}(D)$ is a field containing D so that $\mathrm{QF}(D) = \{ab^{-1} \mid a, b \in D \text{ with } b \neq 0\}$ (cf. Definition 3.52), and P is a prime ideal of D. Let $D_P = \{du^{-1} \in \mathrm{QF}(D) \mid d \in D,\ u \in D \backslash P\}$, called the *localization* of D at P. Show D_P is a local subring of $\mathrm{QF}(D)$.

2.5.5. *First Isomorphism Theorem.*

EXERCISE 3.80. Establish the following ring isomorphisms:

(a) $\mathbb{Z}_{mn}/([m]) \cong \mathbb{Z}_m$ for $m, n \in \mathbb{N}$.

(b) $(R \times S)/(I \times J) \cong (R/I) \times (S/J)$ for I an ideal of R and J an ideal of S.

(c) $\begin{pmatrix} \mathbb{Z} & 5\mathbb{Z} \\ 3\mathbb{Z} & \mathbb{Z} \end{pmatrix} / \begin{pmatrix} \mathbb{Z} & 5\mathbb{Z} \\ 3\mathbb{Z} & 15\mathbb{Z} \end{pmatrix} \cong \mathbb{Z}_{15}$. *Hint:* Try $\varphi(A_{i,j}) = [A_{2,2}]$.

(d) $\begin{pmatrix} \mathbb{Q} & \mathbb{Q} \\ 0 & \mathbb{Q} \end{pmatrix} / \begin{pmatrix} 0 & \mathbb{Q} \\ 0 & \mathbb{Q} \end{pmatrix} \cong \mathbb{Q}$.

(e) $\mathcal{F}(\mathbb{R},\mathbb{R})/\{f \in \mathcal{F}(\mathbb{R},\mathbb{R}) \mid f(\pi) = 0\} \cong \mathbb{R}$.

(f) $R[x]/I[x] \cong (R/I)[x]$ for I and ideal of R.

(g) $M_{n,n}(R)/M_{n,n}(I) \cong M_{n,n}(R/I)$ for I and ideal of R.

EXERCISE 3.81 ($\mathbb{Z} \subseteq R$ or $\mathbb{Z}_k \subseteq R$). Let R be a ring with identity. Let $\varphi : \mathbb{Z} \to R$ be given by $\varphi(n) = n \cdot 1_R$, $n \in \mathbb{Z}$.

(a) Show φ is a homomorphism.

(b) Show $\operatorname{Im} \varphi$ is isomorphic to $\mathbb{Z}$ or $\mathbb{Z}_k$, $k \in \mathbb{N}$ (cf. Exercise 3.47). N.B.: It is common to identify $\mathbb{Z}$ or $\mathbb{Z}_k$ with its image under φ and to say that either $\mathbb{Z} \subseteq R$ or $\mathbb{Z}_k \subseteq R$.

(c) Show that $\mathbb{Z} \subseteq R$ if and only if $\operatorname{char} R = 0$ and that $\mathbb{Z}_k \subseteq R$ if and only if $\operatorname{char} R = k$ (cf. Exercise 3.19).

(d) If R is a field with $\operatorname{char} R = 0$, show that φ extends to an injective homomorphism $\varphi : \mathbb{Q} \to R$. N.B.: It is common to identify $\mathbb{Q}$ with its image under φ and to say that $\mathbb{Q} \subseteq R$.

(e) If R is a field with $\operatorname{char} R \neq 0$, show $\operatorname{char} R$ is prime.

EXERCISE 3.82. Let R be a ring with identity and let $|R| = p$, p prime. Show that R is a field and that $R \cong \mathbb{Z}_p$. *Hint:* Lagrange's Theorem and Exercise 3.81.

EXERCISE 3.83. As additive groups, $m\mathbb{Z}/n\mathbb{Z} \cong \mathbb{Z}_{\frac{n}{m}}$ for $m \mid n$, $m, n \in \mathbb{N}$ (Exercise 2.156). The map inducing the isomorphism as groups is $\varphi : m\mathbb{Z} \to \mathbb{Z}_{\frac{n}{m}}$ given by $\varphi(x) = \left[\frac{x}{m}\right]$.

(a) Show φ descends to a ring homomorphism from $m\mathbb{Z}/n\mathbb{Z}$ to $\mathbb{Z}_{\frac{n}{m}}$ if and only if $m = 1$ or $m = n$. *Hint:* Noting $1 \leq m \leq n$, look at $\varphi(m^2)$ versus $\varphi(m)^2$ and reduce to $\operatorname{mod}(n)$.

(b) Show $\psi(x) = [-\frac{x}{2}]$ induces $2\mathbb{Z}/6\mathbb{Z} \cong \mathbb{Z}_3$.

(c) Show $3\mathbb{Z}/9\mathbb{Z}$ is a trivial ring (on the additive group $(\mathbb{Z}_3, +)$) and so is not isomorphic to $\mathbb{Z}_3$ (as a ring).

(d) If $(m, \frac{n}{m}) = 1$, show $m\mathbb{Z}/n\mathbb{Z} \cong \mathbb{Z}_{\frac{n}{m}}$ as rings. *Hint:* Write $mx + \frac{n}{m}y = 1$ and map $mxk + n\mathbb{Z}$ to $[k]$.

EXERCISE 3.84 (Generalized Chinese Remainder Theorem). Suppose I and J are ideals of a ring R with identity so that $I + J = R$ (cf. Exercise 3.51). Let $\varphi : R \to (R/I) \times (R/J)$ be given by $\varphi(r) = (r + I, r + J)$.

(a) Show φ is a homomorphism with $\ker \varphi = I \cap J$.

(b) Show φ is surjective. Conclude that $R/(I \cap J) \cong (R/I) \times (R/J)$. *Hint:* $1 \in I + J$.

(c) As a corollary, show $\mathbb{Z}_{mn} \cong \mathbb{Z}_m \times \mathbb{Z}_n$ when $(n, m) = 1$.

(d) More generally, let $I_1, \ldots, I_n$ be ideals of R so that $I_i + I_j = R$, $i \neq j$. Show $R/\left(\bigcap_i I_i\right) \cong (R/I_1) \times \cdots \times (R/I_n)$. *Hint:* To see surjectivity, first fix i and write $1 = a_j + b_j$ with $a_j \in I_i$ and $b_j \in I_j$. Look at $\prod_{j \neq i}(a_j + b_j)$ to show you can write $1 = c_i + d_i$ with $c_i \in I_i$ and $d_i \in \bigcap_{j \neq i} I_j$. Now given $(r_i + I_i) \in \prod_i (R/I_i)$, look at $\sum_i r_i d_i$.

EXERCISE 3.85 (Second Isomorphism Theorem). Let I be an ideal and let S be a subring of R.

(a) Show I is an ideal of $S + I$ (cf. Exercise 3.50).

(b) Define $\varphi : S \to (S + I)/I$ by $\varphi(s) = s + I$ for $s \in S$. Show φ is a surjective homomorphism with $\ker \varphi = S \cap I$.

(c) Conclude that

$$(S + I)/I \cong S/(S \cap I).$$

EXERCISE 3.86 (Third Isomorphism Theorem). Let I and J be ideals of R with $I \subseteq J$.

(a) Show J/I is an ideal of R/I.

(b) Define $\varphi : R \to (R/I)\,/\,(J/I)$ by $\varphi(r) = (r+I) + J/I$ for $r \in R$. Show φ is a surjective homomorphism with $\ker\varphi = J$.

(c) Conclude that

$$(R/I)\,/\,(J/I) \cong R/J.$$

3. Polynomials and Roots

Polynomial rings turn out to be very powerful gadgets. We begin a careful study of some of their properties here.

DEFINITION 3.24. Let R be a ring.

(1) The *polynomial ring with coefficients in* R is the set of all formal sums

$$R[x] = \{\sum_{k=0}^{\infty} a_k x^k \mid a_k \in R, \text{ all but finitely many } a_k = 0\}$$

where x is an *indeterminate*, i.e., a formal symbol. If $f = \sum_k a_k x^k$ and $g = \sum_k b_k x^k$ are polynomials, addition and multiplication of polynomials are defined by

$$f + g = \sum_k (a_k + b_k)\, x^k,$$

$$fg = \sum_k \left(\sum_{i,j \text{ with } i+j=k} a_i b_j \right) x^k = \sum_k \left(\sum_{j=0}^{k} a_j b_{k-j} \right) x^k.$$

When writing a polynomial, terms of the form $0_R x^k$ are typically omitted, terms of the form $1_R x^k$ (if R has an identity) are typically written as x^k, and terms of the form $a_0 x^0$ are typically written as a_0. The polynomial f is sometimes written as $f(x)$.

(2) If $f = \sum_k a_k x^k$ satisfies $a_n \neq 0_R$ and $a_k = 0_R$ for $k > n$, $n \in \mathbb{N}$, the *degree* of f is

$$\deg f = n.$$

If all $a_k = 0_R$, $k \in \mathbb{Z}_{\geq 0}$, we write $f = 0$ and say

$$\deg(0) = -\infty.$$

(3) If $f = \sum_{k=0}^{n} a_k x^k$ is nonzero with $n = \deg f$, a_k is called a *coefficient* of f for $0 \leq k \leq n$, a_0 is called the *constant coefficient*, and a_n is called the *leading coefficient*. If R has an identity, the polynomial is called *monic* if $a_n = 1_R$.

Since most students hit abstract algebra after spending a lot of time in a calculus oriented life, it is worth noting that polynomials are viewed a bit differently in algebra than in calculus. In calculus, a polynomial is usually viewed as a special type of function designed to be evaluated at some specific value of x. In an algebra oriented life, polynomials are primarily abstract formal sums designed to be manipulated in exciting ways. Much of the time, though not always, we do not even think of them as functions and often have no desire to plug in something specific for the x.

As a further note, defining $\deg(0) = -\infty$ is a technical sleight of hand that allows a number of theorems (e.g., Theorem 3.26) to be stated more cleanly and certain proofs to be simplified by removing special cases. It should be noted that some mathematicians adopt a different convention and define the degree of 0 to be 0.

In any case, the addition and multiplication rules for polynomials are the same ones we have known since our youth—just souped up with fancier coefficients. Verifying that $R[x]$ is a ring is straightforward, though associativity is mildly unpleasant. We leave the details to Exercise 3.87.

From the definitions, it is clear that the map from R to $R[x]$ given by sending $r \to r$, $r \in R$, is an injective homomorphism. Identifying R with its image in $R[x]$, we view R as a subring of $R[x]$ consisting of the *constant polynomials*. Of course, $R[x]$ is commutative if R is and $R[x]$ has an identity if R does. Of special interest will be the case when R is a domain and, especially, a field.

THEOREM 3.25. *Let D be a domain. Then:*
(1) *For $f, g \in D[x]$, $\deg(fg) = \deg f + \deg g$.*
(2) *$D[x]$ is a domain.*
(3) *If D has an identity, $D[x]^\times = D^\times$.*

PROOF. For part (1), the case of $f = 0$ or $g = 0$ is clear so assume $f, g \neq 0$. Write $f = \sum_{k=0}^{n} a_k x^k$ and $g = \sum_{k=0}^{m} b_k x^k$ with $a_n, b_m \neq 0$. Then it follows from the definitions that $fg = a_0 b_0 + (a_0 b_1 + a_1 b_0)x + \cdots + a_n b_m x^{n+m}$. Since D is a domain, $a_n b_m \neq 0$ so that $\deg(fg) = n + m$ as desired. For part (2), note that a polynomial is zero if and only if its degree is strictly less than zero. Thus part (1) implies that when $f, g \neq 0$, then $fg \neq 0$. For part (3), note that $\deg 1 = 0$. Thus part (1) shows that the equation $fg = 1$ forces f and g to be constant polynomials, i.e., that $f, g \in D$. By definition, the set of units in D is $D^\times$. □

When R is not a domain, interesting things can happen that make the results of Theorem 3.25 false. For instance, $\mathbb{Z}_4[x]$ is not a domain since already $\mathbb{Z}_4 \subseteq \mathbb{Z}_4[x]$. Even stranger, consider $f = [1] + [2]x \in \mathbb{Z}_4[x]$. Then

$$f^2 = ([1] + [2]x)^2 = [1] + [4]x + [4]x^2 = [1]$$

so that $[1] + [2]x$ is a unit and $\deg(f \cdot f) = 0 < 2 = \deg(f) + \deg(f)$. Because of such examples, the best we can say for a general ring R about the degree of a product is that

$$\deg(fg) \leq \deg f + \deg g,$$

$f, g \in R[x]$.

Though not focused on here, there are many instances when multiple indeterminates are needed. For example, if R is a ring and x and y are two indeterminates, we can form $(R[x])[y]$ whose elements have the form $\sum_k a_k(x) y^k$ with $a_k(x) \in R[x]$. If we adopt the usual convention of writing $xy = yx$, then $(R[x])[y] = (R[y])[x]$ and is written simply as $R[x, y]$. Its elements then have the form

$$\sum_{i,j} a_{i,j}\, x^i y^j$$

with $a_{i,j} \in R$ and standard polynomial addition and multiplication are used. Similar comments hold for polynomial rings in many variables, $R[x_1, x_2, \ldots, x_n]$. In fact, it is possible to attach an infinite, even uncountable, number of indeterminates as long as each polynomial only uses a finite number of indeterminates at a time.

3.1. Division Algorithm. In a similar spirit to $\mathbb{Z}$ (cf. Theorem 1.3), polynomial rings over fields have a division algorithm.

THEOREM 3.26 (Division Algorithm). *Let F be a field and suppose $f, g \in F[x]$ with $g \neq 0$. There exist unique $q, r \in F[x]$ so that:*

(1) $f = gq + r$.

(2) $\deg r < \deg g$.

PROOF. Consider existence first. If $\deg f < \deg g$, use $q = 0$ and $r = f$. Therefore assume $\deg f \geq \deg g$. If $\deg f = 0$, use $q = fg^{-1}$ and $r = 0$. Now argue by induction on $\deg f$. Write $f = a_0 + \cdots + a_n x^n$ and $g = b_0 + \cdots + b_m x^m$ with $a_n, b_m \neq 0$ and $n \geq m$ and set $\widetilde{f} = f - a_n b_m^{-1} x^{n-m} g$. Since $\deg \widetilde{f} < \deg f$, we can write $\widetilde{f} = g\widetilde{q} + \widetilde{r}$ with $\deg \widetilde{r} < \deg g$ for some $\widetilde{q}, \widetilde{r} \in F[x]$. Thus $f = g\left(\widetilde{q} + a_n b_m^{-1} x^{n-m}\right) + \widetilde{r}$ and we are done.

For uniqueness, suppose $f = gq + r$ and $f = g'q + r'$ with $\deg r, \deg r' < \deg g$. Subtraction shows $g(q - q') = r' - r$ so that

$$\deg g + \deg(q - q') = \deg(g(q - q')) = \deg(r' - r) < \deg g,$$

which can only happen if $\deg(q - q') = -\infty$, i.e., if and only if $q - q' = 0$. In turn, this shows $r' - r = 0$ as desired. □

The Division Algorithm can be easily generalized to the setting of any ring with identity as long as the leading coefficient of g is a unit (Exercise 3.96). In any case, here is an example of the Division Algorithm in $\mathbb{Z}_2[x]$ with $f = x^3 + x^2 + x$ and $g = x + [1]$. As a youth, I had to write out the long division of f by g in the following type of format:

$$\begin{array}{rl} & x^2 \qquad + [1] \qquad\qquad R = [1]. \\ x + [1] & \overline{)\; x^3 + x^2 + x + [0]} \\ & \;\; x^3 + x^2 \\ \hline & \qquad\qquad x + [0] \\ & \qquad\qquad x + [1] \\ \hline & \qquad\qquad\qquad [1] \end{array}$$

This shows that g goes into f exactly $x^2 + [1]$ times with a remainder of $[1]$. In other words, $f = g\left(x^2 + [1]\right) + [1]$.

Our first application of the Division Algorithm gives the following important description of ideals in polynomial rings over fields.

COROLLARY 3.27. *Let F be a field. Ideals of $F[x]$ are principal, i.e., every ideal is of the form (f) for some $f \in F[x]$.*

PROOF. Let I be a nonzero ideal and choose nonzero $f \in I$ of minimal degree. Of course $(f) \subseteq I$. On the other hand, if $g \in I$, use the Division Algorithm to write $g = fq + r$ for $q, r \in F[x]$ and $\deg r < \deg f$. Then $r = g - fq \in I$. If $r \neq 0$, this would violate our choice of f as having minimal degree. Thus $r = 0$ so that $g = fq \in (f)$ and $I \subseteq (f)$ as desired. □

In Section 5 we will call $F[x]$ a *principal ideal domain* since all of its ideals are principal.

3.2. Roots. Many of us spent a considerable portion of our mathematical training attempting to factor polynomials over $\mathbb{Q}$ or $\mathbb{R}$. An important element of that program is the extraction of roots of a polynomial. It turns out that such questions are also very important for more general rings and, especially, fields. Besides, you have not really lived until you have found all the roots of a polynomial in $\mathbb{Z}_4[x]$.

DEFINITION 3.28. Let R be a commutative ring, let $f = \sum_{k=0}^{n} a_k x^k \in R[x]$, and let $r \in R$.

(1) We define the symbol $f(r)$ to be the element of R given by $\sum_{k=0}^{n} a_k r^k$.

(2) We say r is a *root* or a *zero* of f if $f(r) = 0_R$.

(3) Suppose R has an identity. If $f = (x - r)^m g$, $m \in \mathbb{N}$ and $g \in R[x]$ with $g(r) \neq 0$, then the root r is said to have *multiplicity* m. If the multiplicity is 1, the root is called *simple*.

(4) The *formal derivative* of f is $f' = \sum_{k=1}^{n} k a_k x^{k-1}$.

Note that the definition of $f(r)$ requires commutivity of R since $ax = xa$ in $R[x]$, $a \in R$. The fact that the multiplicity is well defined is found in Exercise 3.112.

THEOREM 3.29. *Let R be a commutative ring and let $r \in R$. The* evaluation map *$\varphi_r : R[x] \to R$ given by $\varphi_r(f) = f(r)$, $f \in R[x]$, is a surjective homomorphism.*

PROOF. If $f = \sum_k a_k x^k$ and $g = \sum_k b_k x^k$ are in $R[x]$, then

$$\varphi_r(f+g) = \sum_k (a_k + b_k) r^k = \sum_k a_k r^k + \sum_k b_k r^k = \varphi_r(f) + \varphi_r(g)$$

and

$$\begin{aligned}\varphi_r(fg) &= \sum_k \left(\sum_{j=0}^{k} a_j b_{k-j} \right) r^k \\ &= \left(\sum_k a_k r^k \right) \left(\sum_k b_k r^k \right) = \varphi_r(f)\varphi_r(g)\end{aligned}$$

so that φ_r is a homomorphism. To see surjectivity, use the constant polynomials. □

In general, determining the roots of a polynomial is a very hard problem depending heavily on the ring R. Changing R can drastically affect even the number of roots of a polynomial. For instance, $x^2 + 1$ has no roots in $\mathbb{R}$, but it has two roots in $\mathbb{C}$ (i and $-i$). In any event, based on our experience with $\mathbb{R}$ and $\mathbb{C}$, we naturally look for roots by trying to factor a polynomial into linear terms. It turns out that this always works for fields.

THEOREM 3.30. *Let F be a field, let $f \in F[x]$, and let $a \in F$.*

(1) *The polynomial f can be written as $f = (x - a)q$ for some $q \in F[x]$ if and only if a is a root of f.*

(2) *Let $n \in \mathbb{Z}_{\geq 0}$. If* $\deg f = n$, *then f has at most n distinct roots.*

(3) *Suppose a is a root of f. Then a has multiplicity at least two if and only if a is also a root of f'.*

PROOF. For (1), use the Division Algorithm to uniquely write $f = (x-a)q+c$ for $q, c \in F[x]$ with $\deg c \leq \deg(x-a) = 1$. In particular, $c \in F$. In fact, evaluation at a shows $f(a) = c$ so that $f = (x-a)q + f(a)$. Hence $f = (x-a)q$ if and only if $f(a) = 0$.

For part (2), suppose $a_1, \ldots, a_n$ are distinct roots of f in F. To begin, use part (1) to write $f = (x-a_1)q_1$, $q_1 \in F[x]$. Moreover, $0 = f(a_2) = (a_2 - a_1)q_1(a_2)$ so that $q_1(a_2) = 0$ and so a_2 is a root of q_1. Thus $q_1 = (x-a_2)q_2$, $q_2 \in F[x]$, and $f = (x-a_1)(x-a_2)q_2$. By induction, we eventually get $f = \prod_{i=1}^{n}(x-a_i)q_{n+1}$ which, by Theorem 3.25, shows $\deg f = n + \deg q_{n+1}$. In particular, $n \leq \deg f$ as desired.

For part (3), note that the formal derivative satisfies the same basic properties that the derivative from calculus does (Exercise 3.113). In particular if a has multiplicity m, $m \geq 1$, then $f = (x-a)^m q$, $q \in F[x]$ with $q(a) \neq 0$. Thus

$$f' = m(x-a)^{m-1}q + (x-a)^m q' = (x-a)^{m-1}[mq + (x-a)q'].$$

If $m \geq 2$, clearly a is a root of f'. Conversely if a is a root of f', we claim $m \geq 2$. If not, then $m = 1$ and a must then be a root of $q + (x-a)q'$. However, evaluating $q + (x-a)q'$ at a gives $q(a)$ which is nonzero and therefore a contradiction. □

Part (1) of Theorem 3.30 is actually true if F is replaced by a commutative ring with identity and part (2) is true for any integral domain (Exercise 3.96). However, Theorem 3.30 can easily be false if F is replaced by a ring that is not an integral domain. For instance, consider $f = [2]x^2 - [2]x \in \mathbb{Z}_4[x]$. Then $\deg f = 2$, but there are actually four roots—all of $\mathbb{Z}_4$! This can be checked by directly plugging in or by observing that $f = [2]x(x - [1])$ and that $y(y-1)$ is even for any $y \in \mathbb{Z}$. Perhaps more interestingly, it is easy to check that $x^2 - [1] \in \mathbb{Z}_8[x]$ has four roots, namely [1], [3], [5], and [7].

As already noted, finding roots of polynomials is a very hard problem. However, working with finite fields greatly simplifies matters since it is possible to plug every element of the field into the polynomial. For example, consider the polynomial $f = x^2 + x + 1 \in \mathbb{Z}_2[x]$. There are only two possibilities for roots, namely [0] or [1]. Thus to determine if f has any roots, simply evaluate $f([0]) = [1]$ and $f([1]) = [1]$ to see that f has no roots in $\mathbb{Z}_2$.

We finish this subsection with a surprising and very powerful application of Theorem 3.30. Recall that if F is a field, $F^\times = F \backslash \{0\}$.

COROLLARY 3.31 (Finite Subgroups of Fields Are Cyclic). **(1)** *Let F be a field and let G be a finite subgroup of $F^\times$. Then G is cyclic.*

(2) *$U_p \cong \mathbb{Z}_{p-1}$ as groups for p a positive prime.*

PROOF. Consider part (1) first. Since G is a finite abelian group, let $p_i^{n_{i,j}}$, $1 \leq i \leq m$, be the elementary divisors of G where $m, k_i, n_{i,j} \in \mathbb{N}$, $p_1 < p_2 < \cdots < p_m$ are positive primes, and $n_{i,1} \geq n_{i,2} \geq \cdots \geq n_{i,k_i}$. Choose $c_i \in G$ to have order $p_i^{n_{i,1}}$ (e.g., the element corresponding to $[1] \in \mathbb{Z}_{p_i^{n_{i,1}}}$) and let $c = \prod_i c_i$. By construction, $|g|$ divides $|c|$ for all $g \in G$. Thus every element of G is a root of the polynomial $x^{|c|} - 1_F \in F[x]$. But since Theorem 3.30 shows there are at most $|c|$ roots, we see $|G| \leq |c|$. Of course, $\langle c \rangle \subseteq G$ so that $|c| = |\langle c \rangle| \leq |G|$. This implies $\langle c \rangle = G$ as desired. For part (2), look at the special case of $G = U_p = \mathbb{Z}_p^\times \subseteq \mathbb{Z}_p$. □

In the above corollary, a generator for G is sometimes called a *primitive element* or *primitive root* of G. In the case of $G = U_p$, there are tricks that make finding a primitive element computationally efficient (cf. Exercise 2.127). The choice of a primitive element $[c] \in U_p$ makes it possible to write down an explicit isomorphism $\phi : \mathbb{Z}_{p-1} \to U_p$ by setting $\phi([k]) = [c]^k$. Modern cryptography, in part, rests on computing the inverse of this map for very large primes (on the order of 2^{1000}), which is computationally very difficult.

3.3. Rational Root Test. Since $\mathbb{Q}$ is such an important field, we give a very useful trick for finding rational roots of polynomials over $\mathbb{Q}$.

THEOREM 3.32 (Rational Roots over $\mathbb{Q}$). **(1)** Reduction to $\mathbb{Z}[x]$: *If $f \in \mathbb{Q}[x]$ is nonzero, there exists a nonzero $c \in \mathbb{Z}$ so that $cf \in \mathbb{Z}[x]$. In that case, f and cf have the same roots.*

(2) Rational Root Test: *If $f = \sum_{k=0}^{n} a_k x^k \in \mathbb{Z}[x]$ with $a_0, a_n \neq 0$, then the only* possible *roots in $\mathbb{Q}$ lie in $\{\frac{r}{s} \mid r, s \in \mathbb{Z}^{\times}$ with $r \mid a_0,\ s \mid a_n\}$.*

PROOF. For part (1), choose c to be the least common multiple (though any common multiple works) of the denominators of the nonzero coefficients of f. Clearly $cf \in \mathbb{Z}[x]$ and, since $\mathbb{Z}$ is an integral domain, $cf(x) = 0$ if and only if $f(x) = 0$.

For part (2), suppose $\frac{r}{s} \in \mathbb{Q}$ with $r, s \in \mathbb{Z}$ is a root of f written in lowest terms, i.e., so that $(r, s) = 1$. Then multiplying both sides of the equation $f(\frac{r}{s}) = 0$ by s^n shows

$$a_0 s^n + a_1 r s^{n-1} + \cdots + a_{n-1} r^{n-1} s + a_n r^n = 0.$$

In particular, $a_0 s^n = -r(a_1 s^{n-1} + \cdots + a_n r^{n-1})$. As $a_0, s \neq 0$, we see $r \neq 0$ and that $r \mid (a_0 s^n)$. Since $(r, s) = 1$, it follows that $r \mid a_0$. Similarly, $a_n r^n = -s(a_0 s^{n-1} + \cdots + a_{n-1} r^{n-1})$ so that $s \mid (a_n r^n)$ and $s \mid a_n$ as desired. □

For example, consider $f = x^3 + \frac{1}{2}x^2 + \frac{1}{2}x - \frac{1}{2} \in \mathbb{Q}[x]$. Multiplying by 2 gives $\widetilde{f} = 2x^3 + x^2 + x - 1 \in \mathbb{Z}[x]$ with the same roots as f. Applying the Rational Root Test shows that the only possible rational roots of $\widetilde{f}$ are $\pm\frac{1}{2}$ and ± 1. Evaluating gives $\widetilde{f}(-\frac{1}{2}) = -\frac{3}{2}$, $\widetilde{f}(\frac{1}{2}) = 0$, $\widetilde{f}(-1) = -3$, and $\widetilde{f}(1) = 3$. Thus the only rational root is $\frac{1}{2}$. Since $\frac{1}{2}$ is a root, we know that $(x - \frac{1}{2})$ divides evenly into $\widetilde{f}$ (Theorem 3.30). Doing long division shows $\widetilde{f} = (x - \frac{1}{2})2(x^2 + x + 1)$. By evaluation (or even by the Rational Root Test again), clearly $\frac{1}{2}$ is not a root of $x^2 + x + 1$ so that $\frac{1}{2}$ is actually a simple root of f.

If some $f \in \mathbb{Q}[x]$ has its constant coefficient equal to zero, then a slight modification is needed before applying the Rational Root Test. For example, consider $f = 3x^6 - 7x^4 + x^3$. Factor out x^3 and write $f = x^3(3x^3 - 7x + 1)$ so that 0 is a root of f with multiplicity 3. All other roots of f are roots of $3x^3 - 7x + 1$. The Rational Root Test then says the only rational possibilities are $\pm\frac{1}{3}$ and ± 1. By direct substitution, none make the cut so that 0 is the only rational root of f.

It is important to emphasize that the Rational Root Test only gives possible rational roots. It does not say that all the possibilities actually are roots and it says nothing about the existence of irrational or complex roots. For example, $x^2 + x + 1$ has no rational roots (e.g., the Rational Root Test says ± 1 are the only possibilities and they clearly fail). However, applying the beloved quadratic equation shows there are two complex roots, namely $-\frac{1}{2} \pm i\frac{\sqrt{3}}{2}$ (Exercise 3.124).

3.4. Fundamental Theorem of Algebra. The quadratic equation is a beautiful formula for the roots of any polynomial $f \in \mathbb{C}[x]$ with $\deg f = 2$. It turns out that analogous formulas exist for the roots of f when $\deg f = 3$ and $\deg f = 4$, due to Tartaglia and Ferrari in the 17th century. Few of us have really worked with these equations in school because, frankly, they are rather hideous (see Exercises 3.125 and 3.126). Even so, the point is that the problem of finding roots of a polynomial f over $\mathbb{C}$ is completely solved when $\deg f \leq 4$.

The story takes a startling turn at this point. It is one of the most famous results of mathematics, due to Abel in the 19th century and made possible by work of Galois, that when $\deg f = 5$, no radical formula for the roots exists (cf. §5.2 in Chapter 4). By radical we mean a formula using only addition, subtraction, multiplication, division, and the extraction of nth roots. Notice that it is not that no one has yet been clever enough to find such a formula, but rather that no such formula exists! This is a truly stunning result and, at least for now, seems completely impossible to prove. In any case, the upshot is that life is hard. Finding roots of general polynomials is extremely difficult work with no simple sure-fire method to success.

Nevertheless, there is one important thing we can say about polynomials over $\mathbb{C}[x]$ that we have seen is not always true over other fields. Namely, nonconstant polynomials over $\mathbb{C}$ always have roots in $\mathbb{C}$—even if we cannot always explicitly write them down. This result is so important that it is called the *Fundamental Theorem of Algebra*. It turns out that pretty much any elementary proof of the first part of the following theorem relies on results from analysis. After all, the difference between $\mathbb{Q}$ and $\mathbb{R}$ mostly has to do with limits and Cauchy sequences so it is no surprise that analysis is needed. Since we are not assuming everyone knows that a continuous function achieves a maximum on a compact set, we relegate the proof to Exercise 3.129 where it can safely be omitted by "analysisphobes" (cf. Exercise 4.135).

THEOREM 3.33 (Fundamental Theorem of Algebra). **(1)** *Let $f \in \mathbb{C}[x]$ be nonconstant. Then f has a root in $\mathbb{C}$.*

(2) *If $\deg f = n$, $n \in \mathbb{N}$, there exist $z_i \in \mathbb{C}$ so that*

$$f = c\prod_{i=1}^{n}(x - z_i)$$

where $c \in \mathbb{C}^{\times}$ is the leading coefficient of f and z_i is a root of f.

PROOF. For part (1), see Exercise 3.129. For part (2), inductively apply Theorem 3.30 to part (1). By construction, c is clearly the leading coefficient. □

COROLLARY 3.34. *Let $f \in \mathbb{R}[x]$ be nonconstant. Then*

$$f = c\prod_{i=1}^{k}(x - r_i)\prod_{j=1}^{m} f_j$$

where $k, m \in \mathbb{Z}_{\geq 0}$ with $k + 2m = \deg f$, $c \in \mathbb{R}^{\times}$ is the leading coefficient of f, $r_i \in \mathbb{R}$ is a real root of f, and $f_j \in \mathbb{R}[x]$ is monic of degree 2 with no real roots.

PROOF. Viewing $\mathbb{R}[x]$ as a subring of $\mathbb{C}[x]$, use the Fundamental Theorem of Algebra to write $f = c\prod_{i=1}^{k}(x-r_i)\prod_{i=1}^{n-k}(x-z_i)$ where r_i are real roots (if there are any) and z_i are nonreal roots in $\mathbb{C}$ (if there are any). By repeated use of Theorem 3.30, $f_0 = \prod_{i=1}^{n-k}(x - z_i) \in \mathbb{R}[x]$.

If we write m_i for the multiplicity of z_i, we have $f_0 = (x - z_i)^{m_i} q_i$ for some $q_i \in \mathbb{C}[x]$ with $q_i(z_i) \neq 0$. Now for any $g = \sum_k a_k x^k \in \mathbb{C}[x]$, define $\overline{g} = \sum_k \overline{a_k} x^k$. It follows from the fact that complex conjugation is a homomorphism of $\mathbb{C}$ that the map $g \to \overline{g}$ is a homomorphism of $\mathbb{C}[x]$ (Exercise 3.119). Thus $f_0 = \overline{f}_0 = (x - \overline{z}_i)^{m_i} \overline{q}_i$ so that $\overline{z}_i$ is a root of f_0 (in fact, of q_i) with multiplicity at least m_i. Since conjugating twice is the identity map, the multiplicity is exactly m_i. Thus $f_0 = [(x - z_i)(x - \overline{z}_i)]^{m_i} \widetilde{q}_i$ for some $\widetilde{q}_i \in \mathbb{C}[x]$ with $\widetilde{q}_i(z_i), \widetilde{q}_i(\overline{z}_i) \neq 0$. By construction $(x - z_i)(x - \overline{z}_i) = x^2 - 2(\operatorname{Re} z_i)\, x + |z|^2 \in \mathbb{R}[x]$ is monic of degree 2 with no real roots (since the roots are z_i and $\overline{z}_i$). Induction on the number of distinct z_i finishes the proof. $\square$

3.5. Exercises 3.87–3.129.

EXERCISE 3.87. If R is a ring, show $R[x]$ is a ring.

EXERCISE 3.88. Add the following polynomials:
(a) $([2]x^3 + [1]) + (x^3 + x^2 + [3]x + [4])$ in $\mathbb{Z}_5[x]$,
(b) $([2]x + [1]) + (x^2 + x + [1])$ in $\mathbb{Z}_3[x]$.

EXERCISE 3.89. Multiply the following polynomials:
(a) $([2]x^3 + [1])(x^2 + [3]x + [4])$ in $\mathbb{Z}_5[x]$,
(b) $([2]x + [1])^3$ in $\mathbb{Z}_3[x]$,
(c) $(x + [1])^5$ in $\mathbb{Z}_2[x]$.

EXERCISE 3.90. Let R be a ring and let $f, g \in R[x]$.
(a) Show $\deg(f + g) \leq \max\{\deg f, \deg g\}$.
(b) Show the inequality in part (a) can be strict.

EXERCISE 3.91. Write down all polynomials of degree at most 3 in $\mathbb{Z}_2[x]$.

EXERCISE 3.92. **(a)** Let $\varphi : R_1 \to R_2$ be a homomorphism of rings. For $f = \sum_k c_k x^k \in R_1[x]$, define $\Phi : R_1[x] \to R_2[x]$ by $\Phi(f) = \sum_k \varphi(c_k)\, x^k$. Show Φ is a homomorphism.

(b) If φ is an isomorphism, show Φ is an isomorphism.

(c) If φ is an isomorphism, show f is a product of two polynomials of degree n and m in $R_1[x]$ if and only if $\Phi(f)$ is a product of two polynomials of degree n and m in $R_2[x]$.

EXERCISE 3.93. For the polynomials f, g indicated below, find q, r so that (i) $f = gq + r$ and (ii) $\deg r < \deg g$.
(a) $f = x^5 + 2x^2 + 1$ and $g = x^2 + 1$ in $\mathbb{Q}[x]$.
(b) $f = x^5 + [2]x^2 + [1]$ and $g = x^2 + [1]$ in $\mathbb{Z}_5[x]$.
(c) $f = x^4 + x^3 + [2]x$ and $g = x + [2]$ in $\mathbb{Z}_3[x]$.
(d) $f = x^7 + x^5 + x^3 + [1]$ and $g = x^2 + x + [1]$ in $\mathbb{Z}_2[x]$.

EXERCISE 3.94. Let F be a field and let $I = \{\sum_k a_k x^k \in F[x] \mid \sum_k a_k = 0\}$.
(a) Show that I is an ideal and find $f \in F$ so that $I = (f)$. *Hint:* Look at $x = 1$.
(b) Show $F[x]/I \cong F$.
(c) In general, for $a \in F$, show $\{f \in F[x] \mid f(a) = 0\} = (x - a)$.

EXERCISE 3.95. Let F be a field. Show $F^\times$ contains no subgroup isomorphic to $\mathbb{Z}_2 \times \mathbb{Z}_4$.

EXERCISE 3.96 (Generalized Division Algorithm). **(a)** Let R be a ring with identity and suppose $f, g \in R[x]$ with $g \neq 0$ and the leading coefficient of g a unit. Show there exist unique $q, r \in R[x]$ so that (i) $f = qg + r$ and (ii) $\deg r < \deg g$.

(b) If R is also commutative with $a \in R$, show $f = (x-a)q$ for some $q \in R[x]$ if and only if a is a root of f.

(c) If R is also an integral domain, show that a nonzero f has at most $\deg f$ roots in R.

EXERCISE 3.97 (Independence of Field Extensions). Let $F \subseteq K$ be fields with $f, g \in F[x]$ and $g \neq 0$. If there is a $q \in K[x]$ so that $f = gq$, show $q \in F[x]$. *Hint:* Use uniqueness of the Division Algorithm in $F[x]$ and then in $K[x]$.

EXERCISE 3.98. Show $x^2 \in (M_{2,2}(\mathbb{R}))[x]$ has an infinite number of roots in $M_{2,2}(\mathbb{R})$.

EXERCISE 3.99. Let $\varphi : \mathbb{Q}[x] \to \mathbb{Q}(\sqrt{2})$ be given by $\varphi(f) = f(\sqrt{2})$, $f \in \mathbb{Q}[x]$.

(a) Show φ is a surjective homomorphism.

(b) Show $\ker \varphi = (x^2 - 2)$ and conclude $\mathbb{Q}[x]/(x^2-2) \cong \mathbb{Q}(\sqrt{2})$ (cf. Exercise 3.65). *Hint:* Corollary 3.27.

EXERCISE 3.100. For each polynomial in the given polynomial ring $F[x]$ indicated below, find all roots in F and determine their multiplicity. *Hint:* These are finite fields—every possibility can be tried.

(a) $x^5 + x^2 + x + [1]$ in $\mathbb{Z}_2[x]$,
(b) $x^5 + x + [1]$ in $\mathbb{Z}_2[x]$,
(c) $x^4 + [2]x^3 + x$ in $\mathbb{Z}_3[x]$,
(d) $[2]x^4 + [3]x^2 + [x] + [4]$ in $\mathbb{Z}_5[x]$,
(e) $x^5 + [1]$ in $\mathbb{Z}_5[x]$.

EXERCISE 3.101. For each polynomial in $\mathbb{Q}[x]$, find all rational roots and their multiplicity. *Hint:* Rational Root Test.

(a) $x^3 - 2x^2 + x - 2$,
(b) $x^4 - 3x^3 + x^2 + 4$,
(c) $x^3 - 6x^2 + 11x - 6$,
(d) $x^4 - \frac{4}{3}x^3 + \frac{4}{3}x^2 - \frac{4}{3}x + \frac{1}{3}$.

EXERCISE 3.102. Show that the only rational roots of a monic polynomial in $\mathbb{Z}[x]$ are integers.

EXERCISE 3.103. Let $f_c = x^{17} + 2x^{11} - 3x^8 + x^5 + 2x + c \in \mathbb{Q}[x]$ for $c \in \mathbb{Q}$. For what value of c can you write $f_c = (x-1)q$ for some $q \in \mathbb{Q}[x]$?

EXERCISE 3.104. Let p be a positive prime. Factor $x^p - x \in \mathbb{Z}_p[x]$ as a product of degree one polynomials. *Hint:* Fermat's Little Theorem.

EXERCISE 3.105. Let F be a field and suppose $f, g \in F[x]$ with $\deg f, \deg g \leq n$. If there are distinct $r_1, r_2, \ldots, r_{n+1} \in F$ so that $f(r_i) = g(r_i)$, show $f = g$. *Hint:* Look at the roots of $f - g$.

EXERCISE 3.106. Let F be a field and consider the set of functions $\mathcal{F}(F) = \{f : F \to F\}$. If $p \in F[x]$, define $\varphi(p) \in \mathcal{F}(F)$ by $(\varphi(p))(r) = p(r)$, $r \in F$. We say

$p_1, p_2 \in F[x]$ *induce the same function* on F if $\varphi(p_1) = \varphi(p_2)$, i.e., if $p_1(r) = p_2(r)$ for all $r \in F$.

(a) For finite fields, show by example that different polynomials can induce the same function on F. *Hint:* Try $F = \mathbb{Z}_2$.

(b) For infinite fields, show this cannot happen. *Hint:* Exercise 3.105.

EXERCISE 3.107. Let F be a field and let $f \in F[x]$ be nonzero. Let $a_1, \ldots, a_n$ be the distinct roots of f in F with multiplicity $m_1, \ldots, m_n$. Show $\sum_{i=1}^{n} m_i \leq \deg f$.

EXERCISE 3.108. Let F be a field, let $f \in F[x]$, and let $a_1, \ldots, a_n$ be distinct elements of F. If $f(a_1) = f(a_2) = \cdots = f(a_n)$, show f is either constant or $\deg f \geq n$. *Hint:* $f - f(a_1)$.

EXERCISE 3.109. Show every polynomial of odd degree in $\mathbb{R}[x]$ has a root in $\mathbb{R}$.

EXERCISE 3.110. Let $m, n \in \mathbb{N}$. Show $\sqrt[n]{m} \in \mathbb{Q}$ if and only if $\sqrt[n]{m} \in \mathbb{N}$, i.e., it is either an integer or irrational. *Hint:* Rational Root Test.

EXERCISE 3.111 (Taylor's Theorem). Let F be a field and suppose $p \in F[x]$ is nonconstant.

(a) Given $f \in F[x]$, show there exist unique $r_0, \ldots, r_m \in F[x]$ so that (i) $f = \sum_{k=0}^{m} r_k p^k$ and (ii) $\deg r_k < \deg p$. *Hint:* Write $\deg f = m \deg p + s$ with $s < \deg p$, use the Division Algorithm with f and p^m, and induct on $\deg f$.

(b) If $a \in F$, consider the special case of $p = x - a$. In this case, prove *Taylor's Theorem*: If $\operatorname{char} F = 0$ (cf. Exercise 3.19), show $r_k = \frac{1}{k!} f^{(k)}(a)$ where $f^{(k)}$ is the kth derivative of f.

(c) Show that part (a) remains true if F is replaced by a ring R with identity for which the leading coefficient of p is a unit. *Hint:* Exercise 3.96.

(d) Show that part (c) can fail if the leading coefficient of p is not a unit.

EXERCISE 3.112. Show that the definition of the multiplicity of a root in Definition 3.28 is well defined. *Hint:* Exercise 3.111.

EXERCISE 3.113. Let R be a commutative ring, and let $f, g \in R[x]$, $a, b \in R$, and $n \in \mathbb{N}$. Show that the formal derivative satisfies the following properties:

(a) $(af + bg)' = af' + bg'$.

(b) $(fg)' = f'g + fg'$ (the *product rule*, also called *Leibniz's rule*).

(c) $((x - a)^n)' = n(x - a)^{n-1}$ when R has an identity.

EXERCISE 3.114. Let R be a ring. Show $\operatorname{char} R = \operatorname{char} R[x]$ (cf. Exercise 3.19).

EXERCISE 3.115. Let F be a field with $f \in F[x]$ and $\deg f = 1$. Show (f) is a maximal ideal in $F[x]$.

EXERCISE 3.116. Let $n \in \mathbb{N}$ and $k \in \mathbb{Z}_{\geq 0}$. Show there are $n^{k+1} - n^k$ polynomials of degree k in $\mathbb{Z}_n[x]$.

EXERCISE 3.117. Find all finite subgroups of $\mathbb{C}^\times$. *Hint:* Show such subgroups are cyclic and that the generator must have norm 1.

EXERCISE 3.118. If F is a field and G and H are finite subgroups of $F^\times$ with $|G| = |F|$, show $G = F$. *Hint:* $x^{|G|} - 1$.

EXERCISE 3.119. For $g = \sum_k a_k x^k \in \mathbb{C}[x]$, let $\overline{g} = \sum_k \overline{a_k} x^k$. Show the map $g \to \overline{g}$ is a homomorphism of $\mathbb{C}[x]$.

EXERCISE 3.120. Let F be a field and let $r_1, r_2 \in F$. If $x^2 + bx + c \in F[x]$ has roots r_i, show $b = -r_1 - r_2$ and $c = r_1 r_2$.

EXERCISE 3.121 (Lagrange Interpolation). Let F be a field with $a_i \in F$, $0 \le i \le k$, $k \in \mathbb{N}$, and the a_i distinct.

(a) Let $l_i \in F[x]$ be given by

$$l_i = \prod_{\substack{j=0 \\ j \neq i}}^{k} \frac{x - a_j}{a_i - a_j}.$$

Show that $\deg l_i = k$ and that $l_i(a_j) = 0$ for $i \neq j$ and $l_i(a_i) = 1$.

(b) If $b_i \in F$ and if $L \in F[x]$ is given by $L = \sum_{i=0}^{k} b_i l_i$, show $\deg L \le k$ and $L(a_i) = b_i$.

(c) Show L is the unique polynomial in $F[x]$ satisfying $L(a_i) = b_i$ of degree at most k. *Hint:* If $\widetilde{L}$ is another solution, show $L - \widetilde{L}$ has at least $k+1$ distinct roots and so it is 0.

(d) Use this technique to find a polynomial $f \in \mathbb{Q}[x]$ of degree at most 3 so that $f(0) = 2$, $f(1) = 3$, $f(2) = 5$, and $f(3) = 7$.

EXERCISE 3.122 (Formal Power Series). Let R be a ring. The *formal power series* ring is

$$R[[x]] = \{\sum_{k=0}^{\infty} a_k x^k \mid a_k \in R\}$$

with addition and multiplication defined just as was done for polynomials. N.B.: These are called *formal* power series since notions of convergence are ignored.

(a) Prove $R[[x]]$ is a ring containing $R[x]$ as a subring.

(b) If R is an integral domain, show $R[[x]]$ is as well.

(c) If R has an identity, show $1-x$ is a unit in $R[[x]]$ with inverse $1+x+x^2+\cdots$ but that $1-x$ is not a unit in $R[x]$.

(d) More generally if $a_0 \in R$ is a unit, show $\sum_{k=0}^{\infty} a_k x^k$ is a unit in $R[[x]]$.

EXERCISE 3.123 (Formal Laurent Series). Let R be a ring. The *formal Laurent series* ring is

$$R((x)) = \{\sum_{k=k_0}^{\infty} a_k x^k \mid k_0 \in \mathbb{Z},\ a_k \in R\}$$

with addition and multiplication defined similarly to the definitions for polynomials.

(a) Prove $R((x))$ is a ring containing $R[[x]]$ as a subring.

(b) If F is a field, show $F((x))$ is a field.

(c) If $f = \sum_{k=k_0}^{\infty} a_k x^k \in F((x))$ with $a_{k_0} \neq 0$, let $\nu(f) = k_0$. Show ν is a discrete valuation of $F((x))$ with discrete valuation ring $F[[x]]$ (cf. Exercise 3.38).

EXERCISE 3.124 (Quadratic Equation). Let $f = ax^2+bx+c \in \mathbb{C}[x]$ with $a \neq 0$.

(a) Factor out a and complete the square to show

$$ax^2 + bx + c = a\left[\left(x + \frac{b}{2a}\right)^2 + \frac{4ac - b^2}{4a^2}\right].$$

(b) Show that the roots of ax^2+bx+c are given by

$$\frac{-b\pm\sqrt{b^2-4ac}}{2a}.$$

N.B.: $\Delta=b^2-4ac$ is called the *discriminant* and a square root for Δ can be found by writing Δ in polar form as $re^{i\theta}$, $r,\theta\in\mathbb{R}$ with $r\geq 0$, and using $\sqrt{r}e^{i\frac{\theta}{2}}$.

(c) If $f\in\mathbb{R}[x]$, show f has no real roots if and only if $\Delta<0$.

EXERCISE 3.125 (Tartaglia's Formula). Let $f\in\mathbb{C}[x]$ with $\deg f=3$. This exercise gives a formula for the roots of f.

(a) Show that we may assume f is of the form $f=x^3+bx^2+cx+d$.

(b) Show that substituting

$$x=y-\frac{b}{3}$$

changes f to $\widetilde{f}=y^3+py+q\in\mathbb{C}[y]$ where

$$p=c-\frac{1}{3}b^2 \quad\text{and}\quad q=d-\frac{1}{3}bc+\frac{2}{27}b^3.$$

The polynomial $\widetilde{f}$ is called the *depressed cubic.*

(c) If we find $u,v\in\mathbb{C}$ so that $u^3-v^3=q$ and $uv=\frac{1}{3}p$, show $y=v-u$ is a root of $\widetilde{f}$. *Hint:* Substitute in and use the binomial formula to show

$$(v-u)^3+3uv(v-u)+u^3-v^3=0.$$

(d) Show that solving (c) reduces to solving $27u^6-27qu^3-p^3=0$ with $v=\frac{p}{3u}$ or $27v^6+27qv^3-p^3=0$ with $u=\frac{p}{3v}$ when $p\neq 0$.

(e) Show that (c) requires $u^3=\frac{q}{2}+\sqrt{\frac{q^2}{4}+\frac{p^3}{27}}$ and $v^3=-\frac{q}{2}+\sqrt{\frac{q^2}{4}+\frac{p^3}{27}}$. *Note:* Of course there are generically two choices for the square root of $\frac{q^2}{4}+\frac{p^3}{27}$. Write and fix $\sqrt{\frac{q^2}{4}+\frac{p^3}{27}}$ for one of the choices. Either choice, used consistently, will give the same set of roots for f in the end.

(f) Write $\omega=\frac{-1+i\sqrt{3}}{2}$. Show that $\{1,\omega,\omega^2=\frac{-1-i\sqrt{3}}{2}\}$ are the three cube roots of 1 in $\mathbb{C}$.

(g) There are generically three cube roots of $-\frac{q}{2}+\sqrt{\frac{q^2}{4}+\frac{p^3}{27}}$. Write and fix $v=\sqrt[3]{-\frac{q}{2}+\sqrt{\frac{q^2}{4}+\frac{p^3}{27}}}$ for one of the choices. Bearing in mind that $u^3v^3=\frac{p^3}{27}$ from part (e), write and fix $u=\sqrt[3]{\frac{q}{2}+\sqrt{\frac{q^2}{4}+\frac{p^3}{27}}}$ for the choice of a cube root of $\frac{q}{2}+\sqrt{\frac{q^2}{4}+\frac{p^3}{27}}$ satisfying $uv=\frac{p}{3}$. Show that the roots of $\widetilde{f}$ are given by

$$y_1=\sqrt[3]{-\frac{q}{2}+\sqrt{\frac{q^2}{4}+\frac{p^3}{27}}}-\sqrt[3]{\frac{q}{2}+\sqrt{\frac{q^2}{4}+\frac{p^3}{27}}},$$

$$y_2=\omega\sqrt[3]{-\frac{q}{2}+\sqrt{\frac{q^2}{4}+\frac{p^3}{27}}}-\omega^2\sqrt[3]{\frac{q}{2}+\sqrt{\frac{q^2}{4}+\frac{p^3}{27}}},$$

$$y_3=\omega^2\sqrt[3]{-\frac{q}{2}+\sqrt{\frac{q^2}{4}+\frac{p^3}{27}}}-\omega\sqrt[3]{\frac{q}{2}+\sqrt{\frac{q^2}{4}+\frac{p^3}{27}}}$$

and that the roots of f are given by $x_i = y_i - \frac{b}{3}$. The expression $\Delta = \frac{q^2}{4} + \frac{p^3}{27}$ is called the *discriminant.*

(h) Use this technique to find the roots of $x^3 + 2x^2 + 3x + 4$.

EXERCISE 3.126 (Ferrari's Formula). Let $f \in \mathbb{C}[x]$ with $\deg f = 4$. This exercise gives a formula for the roots of f. See Exercise 3.127 for an alternate method.

(a) Show that we may assume f is of the form $f = x^4 + bx^3 + cx^2 + dx + e$ and show that the substitution $x = y - \frac{a}{4}$ transforms f into $\widetilde{f} = y^4 + py^2 + qy + r$, called the *depressed quartic*, where $p = -\frac{3b^2}{8} + c$, $q = \frac{b^3}{8} - \frac{bc}{2} + d$, and $r = -\frac{3b^4}{256} + \frac{b^2c}{16} - \frac{bd}{4} + e$.

(b) Show $\widetilde{f}(y) = 0$ if and only if $(y^2 + p)^2 + qy + r = py^2 + p^2$ if and only if $(y^2 + p + z)^2 = (p + 2z)y^2 - qy + (z^2 + 2zp + p^2 - r)$ for any fixed $z \in \mathbb{C}$.

(c) For $\beta \neq 0$, show $(p + 2z)y^2 - qy + (z^2 + 2zp + p^2 - r)$ is a perfect square if and only if $2z^3 + 5pz^2 + (4p^2 - 2r)z + (p^3 - pr - \frac{q^2}{4}) = 0$ and that $z_0 = -\frac{5}{6}p + \sqrt[3]{-\frac{Q}{2} + \sqrt{\frac{Q^2}{4} + \frac{P^3}{27}}} - \sqrt[3]{\frac{Q}{2} + \sqrt{\frac{Q^2}{4} + \frac{P^3}{27}}}$ is a solution where $P = -\frac{p^2}{12} - r$ and $Q = -\frac{p^3}{108} + \frac{pr}{3} - \frac{q^2}{8}$ (cf. Exercise 3.125).

(d) Show that roots of $\widetilde{f}$ are solutions to

$$(y^2 + p + z_0)^2 = \left(\left(\sqrt{p + 2z_0}\right) y - \frac{q}{2\sqrt{p + 2z_0}}\right)^2, \qquad \beta \neq 0.$$

(e) Show that the roots of f are

$$x = -\frac{b}{4} + \frac{\varepsilon_1\sqrt{p + 2z_0} + \varepsilon_2\sqrt{-\left(3p + 2z_0 + \varepsilon_1\frac{2q}{\sqrt{p+2z_0}}\right)}}{2}$$

for $\varepsilon_1, \varepsilon_2 \in \{\pm 1\}$, $\beta \neq 0$.

(f) If $\beta = 0$, show that the roots of f are

$$x = -\frac{b}{4} + \varepsilon_1\sqrt{\frac{-p + \varepsilon_1\sqrt{p^2 - 4r}}{2}}.$$

(g) Use this technique to find the roots of $x^4 + 3x^3 + 10x^2 + 9x + 7$.

EXERCISE 3.127. This exercise gives another method for solving the quintic equation. Let $f \in \mathbb{C}[x]$ with $\deg f = 4$.

(a) Show that we may reduce to the case of $f = x^4 + cx^2 + dx + e$ (cf. Exercise 3.126, part (a)).

(b) Attempt to find $p, q, r, s \in \mathbb{C}$ so that $f = (x^2 + px + q)(x^2 + rx + s)$. If this is done, the quadratic equation finishes the job. Show p, q, r, s must satisfy

$$r = -p, \quad c + p^2 = s + q, \quad \frac{d}{p} = s - q, \quad \text{and} \quad e = sq$$

when $d \neq 0$. N.B.: If $d = 0$, solve the equation by setting $X = x^2$.

(c) Let $P = p^2$ and eliminate s and q from (b) to get $P^3 + 2cP^2 + (x^2 - 4e)P - d^2 = 0$. This can be solved by Exercise 3.125. Choose p to be a square root of P and show that

$$r = -p, \quad s = \frac{c + P + \frac{d}{p}}{2}, \quad q = \frac{c + P - \frac{d}{p}}{2}$$

can be used to factor the quintic into two quadratics. *Hint:* Start with $\left(c+p^2\right)^2 - \left(\frac{d}{p}\right)^2$.

(d) Use this technique to find the roots of $x^4+3x^3+10x^2+11x+15$.

EXERCISE 3.128 (Descartes' Rule of Signs). Let $f = \sum_k a_k x^k \in \mathbb{R}[x]$. A *sign change* in the coefficients of f occurs when, ignoring the zero coefficients, two successive coefficients have opposite sign. More precisely, a sign change occurs at $k_1 \in \mathbb{Z}_{\geq 0}$ when there is $k_1 < k_2 \in \mathbb{N}$ so that $a_{k_1}a_{k_2} < 0$ (i.e., one is negative and the other positive) and $a_j = 0$ for $k_1 < j < k_2$. Let S_f be the number of sign changes in f and let P_f be the number of real positive roots of f counted with multiplicity. This exercise shows $P_f \leq S_f$ with $P_f \equiv S_f \bmod(2)$. N.B.: To get information on the number of negative real roots, apply the same result to the polynomial $f(-x)$.

(a) Let $g = \sum_{k=k_0}^n b_k x^k \in \mathbb{R}[x]$ with $b_{k_0}, b_n \neq 0$. Show

$$S_g \equiv S_{b_{k_0}x^{k_0}+b_n x^n} \bmod(2).$$

Hint: If $r_0, \ldots, r_m \in \mathbb{R}$ are nonzero with S sign changes, show $(-1)^S = \prod_{j=0}^{m-1} \frac{r_j}{r_{j+1}} = \frac{r_0}{r_m}$.

(b) If g has no positive real roots, show $S_g \equiv 0$. *Hint:* Factor out x^{k_0} and use the Intermediate Value Theorem to show $S_g \not\equiv 1$.

(c) Let $r \in \mathbb{R}^+$ and write $h = (x-r)g = \sum_{k=k_0}^{n+1} c_k x^k$. Show (i) $\operatorname{sgn} c_{k_0} = -\operatorname{sgn} b_{k_0}$, (ii) if g has a sign change at $k-1$ for $k > k_0$, then $\operatorname{sgn} c_k = \operatorname{sgn} b_{k-1}$, and (iii) $\operatorname{sgn} c_{n+1} = \operatorname{sgn} b_n$.

(d) Show $S_h \geq S_g + 1$ with $S_h \equiv S_g + 1 \bmod(2)$.

(e) Induct on the number of positive roots to show $P_f \leq S_f$ and $P_f \equiv S_f \bmod(2)$.

(f) If all coefficients of f are positive, show there are no positive real roots.

(g) If there is exactly one sign change in f, show there is exactly one positive real root.

(h) Consider $x^{12}+3x^{10}-145x^9+13x^7+x^6-16x^4-6x^3+2x+3$. Find the number of possible (i) real positive roots, (ii) real negative roots, and (iii) purely complex roots.

EXERCISE 3.129 (Fundamental Theorem of Algebra). Let $f \in \mathbb{C}[x]$ be nonconstant. This exercise shows f has a root in $\mathbb{C}$ and requires analysis.

(a) Suppose $z_0 \in \mathbb{C}$ satisfies $f(z_0) \neq 0$. Without changing the possible existence of a root of f, show that we may modify f and assume that $z_0 = 0$, $f(z_0) = 1$, and that f satisfies

$$f = 1 - x^k + gx^{k+1}$$

for some $k \in \mathbb{N}$ and $g \in \mathbb{C}[x]$. *Hint:* Try $x \to x + z_0$ and multiply by $f(0)^{-1}$ to get $f = 1 + ax^k + \cdots$. Then try $x \to \sqrt[k]{-a^{-1}}x$.

(b) For $x > 0$, show $|f(x)| \leq 1 - x^k\left(1 - x\,|g(x)|\right)$. Conclude $|f(x)| < |f(0)|$ for sufficiently small x so that $|f(0)|$ is not a minimum of $|f(z)|$.

(c) Show that the only possible minimum value of $|f(z)|$ is 0.

(d) Use induction on degree to show $|f(z)| \to \infty$ as $|z| \to \infty$.

(e) Show that the greatest lower bound for $|f(z)|$ on $\mathbb{C}$ is the same as the greatest lower bound for $|f(z)|$ on some sufficiently large disk.

(f) Use continuity of $|f(z)|$ and the fact that a disk is compact to show $|f(z)|$ achieves its minimum value.

4. Polynomials and Irreducibility

4.1. Irreducibility. Just as every integer factors into a product of primes (Theorem 1.14), polynomials over a field will also factor into "indivisible" pieces. In fact, the factorization theories are strikingly similar. Essentially the only differences in passing from the integers to a polynomial ring over a field, $F[x]$, is that we use the word *irreducible* instead of the word *prime* and that $F[x]$ can have more units than $\mathbb{Z}$, namely $F[x]^\times = F^\times$ while $\mathbb{Z}^\times = \{\pm 1\}$ (Theorem 3.25). Below we give a general definition suitable for any integral domain and so, in particular, for a polynomial ring over a field. It is instructive to compare the following to Definition 1.5 and to recall that $p \in \mathbb{Z}$, $p \neq \pm 1, 0$, is prime if its only divisors are $\pm 1, \pm p$ (Definition 1.5).

DEFINITION 3.35. Let D be an integral domain with $f, g \in D$.

(1) For $g \neq 0$, we say g *divides* f or that g is a *divisor* of f, written $g \mid f$, if

$$f = gh$$

for some $h \in D$.

(2) For $f \notin D^\times \cup \{0\}$, we say f is *irreducible* if its only divisors are $\{u, uf \mid u \in D^\times\}$. Otherwise we say f is *reducible.*

An element of the form uf, $u \in D^\times$, is called an *associate* of f. Thus we can say that $f \in D$ is irreducible if its only divisors are units and associates of f. In general, determining irreducibility can be an extraordinarily hard problem. We will concentrate most of our effort on polynomial rings over a field.

4.2. Polynomials over a Field. As already noted, determining when a polynomial is irreducible can be very difficult. However, we do have one general theorem.

THEOREM 3.36. *Let F be a field and let $f \in F[x]$.*

(1) *For $f \neq 0$, f is reducible if and only if it is the product of two polynomials of lower degree.*

(2) *If $\deg f = 1$, then f is irreducible.*

(3) *If $\deg f = 2$ or 3, then f is irreducible if and only if f has no roots in F.*

PROOF. For part (1), suppose first that $f \neq 0$ is reducible. Then we can write $f = gh$ with $g, h \in F[x]$ and $g \notin \{u, uf \mid u \in F^\times\}$. Then $\deg f = \deg g + \deg h$ and $\deg g \geq 1$ (since $g \notin F$) so that $\deg h < \deg f$. If we had $\deg g = \deg f$, then we would have $\deg h = 0$ so that $h \in F^\times$. This would imply that $g = h^{-1}f$ and would contradict the fact that $g \notin F^\times f$. Thus we also have $\deg g < \deg f$ as needed.

Conversely, suppose $f = gh$ with $\deg g, \deg h < \deg f$. From the first inequality we get $g \notin F^\times f$. From the second inequality coupled with $\deg f = \deg g + \deg h$ we get $\deg g \geq 1$ so that $g \notin F^\times$. Thus g is a divisor not in $\{u, uf \mid u \in F^\times\}$.

Part (2) follows immediately from part (1) and the equality $\deg f = \deg g + \deg h$ whenever $f = gh$. For part (3), observe that it is possible to solve $\deg f = \deg g + \deg h$ with $\deg g, \deg h < \deg f$ if and only if either $\deg g = 1$ or $\deg h = 1$. The statement of the theorem is finished by appealing to Theorem 3.30. ☐

For example, it is trivial to check that $x^2 + x + [1] \in \mathbb{Z}_2[x]$ has no roots in $\mathbb{Z}_2$ by plugging in $[0]$ and then $[1]$. Thus $x^2 + x + [1]$ is irreducible in $\mathbb{Z}_2[x]$. In a similar vein, consider $x^2 + x + 1 \in \mathbb{Q}[x]$. By the Rational Root Test, the only possible rational roots are ± 1. By inspection, neither is a root. Thus x^2+x+1 is irreducible in $\mathbb{Q}[x]$. On the other hand if we consider $x^2+x+1 \in \mathbb{C}[x]$, then the quadratic equation shows there are two complex roots and so x^2+x+1 is reducible in $\mathbb{C}[x]$ (cf. Theorem 3.33).

Similar techniques can be used for polynomials of degree three. Unfortunately, there is no simple analogous condition for polynomials of degree 4 or higher. Certainly the existence of a root forces reducibility (Theorem 3.30), but the converse can easily fail. For instance, $x^4+2x^2+1=(x^2+1)^2 \in \mathbb{Q}[x]$ clearly has no rational roots, yet it is still reducible.

For finite fields, there are only a finite number of polynomials of each degree. Therefore it is possible to list all possible products of polynomials of degree n or less. Using this list, one can easily write down all the irreducible polynomials of degree $n+1$ or less. Inductively, this at least gives an algorithm for determining if any particular polynomial is irreducible.

Of course the above procedure is not of much use in establishing general theorems and is of little use in infinite fields. However, at least with $\mathbb{Q}$, $\mathbb{C}$, and $\mathbb{R}$, there are a number of very useful tricks.

4.2.1. *Tricks for* $\mathbb{Q}$. It will turn out that irreducibility questions for polynomials in $\mathbb{Q}[x]$ can be reduced to questions about polynomials with integer coefficients. As a result, the following definition and lemma will be useful.

DEFINITION 3.37. For $f=\sum_k a_k x^k \in \mathbb{Z}[x]$ and $n \in \mathbb{N}$, define $\overline{f} \in \mathbb{Z}_n[x]$ by $\overline{f}=\sum_k [a_k]\, x^k$. The map from $\mathbb{Z}[x] \to \mathbb{Z}_n[x]$ given by mapping $f \to \overline{f}$ is called the *reduction homomorphism.*

Of course, the map $f \to \overline{f}$ really is a homomorphism. This fact follows immediately from the observation that the map $\mathbb{Z} \to \mathbb{Z}_n$ given by $a \to [a]$, $a \in \mathbb{Z}$, is a homomorphism (Exercise 3.150).

LEMMA 3.38 (Gauss's Lemma). **(1)** *Suppose $f=gh$ for $f,g,h \in \mathbb{Z}[x]$ and that p is a prime dividing every coefficient of f. Then p divides every coefficient of g or of h.*

(2) *If $f \in \mathbb{Z}[x]$ and $f=g_1h_1$ for $g_1,h_1 \in \mathbb{Q}[x]$, then there exist $g_2,h_2 \in \mathbb{Z}[x]$ with $g_2=rg_1$ and $h_2=sh_1$ for some $r,s \in \mathbb{Q}^\times$ so that $f=g_2h_2$.*

PROOF. For part (1), use the reduction homomorphism to $\mathbb{Z}_p[x]$. Of course $0=\overline{f}$ as p divides every coefficient of f. Thus $0=\overline{f}=\overline{g}\overline{h}$ in $\mathbb{Z}_p[x]$. Since $\mathbb{Z}_p[x]$ is an integral domain, either $\overline{g}=0$ or $\overline{f}=0$, which is equivalent to the statement of part (1).

For part (2), choose $n,m \in \mathbb{N}$ so $ng_1, mh_1 \in \mathbb{Z}[x]$ (cf. Theorem 3.32). Then $nmf=(ng_1)(mh_1)$. If $nm=\pm 1$, multiply by ± 1 and we are done. Otherwise, choose a prime p with $p \mid nm$. By part (1), p divides the coefficients of, say, ng_1 so that $ng_1=pg_2$ for some $g_2 \in \mathbb{Z}[x]$. Thus $\frac{nm}{p}f=(g_2)\,(mh_1)$. If $\frac{nm}{p}=\pm 1$, we are again done. If not, inductively repeat until ± 1 is achieved. □

Reducing $\mathbb{Q}[x]$ questions of irreducibility to $\mathbb{Z}[x]$ questions of irreducibility has one fly in the ointment. Namely, $\mathbb{Z}[x]$ has lots of irreducible constant polynomials while $\mathbb{Q}[x]$ has none. For instance, 5 is irreducible in $\mathbb{Z}[x]$ since it is prime. However, 5 is a unit in $\mathbb{Q}$ and so it is not even eligible to be irreducible. To take care of this difficulty, we basically ignore factorization in $\mathbb{Z}[x]$ where one of the terms has degree 0.

DEFINITION 3.39. If $f \in \mathbb{Z}[x]$ is nonconstant, we say f has a *proper factorization* if $f = gh$ for $g, h \in \mathbb{Z}[x]$ with $\deg g, \deg h < \deg f$.

COROLLARY 3.40 (Reduction to $\mathbb{Z}[x]$). *Let $f \in \mathbb{Q}[x]$ be nonconstant and choose nonzero $c \in \mathbb{Z}$ so that $\widetilde{f} = cf \in \mathbb{Z}[x]$.*

(1) *f is irreducible in $\mathbb{Q}[x]$ if and only if $\widetilde{f}$ is irreducible in $\mathbb{Q}[x]$.*

(2) *$\widetilde{f}$ is irreducible in $\mathbb{Q}[x]$ if and only if $\widetilde{f}$ has no proper factorization in $\mathbb{Z}[x]$.*

PROOF. Since the set of divisors of f and $\widetilde{f}$ coincide, part (1) follows. For part (2), look at the contrapositive. Clearly any proper factorization of $\widetilde{f}$ in $\mathbb{Z}[x]$ shows $\widetilde{f}$ is reducible in $\mathbb{Q}[x]$ by Theorem 3.36. Conversely, if $\widetilde{f} = gh$ for $g, h \in \mathbb{Q}[x]$ with $\deg g, \deg h < \deg f$, Gauss's Lemma shows $\widetilde{f}$ has a proper factorization in $\mathbb{Z}[x]$. □

Now that we have reduced to $\mathbb{Z}[x]$, there are essentially two good general purpose theorems useful for detecting irreducibility in $\mathbb{Q}[x]$.

THEOREM 3.41 (Irreduciblity Aids in $\mathbb{Q}[x]$). *Let $f = \sum_{k=0}^{n} a_k x^k \in \mathbb{Z}[x]$ be nonconstant with $a_n \neq 0$.*

(1) Eisenstein's Criterion: *If there exists a prime p so that* (i) $p \mid a_i$, $0 \leq i \leq n-1$, (ii) $p^2 \nmid a_0$, *and* (iii) $p \nmid a_n$, *then f is irreducible in $\mathbb{Q}[x]$.*

(2) Modular Reduction: *If p is a positive prime so that* (i) $p \nmid a_n$ *and* (ii) $\overline{f}$ *is irreducible in $\mathbb{Z}_p[x]$, then f is irreducible in $\mathbb{Q}[x]$.*

PROOF. Start with part (1). If f is reducible, write a proper factorization $f = gh$ for $g, h \in \mathbb{Z}[x]$ with $g = \sum_k b_k x^k$ and $h = \sum_k c_k x^k$. Since $p \mid a_0$ and $a_0 = b_0 c_0$, we see that, say, $p \mid b_0$. Since $p^2 \nmid a_0$, we also see that $p \nmid c_0$. Continuing on, as $p \mid a_1, b_0$ and $a_1 = b_0 c_1 + b_1 c_0$, it follows that $p \mid b_1 c_0$. Since $p \nmid c_0$, $p \mid b_1$. In general for $0 \leq i \leq n-1$, we have $p \mid a_i$ with

$$a_i = (b_0 c_i + b_1 c_{i-1} + \cdots + b_{i-1} c_1) + b_i c_0.$$

Applying induction on i, we may assume $p \mid b_0, b_1, \ldots, b_{i-1}$. We therefore see that $p \mid b_i c_0$ and thus $p \mid b_i$. Since $\deg g \leq n-1$, it follows that p divides every coefficient of g. This implies that p divides every coefficient of f, which is a contradiction.

For part (2), suppose f is reducible and write $f = gh$ as above. Since $[a_n] \neq [0]$, $\deg \overline{f} = \deg f$. In particular, $\overline{f} = \overline{g}\overline{h}$ is therefore a factorization of $\overline{f}$ by polynomials of lower degree since $\deg \overline{g} \leq \deg g$ and $\deg \overline{h} \leq \deg h$. This shows $\overline{f} \in \mathbb{Z}_p[x]$ is reducible, which is a contradiction. □

Let us look at some examples. When applicable, Eisenstein's Criterion is a powerful and easy test. Therefore we usually try to use it first. For example, the polynomial

$$6 + 8x + 72x^{29} + 18x^{99} + 7x^{100}$$

is seen to be irreducible in $\mathbb{Q}[x]$ by using $p = 2$ and

$$11 + x^{543}$$

is seen to be irreducible in $\mathbb{Q}[x]$ by using $p = 11$.

Of course there are plenty of irreducible polynomials to which Eisenstein's Criterion does not apply. In that case, we usually turn to modular reduction. For example, consider the polynomial

$$f = 1 + 2x^2 + 3x^3 + 4x^4 + 5x^5.$$

Look at reduction to $\mathbb{Z}_2$ so that $\overline{f} = [1]+x^3+x^5$. We hope to see that $\overline{f}$ is irreducible in $\mathbb{Z}_2[x]$. Since $\overline{f}([0]) = \overline{f}([1]) \neq [0]$, $\overline{f}$ has no roots in $\mathbb{Z}_2$ and so no linear terms factor into $\overline{f}$. Thus the only way $\overline{f}$ can be reducible is if it is a product of a degree 2 polynomial times a degree 3 polynomial. Neither of these two polynomials can have a root in $\mathbb{Z}_2$; i.e., both are irreducible by Theorem 3.36. However, by checking roots, it is trivial to see that the only irreducible polynomial of degree 2 in $\mathbb{Z}_2[x]$ is $x^2+x+[1]$ (the other degree 2 polynomials are x^2, x^2+x, and $x^2+[1]$; cf. Exercise 3.136). Using long division, we check that $\overline{f} = (x^2+x+[1])(x^3+x^2+x)+(x+[1])$ so that $(x^2+x+[1]) \nmid \overline{f}$. Hence $\overline{f}$ is irreducible in $\mathbb{Z}_2[x]$ and thus f is irreducible in $\mathbb{Q}[x]$.

As the reader can see, using modular reduction can work very well, but it takes effort. As a cautionary note, the converse of the modular reduction criterion is false. It can easily happen that an irreducible polynomial over $\mathbb{Q}$ produces a reducible polynomial over $\mathbb{Z}_p$ for some prime p. When this happens, we cannot say anything about the irreducibility/reducibility of the original polynomial. In practice, the only option is to try another prime and hope to get lucky. Unfortunately, we are not always guaranteed success by trying other primes since it can happen that an irreducible polynomial over $\mathbb{Q}$ yields a reducible polynomial for every $\mathbb{Z}_p$. For example, it is a fact that x^4+1 is such a polynomial (Exercise 4.73).

4.2.2. *Irreducibles in $\mathbb{C}$ and $\mathbb{R}$.* As we have seen, irreducibility in $\mathbb{Q}$ can be rather subtle and difficult. None of these problems persist when using $\mathbb{R}$ or $\mathbb{C}$.

THEOREM 3.42 ($\mathbb{C}$). *Let $f \in \mathbb{C}[x]$. Then f is irreducible if and only if* $\deg f = 1$.

PROOF. This follows immediately from the Fundamental Theorem of Algebra. □

THEOREM 3.43 ($\mathbb{R}$). *Let $f \in \mathbb{R}[x]$. Then f is irreducible if and only if either* (i) $\deg f = 1$ *or* (ii) $\deg f = 2$ *with $f = ax^2+bx+c$ satisfying $b^2-4ac<0$, i.e., f has no real roots.*

PROOF. This follows immediately from Corollary 3.34, Theorem 3.36, and the quadratic equation (Exercise 3.124). □

4.3. Exercises 3.130–3.150.

EXERCISE 3.130. Let D be an integral domain. Show that the condition of being associate is an equivalence relation.

EXERCISE 3.131. Let D be an integral domain with nonzero $a, b \in D$. Show $a \mid b$ and $b \mid a$ if and only if they are associates.

EXERCISE 3.132. Show the following polynomials are irreducible by checking for roots:

(a) $x^2-5 \in \mathbb{Q}[x]$.
(b) $2x^3+x^2+2x+5 \in \mathbb{Q}[x]$.
(c) $x^3+x+[1] \in \mathbb{Z}_2[x]$.
(d) $[2]\,x^3+x^2+[2]\,x+[2] \in \mathbb{Z}_3[x]$.
(e) $x^3+x+[1] \in \mathbb{Z}_5[x]$.
(f) $2x^2+x+1 \in \mathbb{R}[x]$.

EXERCISE 3.133. For each positive prime $p \leq 19$, determine whether $x^2+x+1 \in \mathbb{Z}_p[x]$ is irreducible.

EXERCISE 3.134. Let p be a positive prime and let $a \in \mathbb{Z}_p$. Show $x^p - a \in \mathbb{Z}_p[x]$ is reducible (cf. Exercise 3.104).

EXERCISE 3.135. Show $x^4+2x^3+3x^2+2x+1$ has no roots in $\mathbb{Q}$ but is reducible.

EXERCISE 3.136. For each given degree, show that the irreducible polynomials in $\mathbb{Z}_2[x]$ are as listed:

(a) degree 2: $1+x+x^2$,
(b) degree 3: $1+x+x^3$ and $1+x^2+x^3$,
(c) degree 4: $1+x+x^4$, $1+x+x^2+x^3+x^4$, and $1+x^3+x^4$,
(d) degree 5: $1+x^2+x^5$, $1+x+x^2+x^3+x^5$, $1+x^3+x^5$, $1+x+x^3+x^4+x^5$, $1+x^2+x^3+x^4+x^5$, and $1+x+x^2+x^4+x^5$.

EXERCISE 3.137. For each given degree, show that the irreducible polynomials in $\mathbb{Z}_3[x]$, up to multiplication by U_3, are as listed:

(a) degree 2: $1+x^2$, $1-x-x^2$, and $1+x-x^2$,
(b) degree 3: $1-x+x^2+x^3$, $1+x-x^2+x^3$, $1-x^2+x^3$, $1-x+x^3$, $1+x+x^2-x^3$, $1-x-x^2-x^3$, $1-x^2-x^3$, and $1+x-x^3$.

EXERCISE 3.138. Show that the irreducible polynomials of degree 2 in $\mathbb{Z}_5[x]$, up to multiplication by U_5, are $1+x+x^2$, $1+4x+x^2$, $1+x+2x^2$, $1+4x+2x^2$, $1+2x^2$, $1+2x+3x^2$, $1+3x+3x^2$, $1+x^3$, $1+2x+4x^2$, and $1+3x+4x^2$.

EXERCISE 3.139. Show $f = x^4+1 \in \mathbb{Q}[x]$ is irreducible. *Hint:* Reduce to trying to solve $f = (x^2+ax+b)(x^2+cx+d)$ in $\mathbb{Z}[x]$ and showing $a \notin \mathbb{Z}$ to get a contradiction.

EXERCISE 3.140. Use Eisenstein's Criterion to show the following polynomials in $\mathbb{Q}[x]$ are irreducible:

(a) $14+7x+x^5$,
(b) $6+18x^8+12x^9+9x^{10}+4x^{13}$,
(c) $12+30x^4+12x^5+5x^7$,
(d) x^n+pm for $n \in \mathbb{N}$, p a prime, and $m \in \mathbb{Z}\backslash\{0\}$ with $(m,p)=1$.

EXERCISE 3.141. Use modular reduction to show the following polynomials in $\mathbb{Q}[x]$ are irreducible:

(a) $5+3x+x^2$,
(b) $3+7x^2+x^3$,
(c) $3-x+4x^2+2x^3+x^4$,
(d) $9+2x+3x^2+6x^4+x^5$,
(e) $5+x+x^2+x^3$.

EXERCISE 3.142 (Linear Change of Variables). Let F be a field with $u \in F^\times$ and $b \in F$. If $f(x) = \sum_k a_k x^k \in F[x]$, define $\varphi_{u,b} : F[x] \to F[x]$ by $\varphi_{u,b}(f) = \sum_k a_k (ux+b)^k$. Alternately, it is most common to write $f(ux+b)$ for $\varphi_{u,b}(f)$.

(a) Show the inverse to $\varphi_{u,b}$ is $\varphi_{u^{-1},-u^{-1}b}$.
(b) Show $\varphi_{u,b}$ is an isomorphism that preserves degree.
(c) Conclude f is irreducible if and only if $\varphi_{u,b}(f)$ is irreducible.
(d) Consider $f(x) = x^4+1 \in \mathbb{Q}[x]$. Note that Eisenstein's Criterion does not apply to f. Nevertheless, show $f(x+1) = 2+4x+6x^2+4x^3+x^4$ and conclude f is irreducible (cf. Exercise 3.139).

(e) Show $-5 + 11x - 3x^2 - x^3 + x^4$ is irreducible in $\mathbb{Q}[x]$.

EXERCISE 3.143. Let D be an integral domain and let $p \in D$. Show p is irreducible if and only if whenever $(p) \subseteq (q)$ for some $q \in D \backslash D^\times$, then $(p) = (q)$.

EXERCISE 3.144. Let $F \subseteq K$ be fields with $f \in F[x]$.

(a) If f is irreducible when viewed as an element of $K[x]$, show it is irreducible as an element of $F[x]$.

(b) Show the converse to part (a) can be false.

EXERCISE 3.145 (Cyclotomic Polynomials). Let p be a positive prime. The pth *cyclotomic polynomial* is

$$\Phi_p(x) = 1 + x + x^2 + \cdots + x^{p-1}.$$

This exercise shows Φ_p is irreducible in $\mathbb{Q}[x]$.

(a) Show $x^p - 1 = (x-1)\Phi_p$.

(b) Show $\Phi_p(x)$ is irreducible if and only if $\Phi_p(x+1)$ is irreducible (cf. Exercise 3.142).

(c) Show $x\Phi_p(x+1) = (x+1)^p - 1$ and use the Binomial Theorem to conclude that

$$\Phi_p(x+1) = \sum_{k=1}^{p} \binom{p}{k} x^{k-1} = p + \frac{p(p-1)}{2}x + \cdots + px^{p-2} + x^{p-1}.$$

(d) Show $\Phi_p(x)$ is irreducible (cf. Exercise 3.43 part (a)).

(e) If p is odd, show $1 - x + x^2 - x^3 + \cdots + x^{p-1}$ is irreducible in $\mathbb{Q}[x]$.

(f) Show $1 + x + x^2 + x^3 \in \mathbb{Q}[x]$ is not irreducible (cf. Theorem 4.58).

EXERCISE 3.146. Let p be a positive prime and let $n \in \mathbb{N}$. The $p^{n^{\text{th}}}$ *cyclotomic polynomial* (cf. Exercise 3.145) is

$$\Phi_{p^n}(x) = 1 + x^{p^{n-1}} + x^{2p^{n-1}} + \cdots + x^{(p-1)p^{n-1}}.$$

(a) Show $x^{p^n} - 1 = (x^{p^{n-1}} - 1)\Phi_{p^n}$.

(b) Show $\Phi_{p^n}(x)$ is irreducible in $\mathbb{Q}[x]$.

EXERCISE 3.147. Let F be a field. Show that nonzero $f \in F[x]$ satisfies $f \mid g$ for all $g \in F[x]$ if and only if $f \in F^\times$.

EXERCISE 3.148. Let p be a positive prime. Show there are exactly $\frac{p(p-1)}{2}$ monic irreducible polynomials of degree 2 in $\mathbb{Z}_p[x]$. *Hint:* Count the total number of monic quadratics and the number of reducible quadratics using roots.

EXERCISE 3.149. Find an $f \in \mathbb{Z}[x]$ that is irreducible over $\mathbb{Q}$ but whose reduction $\bmod p$ is reducible over the first three positive primes.

EXERCISE 3.150. Verify that the reduction homomorphism is indeed a homomorphism.

5. Factorization

Let D be an integral domain. In this section we will frequently use the *cancellation law* that says if $ab = ac$ for $a, b, c \in D$ with $a \neq 0$, then $b = c$. To see this, write $ab = ac$ as $a(b - c) = 0$ and use the definition of an integral domain (Exercise 3.4).

Recall from §4.1 the notions of a divisor and an irreducible element in D (applied primarily to polynomial rings over fields). The ideas were closely modeled on the definitions of a divisor and a prime in $\mathbb{Z}$. In fact, many of the theorems about divisors from $\mathbb{Z}$ carry over with little change. For instance, the proof of the following result is nearly identical to that of Lemma 1.6 (Exercise 3.151).

LEMMA 3.44. *Let D be an integral domain with $a, b, c \in D$, $a \neq 0$:*
(1) *If $a \mid b$ and $b \mid c$ (with $b \neq 0$), then $a \mid c$.*
(2) *If $a \mid b$ and $a \mid c$, then $a \mid (bx + cy)$ for all $x, y \in D$.*
(3) *If $a \mid b$ and $b \mid a$ (with $b \neq 0$), then $a = ub$ for some $u \in D^\times$.*

However, we run into some unexpected problems with generalizing the Fundamental Theorem of Arithmetic. For example, consider the integral domain

$$\mathbb{Z}[\sqrt{-5}] = \{a + ib\sqrt{5} \mid a, b \in \mathbb{Z}\}$$

which is a subring of $\mathbb{C}$. Let $z = a + ib\sqrt{5}$ and define the *norm* of z as $N(z) = a^2 + 5b^2 \in \mathbb{Z}_{\geq 0}$. Since $N(z) = |z|^2 = z\overline{z}$, it follows that $N(z_1 z_2) = N(z_1)N(z_2)$. If z is a unit, then $N(z)N(z^{-1}) = N(1) = 1$ so that $N(z) = 1$. Of course, if $N(z) = 1$, then $z^{-1} = \overline{z}$. Thus the units in $\mathbb{Z}[\sqrt{-5}]$ are the elements with norm 1. From this it is easy to see that $\mathbb{Z}[\sqrt{-5}]^\times = \{\pm 1\}$.

Similarly, if z is reducible with $z = uv$, neither units, then $N(z) = N(u)N(v)$. Besides 0 and 1, the smallest possible value of N is 4. In particular, if z is reducible, then certainly $N(z) \geq 16$. Thus 2, 3, and $1 \pm i\sqrt{5}$ are all irreducible. Since simple multiplication shows

$$\begin{aligned} 6 &= 2 \cdot 3 \\ &= \left(1 + i\sqrt{5}\right)\left(1 - i\sqrt{5}\right), \end{aligned}$$

we see that 6 can be factored into irreducible products in two different (non-associate) ways! In other words, the uniqueness part of the analogue of the Fundamental Theorem of Arithmetic does not hold in $\mathbb{Z}[\sqrt{-5}]$. Nevertheless, the Fundamental Theorem of Arithmetic will generalize to certain other important rings. We take up that study now.

5.1. Unique Factorization. The following definition officially gives the generalization of the Fundamental Theorem of Arithmetic.

DEFINITION 3.45. Let D be an integral domain. We say D is a *unique factorization domain* or *UFD* if every $r \in R$ that is nonzero and not a unit satisfies the following:

(1) r is a finite product of irreducible elements; i.e., $r = \prod_{i=1}^n p_i$ with $n \in \mathbb{N}$ and $p_i \in D$ irreducible.

(2) The above decomposition is unique up to multiplication by units and reordering; i.e., if also $r = \prod_{i=1}^m q_i$ with $m \in \mathbb{N}$ and $q_i \in D$ irreducible, then $n = m$ and, after possibly reindexing the subscripts, $q_i = u_i p_i$ for some $u_i \in D^\times$.

The Fundamental Theorem of Arithmetic says that $\mathbb{Z}$ is a UFD. We will soon see that $F[x]$ is also a UFD for any field F (Theorem 3.57). On the other hand, our above discussion of $\mathbb{Z}[\sqrt{-5}]$ shows $\mathbb{Z}[\sqrt{-5}]$ is not a UFD.

We need to develop a means for checking when an integral domain is a UFD. To this end, we introduce three definitions.

Definition 3.46. Let D be an integral domain and let $p \in D$, $p \neq 0$. We say p is *prime* if (p) is a prime ideal, i.e., if when $p \mid ab$ for $a, b \in D$, then $p \mid a$ or $p \mid b$.

Of course, for $\mathbb{Z}$, the notions of irreducible and prime coincide. This is not true in general, but we do have the following.

Lemma 3.47. *Let D be an integral domain and suppose $p \in D$ is prime. Then p is irreducible.*

Proof. If $p = ab$ for some $a, b \in D$, then primeness forces, say, $p \mid a$. Since clearly $a \mid p$, Lemma 3.44 shows $a = up$ for some $u \in D^\times$. Hence $p = ab = pub$ so that $1 = ub$ and $b = u^{-1}$ is a unit. Thus p is irreducible. □

However, the converse can be false. For instance $1 + i\sqrt{5} \in \mathbb{Z}[\sqrt{-5}]$ is irreducible. Yet $1+i\sqrt{5}$ is not prime. To see this, observe that $\left(1 + i\sqrt{5}\right) \mid (2 \cdot 3)$ but $\left(1 + i\sqrt{5}\right) \nmid 2$ and $\left(1 + i\sqrt{5}\right) \nmid 3$ (this can be seen by noting that $N\left(1 + i\sqrt{5}\right) = 6$ does not divide $N(2) = 4$ or $N(3) = 9$). With that said, it turns out that the notions of irreducible and prime coincide in UFD's.

Lemma 3.48. *Let D be a UFD.*

(1) *Then $p \in D$ is irreducible if and only if it is prime.*

(2) *If distinct $p_i \in D$ are irreducible and $m_i \in \mathbb{N}$, the only divisors of the finite product $\prod_j p_j^{m_j}$ are of the form $u \prod_j p_j^{n_j}$ for $u \in D^\times$ and $n_j \in \mathbb{Z}$ with $0 \leq n_j \leq m_j$.*

Proof. For part (1), suppose $p \in D$ is irreducible and $p \mid ab$ for some $a, b \in D$. Write $ab = pc$ for some $c \in D$ and factor each element into irreducibles in D as $a = \prod_j p_j$, $b = \prod_j q_j$, and $c = \prod_j r_j$. Thus $\prod_j p_j \prod_j q_j$ and $p \prod_j r_j$ are two factorizations of ab into irreducibles. By definition, we may reindex and assume, say, that $p = p_1 u$ for some $u \in D^\times$. Thus $a = p\left(u^{-1} \prod_{j>1} p_j\right)$ so that $p \mid a$. Hence p is prime as desired.

For part (2), let $a = \prod_j p_j^{m_j}$ and suppose $b \mid a$ with $a = bc$ for some $b, c \in D$. Write $b = \prod_j q_j^{n_j}$ and $c = \prod_j q_j^{k_j}$ as products of irreducibles in D with the q_j distinct and $n_j, k_j \in \mathbb{Z}_{\geq 0}$. Then $\prod_j q_j^{n_j + k_j}$ is another factorization of a by irreducibles. By uniqueness, the irreducibles agree up to reindexing and multiplication by a unit. Thus we may assume that $q_j = p_j u_j$ for $u_j \in D^\times$ and that $n_j + k_j = m_j$. Writing $u = \prod_j u_j^{n_j}$, we see $b = u \prod_j p_j^{n_j}$ with $0 \leq n_j \leq m_j$. The converse is obvious. □

For the next definition, recall Corollary 1.12.

Definition 3.49. Let D be an integral domain with $a_i \in D$, not all zero, $1 \leq i \leq n$, $n \in \mathbb{N}$. We say $d \in D$ is a *greatest common divisor* of the a_i, written (nonuniquely) as $d = \gcd(a_1, \ldots, a_n)$, if the following hold:

(1) $d \mid a_i$, $1 \leq i \leq n$.

(2) If $c \mid a_i$, $1 \leq i \leq n$, for some $c \in D$, then $c \mid d$.

When the only greatest common divisors of $a_1, \ldots, a_n$ are units, we say $a_1, \ldots, a_n$ are *relatively prime.*

According to the above definition, clearly greatest common divisors exist in $\mathbb{Z}$ and are unique up to multiplication by ± 1. Because of this ambiguity, our definition here is slightly different than Corollary 1.12 where we always chose the *positive* greatest common divisor.

In general, greatest common divisors need not exist. For example, $\gcd(6, 2 + 2i\sqrt{5})$ fails to exist in $\mathbb{Z}[\sqrt{-5}]$. To see this, check that $N(6) = 36$ and $N(2+2i\sqrt{5}) = 24$ so that any common divisor must have norm dividing 12. Using this restriction, it is then easy to verify that the only common divisors of 6 and $2 + 2i\sqrt{5}$ are $\pm\{1, 2, 1+i\sqrt{5}\}$ and to check that none of these are divisible by all the rest (Exercise 3.163).

Nevertheless, when a greatest common divisor exists, it is unique up to multiplication by a unit (Exercise 3.153) and we will soon see that greatest common divisors always exist in UFD's. As a final cautionary note, it is worth mentioning that even when $\gcd(a, b)$ exists, it does not have to be of the form $ax + by$ for some $x, y \in D$ (Exercise 3.155) as was the case for $\mathbb{Z}$ (cf. Exercise 3.178).

The final definition is needed to show that a factorization exists.

DEFINITION 3.50. An integral domain D is said to satisfy the *ascending chain condition on principal ideals* or the *ACCP* if whenever there is $a_i \in D$, $i \in \mathbb{N}$, so that $(a_i) \subseteq (a_{i+1})$ for all i, then there exists $k \in \mathbb{N}$ so that $(a_i) = (a_{i+1})$ for $i \geq k$.

In $\mathbb{Z}$ we have $(n) \subseteq (m)$ if and only if $m \mid n$ (Exercise 3.48). Since n has only a finite number of divisors, every ascending chain of ideals containing (n) eventually stabilizes and stops growing larger. Thus $\mathbb{Z}$ satisfies the ACCP. For an example of something failing to satisfy the ACCP, start with $D = \{f \in \mathbb{Q}[x] \mid f(0) \in \mathbb{Z}\}$ and look at the ideals $(x) \subseteq (\frac{1}{2}x) \subseteq (\frac{1}{4}x) \subseteq \cdots$. Roughly speaking, when rings fail to satisfy the ACCP, it is possible to continue breaking an element into smaller and smaller reducible pieces without ever arriving at an irreducible.

With the above preliminary work done, we are now able to describe when an integral domain is really a UFD.

THEOREM 3.51. *Let D be an integral domain. The following are equivalent.*

(1) *D is a UFD.*

(2) *D satisfies the ACCP and all irreducible elements are prime.*

(3) *D satisfies the ACCP and a greatest common divisor exists for all $a_1, \ldots, a_n \in D$, not all zero.*

PROOF. Look at (1) $\Longrightarrow$ (2) first. If D is a UFD, we already know that irreducibles are prime. Thus suppose $(a_i) \subseteq (a_{i+1})$ for $a_i \in D$, $i \in \mathbb{N}$. Then $a_{i+1} \mid a_i$. To avoid trivialities, assume a_1 is neither 0 nor a unit and write $a_1 = \prod p_j^{m_{1,j}}$ as a finite product of irreducibles with p_j distinct. Since $a_i \mid a_1$, Lemma 3.48 shows $a_i = u_i \prod_j p_j^{m_{i,j}}$ with $m_{i,j} \in \mathbb{Z}$ and $0 \leq m_{i,j} \leq m_{1,j}$ and $u_i \in D^\times$. Using $a_{i+1} \mid a_i$, we also see $0 \leq m_{i+1,j} \leq m_{i,j}$. Since there are only a finite number of choices for the decreasing sequence of $m_{i,j}$'s, there exists $N \in \mathbb{N}$ so that $m_{i,j} = m_{i+1,j}$ for all $i \geq N$. Thus $a_i = (u_i u_{i+1}^{-1})a_{i+1}$ and $(a_i) = (a_{i+1})$ for $i \geq N$ as desired.

Look at (2) $\Longrightarrow$ (1) next. Let $a_0 \in D$ be nonzero and not a unit and let D_F be all elements of D that are a finite product of irreducibles. Arguing by contradiction, assume $a_0 \notin D_F$. Then a_0 is not irreducible so we can write $a_0 = a_1 b_1$ for $a_1, b_1 \in D$, neither zero nor units. If $a_1, b_1 \in D_F$, we would have $a_0 \in D_F$. Thus we must have, say, $a_1 \notin D_F$. Therefore we can write $a_1 = a_2 b_2$ for $a_2, b_2 \in D$, neither units. Applying induction, we get $a_i, b_i \in D$, neither zero nor units, for $i \in \mathbb{N}$ so that $a_i = a_{i+1} b_{i+1}$. By construction, $(a_i) \subseteq (a_{i+1})$. In fact, we must also have $(a_i) \neq (a_{i+1})$. If not, then we would have $a_{i+1} = c_i a_i$, some $c_i \in D$, which would imply $a_i = a_i c_i b_i$. In turn, this would show that $1 = c_i b_i$, which violates the fact that $b_i \notin D^\times$. Thus, if $a_0 \notin D_F$, we arrive at a contradiction to D satisfying the ACCP.

We have just shown that we can write $a_0 = \prod_{j=1}^n p_j$ as a product of irreducible elements of D. Suppose now that there is another such decomposition $a_0 = \prod_{j=1}^m q_j$. Since $p_1 \mid a_0$ and is prime, we may reindex subscripts so that $p_1 \mid q_1$. Since q_1 is irreducible, this can only happen if $q_1 = p_1 u_1$ for $u_1 \in D^\times$. Thus we get $\prod_{j=2}^n p_j = u_1 \prod_{j=2}^m q_j$. If $n < m$, we inductively arrive at the case where $1 = u \prod_{j=n+1}^m q_j$ for some $u \in D^\times$, which violates the fact that q_j are not units. Similarly, we cannot have $m < n$. Thus $n = m$ and the irreducible factors are unique up to reindexing and multiplication by a unit.

Look at (1) $\Longrightarrow$ (3) next. For the sake of clarity, we examine the case of $n = 2$ since the general case is done in the same way. Let D be a UFD with $a, b \in D$. If one of the pair is zero or a unit, the existence of a greatest common divisor is trivial so assume they are nonzero and nonunits. Write $a = \prod_j p_j^{m_j}$ and $b = \prod_j p_j^{n_j}$ with $m, n \in \mathbb{Z}_{\geq 0}$. From the definitions and Lemma 3.48, clearly $\prod_j p_j^{\min\{m_j, n_j\}}$ is a greatest common divisor of a and b.

Finally look at (3) $\Longrightarrow$ (2). Assume D has greatest common denominators. First we prove that if $c \in D$ is nonzero, then

$$\gcd(ac, bc) = cu \gcd(a, b)$$

for some $u \in D^\times$ (bear in mind that greatest common divisors are unique only up to multiplication by units; cf. Exercise 3.153). To see this, let $\gcd(a, b) = d$ and $\gcd(ac, bc) = d'$. Clearly $dc \mid ac, bc$ so that $(dc) \mid d'$. Write $d' = dcu$ for some $u \in D$. Since $(dcu) \mid ac, bc$, it follows that $(du) \mid a, b$ so that $(du) \mid d$. Thus u is a unit as desired.

To finish the proof of this theorem, we show that irreducibles are prime. Thus, suppose $p \in D$ is irreducible and $p \mid (ab)$ for some $a, b \in D$, neither 0 (the case of $a = 0$ is trivial). Write $ab = pc$ for some $c \in D$. If $p \nmid a$, then $\gcd(p, a) = 1$ since p is irreducible. Multiplying by b, we see there exists $u, u' \in D^\times$ so that $b = u \gcd(pb, ab) = u \gcd(pb, pc) = u' p \gcd(b, c)$. Thus $p \mid b$ as desired and irreducibles are prime. □

Each of these equivalent notions of being a UFD can be very hard to check in any particular case. However, there are at least three types of rings for which general theorems exist: polynomial rings over UFD's (Theorem 3.57), principal ideal domains (Theorem 3.60), and Euclidean Domains (Theorem 3.63).

5.2. Quotient Fields. We want to apply our UFD theory to polynomial rings. As we saw in Section 4, there were many connections between irreducibility in $\mathbb{Q}[x]$ and irreducibility in $\mathbb{Z}[x]$. To generalize this, we need to mimic the construction of $\mathbb{Q}$ from $\mathbb{Z}$ (cf. §2.6 in Chapter 1) in order to create a field of fractions from any integral domain.

Definition 3.52. **(1)** Let D be an integral domain. A *fraction* over D is a formal symbol $\frac{a}{b}$ with $a, b \in D$, $b \neq 0$.

(2) Two fractions $\frac{a_1}{b_1}$ and $\frac{a_2}{b_2}$ are called *equivalent* if $a_1 b_2 = a_2 b_1$.

(3) The set of equivalence classes of fractions is called the *quotient field* or *field of fractions* of D. It is equipped with two binary operations on equivalence classes given by

$$\frac{a_1}{b_1} + \frac{a_2}{b_2} = \frac{a_1b_2 + a_2b_1}{b_1b_2},$$
$$\frac{a_1}{b_1} \cdot \frac{a_2}{b_2} = \frac{a_1a_2}{b_1b_2}.$$

Here, as customary (though mildly ambiguous), we write $\frac{a_1}{b_1}$ for the equivalence class $\left[\frac{a_1}{b_1}\right]$ so that $\frac{a_1}{b_1} = \frac{a_2}{b_2}$ if and only if $a_1b_2 = a_2b_1$.

Two details need to be checked in order for Definition 3.52 to be valid. We need to verify that the above notion of equivalent is an equivalence relation and that the binary operations are well defined on the equivalence classes of fractions. In fact, we of course also want to show that these binary operations make the field of fractions into a commutative ring with identity. All these statements are straightforward and are left to Exercise 3.159. Moreover, it is clear from the definitions that the inverse to $\frac{a}{b}$, $a, b \neq 0$, is $\frac{b}{a}$ so that the field of fractions really is a field. Just as with $\mathbb{Z}$ and $\mathbb{Q}$, we can view D as sitting inside its field of fractions by identifying $a \in D$ with $\frac{a}{1}$. It is obvious from the definitions that this map is an injective homomorphism. In summary, we have the following result.

THEOREM 3.53. *Let D be an integral domain and let F be its field of fractions.*
(1) *F is a field.*
(2) *$D \cong \{\frac{a}{1} \in F \mid a \in D\}$ and so we may view D as sitting inside F.*

As an example, it is clear that the quotient field of $\mathbb{Z}$ is $\mathbb{Q}$. Another example is the following.

DEFINITION 3.54. If D is an integral domain, the set of *rational functions* with coefficients in D, written $D(x)$, is defined to be the quotient field of $D[x]$. Thus $D(x)$ is the set of all equivalence classes of fractions of the form $\frac{f}{g}$ with $f, g \in D[x]$, $g \neq 0$.

5.3. Unique Factorization in Polynomial Rings.

DEFINITION 3.55. Let D be a UFD and let $f = \sum_k a_k x^k \in D[x]$, with $f \neq 0$. We say f is *primitive* if $\gcd(a_k)$ is a unit.

If D is a UFD and $f \in D[x]$, $f \neq 0$, it is apparent that f can be written as $f = cf_0$ where $c \in D$ and $f_0 \in D[x]$ is primitive by taking $c = \gcd(a_k)$. In fact, it is easy to see that c and f_0 are unique up to multiplication by a unit of D (Exercise 3.158). The element c, determined up to multiplication by a unit, is called the *content* of f.

LEMMA 3.56. *Let D be a UFD and let F be its field of fractions.*

(1) *If $f \in F[x]$ is nonzero, there is a $c \in F$ and $f_0 \in D[x]$ primitive so that $f = cf_0$. Moreover, $f \in D[x]$ if and only if $c \in D$.*

(2) (Gauss's lemma) *If $f_0, g_0 \in D[x]$ are primitive, then f_0g_0 is primitive.*

(3) *Let $f_0, g \in D[x]$ with f_0 primitive. If $f_0 \mid g$ in $F[x]$, then actually $f_0 \mid g$ in $D[x]$.*

(4) *Let $h \in D[x]$. Then h is irreducible in $D[x]$ if and only if either* (i) $\deg h = 0$ *and h is irreducible in D or* (ii) $\deg h \geq 1$, *h is primitive, and h is irreducible in $F[x]$.*

PROOF. For part (1), first consider existence. Let $b \in D$ be the product of the denominators of the coefficients of f so that $bf \in D[x]$. Let $a \in D$ be the greatest common divisor of the coefficients of bf. Then $\frac{b}{a}f$ is primitive in $D[x]$ and $f = \frac{a}{b}\left(\frac{b}{a}f\right)$ as desired.

For the second part of part (1), suppose $c \in F$ and $f_0 \in D[x]$ is primitive so that $f = cf_0$. If $c \in D$, clearly $f \in D[x]$. Conversely, suppose $f \in D[x]$ and write $c = \frac{a}{b}$ with $a, b \in D$, $b \neq 0$. Using the fact that D is a UFD, we can cancel and assume that the $\gcd(a, b)$ is a unit. To finish part (1), we need only show that b is a unit. If not, there is an irreducible $p \in D$ so $p \mid b$ with $p \nmid a$. Then $bf = af_0$ shows p divides all the coefficients of f_0, which contradicts the fact that f_0 is primitive.

For part (2), let p be any irreducible element of D. Since p is prime, $D/(p)$ is an integral domain. If $h = \sum_k a_k x^k \in D[x]$, define $\overline{h} = \sum_k (a_k + (p))\, x^k \in (D/(p))\,[x]$. Just as we saw for $\mathbb{Z}$, the map $h \to \overline{h}$ is a homomorphism (Exercise 3.157). To prove (2), it suffices to show that p does not divide all the coefficients of $f_0 g_0$, i.e., to show that $\overline{f_0 g_0} \neq 0_{D/(p)}$. But this follows immediately from the fact that $\overline{f}_0, \overline{g}_0 \neq 0_{D/(p)}$ since they are primitive.

For part (3), write $g = f_0 q$ with $q \in F[x]$. By part (1) there is a $c \in F$ and primitive $q_0 \in D[x]$ so that $q = cq_0$. By Gauss's lemma, $f_0 q_0$ is primitive. Since $g = c(f_0 q_0)$, part (1) shows $c \in D$ so that $q \in D[x]$ as desired.

For part (4), suppose $h \in D[x]$. Recall that the degree of a product of polynomials is the sum of the degrees. Thus if $\deg h = 0$, clearly h is irreducible in $D[x]$ if and only if it is irreducible in D. Therefore assume that $\deg h \geq 1$ and consider the contrapositive. For the first direction, suppose h is reducible in $D[x]$ and that it factors as $h = h_1 h_2$ for $h_i \in D[x]$, neither units in D. If $\deg h_1$ or $\deg h_2$ is 0, then h is not primitive. Otherwise, the factorization shows h is reducible in $F[x]$ and we are done. For the opposite direction, suppose either h is not primitive or that h is reducible in $F[x]$. In the first case, we see h is not irreducible in $D[x]$ by factoring out the gcd of the coefficients. In the second case, there exists $f \in F[x]$ so that $f \mid h$ in $F[x]$ with $0 < \deg f < \deg h$. Write $f = cf_0$ as in part (1). Clearly $f_0 \mid h$ in $F[x]$ as well. Thus part (3) shows $f_0 \mid h$ in $D[x]$ with $0 < \deg f_0 < \deg h$ so that f_0 is reducible in $D[x]$ as desired. □

THEOREM 3.57. *Let D be a UFD. Then $D[x]$ is a UFD.*

PROOF. Recall Theorem 3.51. First we show $D[x]$ satisfies the ACCP. Suppose we have nonzero $f_i \in D[x]$, $i \in \mathbb{N}$, with $(f_i) \subseteq (f_{i+1})$ so that $f_{i+1} \mid f_i$. Write a_i for the leading coefficient of f_i. At the very least, $a_{i+1} \mid a_i$ so that $(a_i) \subseteq (a_{i+1})$. Since D is a UFD, this sequence of principal ideals eventually stabilizes. Thus for sufficiently large i, $u_i a_{i+1} = a_i$ with $u_i \in D^\times$. Since $\deg f_i$ can only decrease a finite number of times, we also eventually have $\deg f_i = \deg f_{i+1}$. When this happens, we get $c_i f_{i+1} = f_i$ with $c_i \in D$ for large i. Looking at leading coefficients for large i shows $c_i u_i^{-1} a_i = a_i$ so that $c_i = u_i$ and $(f_i) = (f_{i+1})$ as desired.

Now let $f \in D[x]$ be irreducible. It suffices to show f is prime. If $\deg f = 0$, we are done since D is a UFD. If $\deg f \geq 1$, then f is primitive and irreducible in $F[x]$ where F is the quotient field of D. Suppose $f \mid (gh)$ for some $g, h \in D[x]$. If $f \nmid g$, consider the ideal $I = \{fp + gq \mid p, q \in F[x]\}$. By Corollary 3.27, $I = (k)$ for some $k \in F[x]$. In particular, $k \mid f$ so that k is either an associate of f or a unit in F by the irreducibility of f. Since also $k \mid g$ but $f \nmid g$, k is not an associate of f and so $(k) = F[x]$. In particular, there exist $p, q \in F[x]$ so $1 = fp + gq$. Hence

$h = (f)\,hp + (gh)q$ so that $f \mid h$ in $F[x]$. Since f is primitive and $g \in D[x]$, $f \mid h$ in $D[x]$ by Lemma 3.56 as desired. □

Theorem 3.57 is a rather powerful and surprising result. As applications, $F[x]$ is a UFD for any field F (since F is a UFD by default), $\mathbb{Z}[x]$ is a UFD, and $D[x_1, x_2, \ldots, x_n]$ is a UFD whenever D is (by repeated use of the theorem).

5.4. Exercises 3.151–3.172.

EXERCISE 3.151. Prove Lemma 3.44.

EXERCISE 3.152. Factor the following polynomials into a product of irreducibles:
(a) $x^3 + 2$ in $\mathbb{Q}[x]$, $\mathbb{R}[x]$, and $\mathbb{C}[x]$,
(b) $x^3 + [2]$ in $\mathbb{Z}_2[x]$, $\mathbb{Z}_3[x]$, and $\mathbb{Z}_5[x]$,
(c) $x^4 - 2x^3 - x + 2$ in $\mathbb{Q}[x]$,
(d) $3x^4 - 7x^3 - 19x^2 - 2x - 8$ in $\mathbb{Q}[x]$,
(e) $x^3 + 4x^2 + 8x + 5$ in $\mathbb{Q}[x]$,
(f) $x^5 + x^4 + x^3 + [1]$ in $\mathbb{Z}_2[x]$,
(g) $x^5 + x^4 + [1]$ in $\mathbb{Z}_2[x]$,
(h) $x^4 + x^3 + x - [1]$ in $\mathbb{Z}_3[x]$.

EXERCISE 3.153. Let D be an integral domain with $a, b \in D$, both not zero. If a greatest common divisor exists, show it is unique up to multiplication by units.

* EXERCISE 3.154. **(a)** Factor $f = x^4 - x^2 + x - 1$ and $g = x^3 - x^2 + x - 1$ into irreducibles in $\mathbb{Q}[x]$.
(b) Find a greatest common divisor for f and g.

* EXERCISE 3.155. In $\mathbb{Z}[x]$, show $\gcd(2, x) = 1$ but $1 \neq 2f + xg$ for any $f, g \in \mathbb{Z}[x]$.

EXERCISE 3.156. Let F be a field and let $p \in F[x]$ be irreducible. If $p \mid \prod_{i=1}^{n} f_i$ for some nonzero $f_i \in F[x]$, show there exists i_0 so that $p \mid f_{i_0}$.

EXERCISE 3.157. Let D be a UFD and let $p \in D$ be irreducible. If $h = \sum_k a_k x^k \in D[x]$, define $\overline{h} = \sum_k (a_k + (p))\, x^k \in (D/(p))\,[x]$. Show the map $h \to \overline{h}$ is a homomorphism.

EXERCISE 3.158. Let D be a UFD with nonzero $f \in D[x]$. Write $f = cf_0$ with $c \in D$ and $f_0 \in F[x]$ primitive. Prove that c and f_0 are unique up to multiplication by units in D.

EXERCISE 3.159. **(1)** Show that the notion of equivalent is an equivalence relation in Definition 3.52.
(2) Show that the binary operations from Definition 3.52 are well defined on the equivalence classes of fractions.
(3) Let D be an integral domain. Show its quotient field is a ring.

EXERCISE 3.160. Show that subrings of UFD's are not always UFD's. *Hint:* $\mathbb{C}$ is a UFD.

EXERCISE 3.161. Let F be a field and let $a_1, \ldots, a_n \in F$ be distinct. Show $F[x]/(\prod_{i=1}^{n}(x - a_i)) \cong F^k$. *Hint:* One method is to use Exercise 3.84.

EXERCISE 3.162. Let D be an integral domain so that $D[x]$ is a UFD. Show D is a UFD.

EXERCISE 3.163. Show that the only common divisors of 6 and $2+2i\sqrt{5}$ in $\mathbb{Z}[\sqrt{-5}]$ are $\pm\{1, 2, 1+i\sqrt{5}\}$ and that none are divisible by all the rest. Conclude that a greatest common divisor does not exist.

EXERCISE 3.164. Let D be an integral domain for which every nonunit can be written as a finite product of irreducible elements. Show D is a UFD if and only if every irreducible element is prime.

EXERCISE 3.165. Let R be a commutative ring with identity in which every ideal is *finitely generated*, i.e., every ideal is of the form $\{\sum_{i=1}^N r_i c_i \mid r_i \in R\}$ for some $N \in \mathbb{N}$ and $c_i \in R$. Show R satisfies the *ascending chain condition on ideals (ACC)*, i.e., that if $I_i \subseteq I_{i+1}$ are ideals for $i \in \mathbb{N}$, then eventually $I_i = I_{i+1}$ for all sufficiently large i. *Hint:* Look at $I = \bigcup_i I_i$.

EXERCISE 3.166. Let D be an integral domain. We say that D satisfies the *descending chain condition on principal ideals (DCCP)* if whenever $(a_i) \supseteq (a_{i+1})$ for $a_i \in D$, $i \in \mathbb{N}$, then eventually $(a_i) = (a_{i+1})$ for all sufficiently large i. Show that D satisfies the DCCP if and only if it is a field. *Hint:* $(a^j) \supseteq (a^{j+1})$.

EXERCISE 3.167. A number $z \in \mathbb{C}$ is called *algebraic* if it is the root of a nonzero $f \in \mathbb{Q}[x]$. Show z is algebraic if and only if it is the root of a nonzero $g \in \mathbb{Z}[x]$.

EXERCISE 3.168 (Algebraic Integers). A number $z \in \mathbb{C}$ is called an *algebraic integer* if it is the root of a nonzero *monic* $f \in \mathbb{Z}[x]$. It turns out that $\mathcal{O} = \{\text{algebraic integers}\}$ is a subring of $\mathbb{C}$ (see Theorem 4.13).

(a) Let $z \in \mathbb{C}$ be algebraic (cf. Exercise 3.167) and let $I = \{g \in \mathbb{Q}[x] \mid g(z) = 0\}$. Show there exists an irreducible $f \in \mathbb{Z}[x]$, called the *irreducible polynomial* for z, so that $I = (f)$ in $\mathbb{Q}[x]$.

(b) In $\mathbb{Z}[x]$, show $(f) = \{h \in \mathbb{Z}[x] \mid h(z) = 0\}$ and conclude z is an algebraic integer if and only if f is monic.

(c) Show $\mathcal{O} \cap \mathbb{Q} = \mathbb{Z}$.

(d) Show that all n^{th} roots of unity are algebraic integers for any $n \in \mathbb{N}$.

EXERCISE 3.169 (Quadratic Number Fields). Fix $\omega \in \mathbb{C}$ so that $\omega^2 \in \mathbb{Z}$ but $\omega \notin \mathbb{Z}$ and set $\mathbb{Q}[\omega] = \{a + \omega b \mid a, b \in \mathbb{Q}\}$.

(a) Show $\mathbb{Q}[\omega]$ is a subfield of $\mathbb{C}$.

(b) If $\omega^2 = p^2 w^2$ for $p \in \mathbb{Z}$ a prime and $w \in \mathbb{C}$, show $\mathbb{Q}[\omega] = \mathbb{Q}[w]$. Conclude that we may assume ω^2 is square free so that, in particular, $\omega^2 \not\equiv 0 \bmod(4)$.

(c) For $z = a + \omega b \in \mathbb{Q}[\omega]$, let $z^* = a - \omega b \in \mathbb{Q}[\omega]$. Show z is an algebraic integer if and only if $2a$ and $a^2 - b^2\omega^2$ are integers (cf. Exercise 3.168).

(d) If $\omega^2 \equiv 2$ or $3 \bmod(4)$, show z is an algebraic integer if and only if $a, b \in \mathbb{Z}$.

(e) If $\omega^2 \equiv 1 \bmod(4)$, show z is an algebraic integer if and only if $a, b \in \mathbb{Z}$ or $a, b \in \frac{1}{2} + \mathbb{Z}$.

EXERCISE 3.170 (Integrally Closed). Let $R \subseteq S$ be commutative rings with the same identity. We say $\alpha \in S$ is *integral* over R if there is a monic $f \in R[x]$ so that $f(\alpha) = 0$. Let D be an integral domain and let F be its field of fractions. We say D is *integrally closed* if $\{\alpha \in F \mid \alpha \text{ is integral over } D\} = D$. If D is a UFD, show D is integrally closed. *Hint:* Write $\alpha = \frac{a}{b}$ with $(a, b) = 1$. If $f(\alpha) = 0$, clear denominators to show $a^n + bc = 0$ for some $c \in D$. If $p \in D$ is prime with $p \mid b$, show there is a contradiction to $(a, b) = 1$.

EXERCISE 3.171 (Least Common Multiples). Let D be an integral domain with $a_i \in D$, not all zero, $1 \le i \le n$, $n \in \mathbb{N}$. We say $l \in D$ is a *least common multiple* of the a_i, written (nonuniquely) as $\operatorname{lcm}(a_1, \ldots, a_n)$, if (i) $a_i \mid l$, $1 \le i \le n$, and (ii) if $a_i \mid c$, $1 \le i \le n$, for some $c \in D$, then $l \mid c$ (cf. Exercise 1.28).

(a) If a least common multiple exists, show it is unique up to multiplication by a unit.

(b) If D is a UFD with nonzero $a, b \in D$ written as a product of nonassociate irreducible elements, $a = u \prod_j p_j^{m_j}$ and $b = v \prod_j p_j^{n_j}$ for some units u and v, show $\prod_j p_j^{\max\{m_j, n_j\}}$ is a least common multiple of a and b and that $\operatorname{lcm}(a,b)\gcd(a,b)$ is associate to ab.

(c) Let D be an integral domain. Show D is a UFD if and only if (i) D satisfies the ACCP and (ii) a least common multiple exists for all $a_1, \ldots, a_n \in D$, not all zero.

EXERCISE 3.172 (Kronecker's Factorization Algorithm). This exercise gives an algorithm that factors any nonconstant polynomial in $\mathbb{Q}[x]$ into a product of irreducibles. We may assume we are starting with a primitive nonconstant $f \in \mathbb{Z}[x]$ of degree n. By induction on n, it suffices to determine whether f is irreducible in $\mathbb{Q}[x]$ and, if not, to find a nonconstant factor.

(a) If f has a nonconstant factor g with $f = gh$, show that we may assume $g, h \in \mathbb{Z}[x]$ are primitive with $\deg(g)$ at most $m = \lfloor \frac{n}{2} \rfloor$.

(b) Show that we can find $x_0, x_1, \ldots, x_m \in \{0, 1, \ldots, 2n\}$ so that $f(x_i) \neq 0$.

(c) Show that $g(x_i) \mid f(x_i)$ so that there are only finitely many possibilities for $g(x_i)$. Write D_i for the set of divisors of $f(x_i)$.

(d) For each choice of $c_i \in D_i$, $0 \le i \le m$, show there is a unique $g_{c_0,\ldots,c_m} \in \mathbb{Q}[x]$ of degree at most m satisfying $g_{c_0,\ldots,c_m}(x_i) = c_i$ (cf. Exercise 3.121 for an explicit formula).

(e) Let $G = \{g_{c_0,\ldots,c_m} \mid \deg g \ge 1\}$. If $g \nmid f$ for all $g \in G$ (which can be determined using long division), show f is irreducible. Otherwise, show f is reducible and there is a $g \in G$ providing us with a nonconstant factor in $\mathbb{Q}[x]$.

(f) Carry out the algorithm for $f = 3x^2 - 11x + 10$.

6. Principal Ideal and Euclidean Domains

6.1. Principal Ideal Domains. Besides being a UFD, $\mathbb{Z}$ has further special properties. Namely, all ideals in $\mathbb{Z}$ are principal (Exercise 3.47). This idea turns out to be rather useful.

DEFINITION 3.58. A *principal ideal domain* or *PID* is an integral domain in which every ideal is principal.

For example, $\mathbb{Z}$ is a PID and $F[x]$ is a PID for any field F (Corollary 3.27). On the other hand, it is easy to see that the ideal $\{xf + yg \mid f, g \in \mathbb{R}[x,y]\}$ in $\mathbb{R}[x,y]$ is not principal so that $\mathbb{R}[x,y]$ is not a PID (Exercise 3.175).

The next result shows that not only do greatest common divisors exist in PID's, but they can be written as a very special linear combination.

LEMMA 3.59. *Let D be a PID with $a_1, \ldots, a_n \in D$, not all zero.*

(1) *There exists $d \in D$ so that $(d) = \{\sum_i a_i b_i \mid b_i \in D\}$ and d is unique up to multiplication by a unit.*

(2) *d is a greatest common divisor for $a_1, \ldots, a_n$.*
(3) *There exist $b_i \in D$ so $d = \sum_i a_i b_i$.*

PROOF. Clearly $I = \{\sum_i a_i b_i \mid b_i \in D\}$ is an ideal. Thus d exists since D is a PID. It is unique up to multiplication by a unit since that is the only way two principal ideals can be equal. In particular, parts (1) and (3) are complete. For part (2), first note that $d \mid a_i$ since $a_i \in I$. Next suppose c is a common divisor of the a_i's. Clearly c then divides $\sum_i a_i b_i = d$ and we are done. $\square$

It turns out that PID's are special cases of UFD's.

THEOREM 3.60. *A PID is a UFD.*

PROOF. Let D be a PID. By Theorem 3.51 and Lemma 3.59, it suffices to show that D satisfies the ACCP. Let $a_i \in D$ so that $(a_i) \subseteq (a_{i+1})$. It follows that $I = \bigcup_i (a_i)$ is an ideal (cf. Exercise 3.51) so we can write $I = (a)$. In particular, $a \in (a_i)$ for some i and therefore for all large i. Thus $(a) \subseteq (a_i) \subseteq (a)$ so that $(a_i) = (a)$ as desired. $\square$

As a corollary, irreducible elements in PID's are prime. In fact, more is true.

COROLLARY 3.61. *Prime ideals in a PID are maximal.*

PROOF. Let D be a PID and suppose (p) is a prime ideal in D. If (b) is an ideal containing (p), then $p = br$ for some $r \in D$. If $p \mid b$, then $(b) = (a)$. Otherwise, $p \mid r$ so $r = cp$ for some $c \in D$. In particular, $p = bcp$ so that $bc = 1$, b is a unit, and $(b) = D$. Hence (p) is maximal. $\square$

As the reader no doubt expects, there exist UFD's that are not PID's. An easy example is $\mathbb{Z}[x]$ and the ideal $\{2f + xg \mid f, g \in \mathbb{Z}[x]\}$ (cf. Exercise 3.155). In fact if R is a commutative ring with identity, then $R[x]$ is a PID if and only if R is a field (Exercise 3.174).

6.2. Euclidean Domains. Finally, $\mathbb{Z}$ has an even more specialized property. Namely, it has a Division Algorithm.

DEFINITION 3.62. Let D be an integral domain.

(1) A *size function* on D is a map $\sigma : D\backslash\{0\} \to \mathbb{Z}_{\geq 0}$. We extend σ to D by setting $\sigma(0) = -\infty$.

(2) D is called a *Euclidean domain* with respect to σ if for all $a, b \in D$, $b \neq 0$, there exists $q, r \in D$ so that

$$a = bq + r$$

with $\sigma(r) < \sigma(b)$.

Of course $\mathbb{Z}$ is our first example with $\sigma(n) = |n|$, $n \in \mathbb{Z}\backslash\{0\}$. In that case, the definition of a Euclidean domain just says that the Division Algorithm works. Another example we have seen is $F[x]$ where F is a field. In this case, take $\sigma(f) = \deg(f)$ for $f \in F[x]$. Again the analogue of the Division Algorithm works (Theorem 3.26) so that $F[x]$ is a Euclidean domain as well. As a final example, we will soon see that the set of Gaussian integers $\mathbb{Z}[i] = \{n + im \mid n, m \in \mathbb{Z}\}$ is a Euclidean domain with size function given by $N(n + im) = n^2 + m^2$ (see §6.3 below). As a cautionary note, observe that we do not require q and r to be unique in Definition 3.62.

THEOREM 3.63. *A Euclidean domain is a PID.*

PROOF. Let D be a Euclidean domain and let I be a nonzero ideal. Choose any nonzero $a \in I$ with $\sigma(a)$ minimal. If nonzero $b \in I$, write $b = aq + r$ for some $q, r \in D$ with $\sigma(r) < \sigma(a)$. If $r \neq 0$, we would get nonzero $r \in I$ and violate our choice of minimal a. Thus $r = 0$ so that $I \subseteq (a)$. Since clearly $(a) \subseteq I$, we are done. $\square$

In particular, we see that Euclidean domains are UFD's.

6.3. Gaussian Integers. If $z = a + ib \in \mathbb{C}$, recall that the *norm* of z is $N(z) = |z|^2 = z\overline{z} = a^2 + b^2$ where $\overline{z} = a - ib$ is *complex conjugation*. In particular, the norm is multiplicative; i.e., $N(z_1 z_2) = N(z_1)N(z_2)$, and $z^{-1} = \frac{\overline{z}}{N(z)}$ for $z \neq 0$.

The *Gaussian integers* are defined as the subring of $\mathbb{C}$ given by $\mathbb{Z}[i] = \{n + im \mid n, m \in \mathbb{Z}\}$. Restricting N to $\mathbb{Z}[i] \backslash \{0\}$ gives a size function on $\mathbb{Z}[i]$.

THEOREM 3.64. *$\mathbb{Z}[i]$ is a Euclidean domain (and therefore a PID and UFD).*

PROOF. Let $z = a + ib$ and $w = c + id$ be in $\mathbb{Z}[i]$ with $w \neq 0$. Then in $\mathbb{C}$, it is easy to check that $\frac{z}{w} = \alpha + i\beta$ where $\alpha = \frac{ac+bd}{c^2+d^2}$ and $\beta = \frac{bc-ad}{c^2+d^2}$ are rational. Choose $n, m \in \mathbb{Z}$ so that $|\alpha - n|, |\beta - m| \leq \frac{1}{2}$ and set $q = n + im$ and $r = z - wq$. We need only show $N(r) < N(w)$. For this calculate

$$N(r) = N(w)N(\frac{z}{w} - q) = N(w)N((\alpha - n) + i(\beta - m)) \leq N(w)\left(\frac{1}{4} + \frac{1}{4}\right) = \frac{1}{2}N(w),$$

which is overkill. $\square$

Notice in the above proof that if α or β is in $\frac{1}{2} + \mathbb{Z}$, then there are multiple choices for q and r.

In order to factor in $\mathbb{Z}[i]$, it is necessary to know which elements are irreducible and which elements are units. The units are easy.

LEMMA 3.65. *The units in $\mathbb{Z}[i]$ are $\{\pm 1, \pm i\}$.*

PROOF. First of all, if $z \in \mathbb{Z}[i]$ has an inverse, then $N(z)N(z^{-1}) = N(1) = 1$ so that $N(z) = 1$. Conversely if $N(z) = 1$, the formula for inverses in $\mathbb{C}$ shows $\overline{z} = z^{-1}$. To finish the proof, observe that $\{a+ib \in \mathbb{Z}[i] \mid a^2+b^2 = 1\} = \pm\{1, i\}$. $\square$

Finding the irreducibles is more difficult and requires a famous result of Fermat. The proof is not trivial.

LEMMA 3.66 (Fermat's Theorem on the Sum of Two Squares). *Let $p \in \mathbb{Z}$ be a positive prime. Then $p = n^2 + m^2$ for some $n, m \in \mathbb{Z}$ if and only if $p = 2$ or if p is odd with $p \equiv 1 \bmod(4)$.*

PROOF. Since $2 = 1^2 + 1^2$, suppose p is an odd prime so that $p \equiv 1$ or $3 \bmod(4)$. Calculating squares in $\mathbb{Z}_4$, it follows that $n^2, m^2 \equiv 0$ or $1 \bmod(4)$ so that $n^2 + m^2 \equiv 0$, 1, or $2 \bmod(4)$. In particular if p is a sum of two squares, then $p \not\equiv 3 \bmod(4)$ which forces $p \equiv 1 \bmod(4)$.

Conversely if $p \equiv 1 \bmod(4)$, write $p = 4n + 1$ so that $U_p \cong \mathbb{Z}_{4n}$ by Corollary 3.31. Since $[n] \in \mathbb{Z}_{4n}$ has order 4, there is some $[k] \in U_p$ of order 4. As $[-1] \in U_p$ is the only element of order 2 (the roots of $x^2 - [1]$ are $\pm[1]$), $[k^2] = [-1]$ and so $p \mid (k^2 + 1)$ in $\mathbb{Z}$. In particular, $p \mid (k + i)(k - i)$ in $\mathbb{Z}[i]$. Obviously $\frac{k}{p} \pm i\frac{1}{p} \notin \mathbb{Z}[i]$ (since $\frac{1}{p} \notin \mathbb{Z}$) so that $p \nmid (k \pm i)$ in $\mathbb{Z}[i]$ which shows that p is not prime in $\mathbb{Z}[i]$. Since $\mathbb{Z}[i]$ is a UFD, p has an irreducible divisor, say $z = a+ib \in \mathbb{Z}[i]$. Thus $p = (a+ib)w$

for some nonunit $w \in \mathbb{Z}[i]$. Taking conjugates shows $p = (a - ib)\overline{w}$ and multiplying gives $p^2 = (a^2 + b^2)N(w)$. As z, w are not units, $N(z) = a^2 + b^2, N(w) \neq 1$. Since p is prime in the UFD $\mathbb{Z}$, looking at the prime factorization of p^2 compared to that of $a^2 + b^2$ and of $N(w)$ shows that the only possibility is $p = a^2 + b^2$ (and $p = N(w)$, not that we care). $\square$

As a note on the proof of Lemma 3.66, examination of w can be used to show that the choice of n and m is unique up to multiplying by ± 1 and switching n and m (Exercise 3.192). This result can also be seen using the fact that $\mathbb{Z}[i]$ is a UFD and Lemma 3.65.

THEOREM 3.67. **(1)** *For each positive prime $p \in \mathbb{Z}$ satisfying $p \equiv 3 \bmod(4)$, p is irreducible in $\mathbb{Z}[i]$.*

(2) *For $p = 2$ and each positive prime $p \in \mathbb{Z}$ satisfying $p \equiv 1 \bmod(4)$, $a \pm ib$ is irreducible in $\mathbb{Z}[i]$ for $a, b \in \mathbb{Z}$ satisfying $a^2 + b^2 = p$.*

Moreover, these are the only irreducible elements of $\mathbb{Z}[i]$ up to multiplication by a unit.

PROOF. For part (1), suppose $p = zw$ for $z, w \in \mathbb{Z}[i]$. Then $p^2 = N(z)N(w)$. Since we cannot solve $p = N(z)$ (Lemma 3.66), it follows that $N(z)$ or $N(w)$ is 1 and therefore is a unit. Thus p is irreducible. For part (2), note that $N(a \pm ib) = p$ is a prime in $\mathbb{Z}$. Thus if we write $a \pm ib = zw$, it follows that $p = N(z)N(w)$ so that again $N(z)$ or $N(w)$ is 1 and therefore is a unit.

To see that these are the only irreducibles up to multiplication by a unit, suppose $z \in \mathbb{Z}[i]$ is prime. It is straightforward to see that $(z) \cap \mathbb{Z}$ is still a prime ideal in $\mathbb{Z}$. Hence $(z) \cap \mathbb{Z} = (p)$ for some positive prime $p \in \mathbb{Z}$ (cf. Exercise 3.76). Write $p = zw$ for some $w \in \mathbb{Z}[i]$ so that $p^2 = N(z)N(w)$. If w is a unit, we may replace z by an associate and assume that $w = 1$ and that $z = p$. Since p is irreducible, we cannot write $p = (a + ib)(a - ib)$ for $a, b \in \mathbb{Z}$. Thus $p \equiv 3 \bmod(4)$ (Lemma 3.66) and we are in case (1). If w is not a unit, then $N(w) \neq 1$ so that we must have $N(z) = p$ (and $N(w) = p$). Thus $p = 2$ or p is odd with $p \equiv 1 \bmod(4)$ (Lemma 3.66) and we are in case (2) as desired. $\square$

As a seemingly small generalization of the Gaussian integers, fix $\omega \in \mathbb{C} \backslash \mathbb{Z}$ with $\omega^2 \in \mathbb{Z}$ and consider

$$\mathbb{Z}[\omega] = \{a + \omega b \mid a, b \in \mathbb{Z}\}$$

as a subring of $\mathbb{C}$ (cf. Exercise 3.169). In this setting, the *conjugate* of $a + \omega b \in \mathbb{Z}[\omega]$ is defined as $(a + \omega b)^* = a - \omega b$ and the *norm* of $a + \omega b$ is

$$N(a + \omega b) = (a + \omega b)(a + \omega b)^* = a^2 - \omega^2 b^2.$$

The exercises will examine some properties of $\mathbb{Z}[\omega]$. However in general, it is very difficult to determine when $\mathbb{Z}[\omega]$ is a UFD, a PID, or a Euclidean domain.

6.4. Exercises 3.173–3.201.

EXERCISE 3.173. Let D be an integral domain with nonzero $a, b \in D$. Show $(a) = (b)$ if and only if a and b are associates.

EXERCISE 3.174. Let R be a commutative ring with identity. Show $R[x]$ is a PID if and only if R is a field. *Hint:* If $R[x]$ is a PID, show R is a domain ($R \subseteq R[x]$), then that (x) is prime, and then look at a quotient (Corollary 3.61 and Theorem 3.22).

EXERCISE 3.175. Let F be a field. Though a UFD, show $F[x, y]$ is not a PID. *Hint:* Consider $\{xf + yg \mid f, g \in F[x, y]\}$.

EXERCISE 3.176. Let D be a PID with $a, b, c \in D$ so that $\gcd(a, b) = 1$ (cf. Theorem 1.13).

(a) If $a, b \mid c$, show $ab \mid c$.

(b) If $a \mid bc$, show $a \mid c$.

EXERCISE 3.177. Let D be a PID with $a_1, \ldots, a_n \in D$, not all zero.

(a) Show d is a greatest common divisor if and only if $(a_1) + \cdots + (a_n) = (d)$ (cf. Exercise 3.51).

(b) Show l is a least common multiple if and only if $(a_1) \cap \cdots \cap (a_n) = (l)$ (cf. Exercise 3.171).

EXERCISE 3.178. Let D be an integral domain. This exercise shows D is a PID if and only if (i) D satisfies the ACCP and (ii) for all nonzero $a, b \in D$, there is a common divisor of a and b of the form $ax + by$ for some $x, y \in D$.

(a) If D is a PID, show it satisfies (i) and (ii).

(b) Conversely, suppose now that D satisfies (i). Let I be a nonzero ideal in D. Show I contains a maximal principal ideal; i.e., show there exists $a \in I$ so that if $(a) \subseteq (c) \subseteq I$ for some $c \in I$, then $(a) = (c)$. *Hint:* Argue by contradiction and construct an infinite chain of ideals.

(c) Suppose D also satisfies (ii). Show D is a PID. *Hint:* For nonzero $b \in I$, let $c = ax + by$ be a greatest common divisor.

(d) If D is a UFD, show D is a PID if and only if (ii).

EXERCISE 3.179. Show that the ideal (x) in $\mathbb{Z}[x]$ is prime but not maximal.

EXERCISE 3.180. Let F be a field. Show that F is a Euclidean domain. *Hint:* For nonzero $a \in F$, let $\sigma(a) = 1$.

EXERCISE 3.181. Let (D, σ) be a Euclidean domain with size function σ. Define $\nu : D\backslash\{0\} \to \mathbb{Z}_{\geq 0}$ by $\nu(a) = \min\{\sigma(ab) \mid b \in D\backslash\{0\}\}$, $a \in D$. Show that (D, ν) is still a Euclidean domain with the additional property that

$$\nu(ab) \geq \nu(a)$$

for $a, b \in D\backslash\{0\}$. *Hint:* For $a, b \in D$, $b \neq 0$, choose $q, r \in D$ with $\sigma(r)$ minimal satisfying $a = bq + r$ with $\sigma(r) < \sigma(b)$. If $\nu(r) \geq \nu(b)$, show $\sigma(r) \geq \sigma(bc)$ for some nonzero $c \in D$, write $a = bcq' + r'$ with $\sigma(r') < \sigma(bc)$, and show $\sigma(r')$ violates minimality.

EXERCISE 3.182. Let F be a field. Show that $F[[x]]$ is a Euclidean domain with $\sigma(\sum_{k \geq n} a_k x^k) = n$ when $a_n \neq 0$, $a_k \in F$ (cf. Exercise 3.122). *Hint:* For $f, g \in F[[x]]$ with $g \neq 0$, try $r = 0$ when $\sigma(f) \geq \sigma(g)$ and try $q = 0$ when $\sigma(f) < \sigma(g)$.

EXERCISE 3.183. Let $a = 4 + 7i$ and $b = 2 + 3i$. Find $q, r \in \mathbb{Z}[i]$ so that $a = bq + r$ with either $N(r) < N(b)$ or $r = 0$.

EXERCISE 3.184. Factor the following into a product of irreducibles in $\mathbb{Z}[i]$:

(a) 91,

(b) $3 + 5i$,

(c) $28 - 49i$.

EXERCISE 3.185. Let $\omega \in \mathbb{C}\backslash\mathbb{Z}$ with $\omega^2 \in \mathbb{Z}$

(a) Show $\omega \notin \mathbb{Q}$. *Hint:* Rational Root Test.

(b) If $a_i + \omega b_i \in \mathbb{Z}[\omega]$, show $a_1 + \omega b_1 = a_2 + \omega b_2$ if and only if $a_1 = a_2$ and $b_1 = b_2$.

(c) For $z \in \mathbb{Z}[\omega]$, show $N(z) = 0$ if and only if $z = 0$.

(d) For $z_i \in \mathbb{Z}[\omega]$, show $N(z_1 z_2) = N(z_1)N(z_2)$.

(e) Show $z \in \mathbb{Z}[\omega]$ is a unit if and only if $N(z) = \pm 1$. If so, show $z^{-1} = N(z)z^*$.

(f) If $N(z)$ is prime in $\mathbb{Z}$, show z is irreducible in $\mathbb{Z}[\omega]$.

(g) Show that the converse to part (f) can be false. *Hint:* $\mathbb{Z}[\sqrt{-5}]$.

EXERCISE 3.186. Let $\omega \in \mathbb{C}\backslash\mathbb{Z}$ with $\omega^2 \in \mathbb{Z}$. Show that every nonunit in $\mathbb{Z}[\omega]$ can be written as a (possibly nonunique) product of irreducibles. *Hint:* If $z \in \mathbb{Z}[\omega]$ is reducible, write as $z = ab$ for nonunits $a, b \in \mathbb{Z}[\omega]$ and show $|N(a)|, |N(b)| < |N(z)|$. As needed, repeat this procedure on the factors and prove that this process eventually gives the result.

EXERCISE 3.187. **(a)** Adapt the proof of Theorem 3.64 to show $\mathbb{Z}[i\sqrt{2}]$ is a Euclidean domain with size function given by $N(z)$.

(b) Show that the only units of $\mathbb{Z}[i\sqrt{2}]$ or $\mathbb{Z}[i\sqrt{3}]$ are $\{\pm 1\}$.

(c) Show $\mathbb{Z}[i\sqrt{3}]$ is not a UFD. *Hint:* Find two irreducible factorizations of 4.

(d) Show $\mathbb{Z}[i\sqrt{p}]$ is not a UFD for any prime $p \geq 3$. *Hint:* Note that $2 \mid (1 + p)$.

(e) If $m \geq 3$ is square free, show $\mathbb{Z}[i\sqrt{m}]$ is not a UFD. *Hint:* Show 2, $i\sqrt{m}$, and $1 + i\sqrt{m}$ are irreducible.

EXERCISE 3.188. **(a)** Adapt the proof of Theorem 3.64 to show $\mathbb{Z}[\sqrt{2}]$ and $\mathbb{Z}[\sqrt{3}]$ are Euclidean domains with size function given by $|N(z)|$.

(b) Show $\mathbb{Z}[\sqrt{2}]$, $\mathbb{Z}[\sqrt{3}]$, and $\mathbb{Z}[\sqrt{5}]$ each have an infinite number of units. *Hint:* Find a unit $u \neq \pm 1$ and look at u^n.

(c) Show $\mathbb{Z}[\sqrt{5}]$ is not a UFD. *Hint:* Find two irreducible factorizations of 4. For irreducibility show $|a^2 - 5b^2| \neq 2$ by looking at mod(5).

EXERCISE 3.189. Let ζ be the cube root of 1 given by $\zeta = -\frac{1}{2} + i\frac{\sqrt{3}}{2}$ and let $\mathbb{Z}[\zeta] = \{a + \zeta b \mid a, b \in \mathbb{Z}\}$, called the *Eisenstein integers*.

(a) Show that every element of $\mathbb{Z}[\zeta]$ can be uniquely written in the form $a + b\zeta$ for some $a, b \in \mathbb{Z}$.

(b) Let $N(a + \zeta b) = (a + \zeta b)(a + \overline{\zeta} b) = a^2 - ab + b^2$. Show that the units of $\mathbb{Z}[\zeta]$ are $\pm\{1, \zeta, \zeta^2\}$.

(c) Show $\mathbb{Z}[\zeta]$ is a Euclidean domain with size function N.

EXERCISE 3.190. Let $\omega \in \mathbb{C}\backslash\mathbb{Z}$ with $\omega^2 \in \mathbb{Z}$. Show $\mathbb{Q}[\omega] = \{a + \omega b \mid a, b \in \mathbb{Q}\}$ is the quotient field of $\mathbb{Z}[\omega]$.

EXERCISE 3.191. Let $\omega \in \mathbb{C}\backslash\mathbb{Z}$ with $\omega^2 \in \mathbb{Z}$. Show that there is an injective homomorphism $\varphi : \mathbb{Z}[\omega] \hookrightarrow M_{2,2}(\mathbb{Z})$ with $N(z) = \det \varphi(z)$ for $z \in \mathbb{Z}[\omega]$.

EXERCISE 3.192. Let $p \in \mathbb{Z}$ be a positive prime satisfying $p = 2$ or $p \equiv 1 \bmod(4)$. Show that the decomposition $p = n^2 + m^2$ with $n, m \in \mathbb{Z}$ is unique up to multiplying by ± 1 and switching n and m.

EXERCISE 3.193. Let p be a prime in $\mathbb{Z}$. Then $p \mid (n^2+1)$ for some $n \in \mathbb{Z}$ if and only if $p = 2$ or if p is odd and $p \equiv 1 \bmod(4)$. *Hint:* See the use of Corollary 3.31 in the proof of Lemma 3.66.

EXERCISE 3.194. Write $n \in \mathbb{N}$ as a product of distinct primes $n = 2^j p_1^{k_1} \cdots p_r^{k_r} q_1^{l_1} \cdots q_s^{l_s}$ with $r, s, j, k_i, l_i \in \mathbb{Z}_{\geq 0}$, $p_i \equiv 3 \bmod(4)$, $p_i \equiv 1 \bmod(4)$. Show n can be written as a sum of two squares if and only if each k_i is even. *Hint:* $N(z_1)N(z_2) = N(z_1 z_2)$.

EXERCISE 3.195 (Euclidean Algorithm). Let D be a Euclidean domain with $a, b \in D$, $b \neq 0$. Repeatedly use the Division Algorithm to write (cf. Theorem 1.9 and Exercise 1.20)

$$\begin{aligned}
a &= bq_0 + r_0 \text{ with } \sigma(b) > \sigma(r_0),\\
b &= r_0 q_1 + r_1 \text{ with } \sigma(r_0) > \sigma(r_1),\\
r_0 &= r_1 q_2 + r_2 \text{ with } \sigma(r_1) > \sigma(r_2),\\
&\ \vdots\\
r_{N-2} &= r_{N-1} q_N + r_N \text{ with } \sigma(r_{N-1}) > \sigma(r_N) > 0,\\
r_{N-1} &= r_N q_{N+1} + 0.
\end{aligned}$$

(a) As claimed, show that the preceding procedure is always finite.
(b) Show r_N is a greatest common divisor of a and b.
(c) Use the Euclidean Algorithm to find a greatest common divisor of $x^4 - x^2 + x - 1$ and $x^3 - x^2 + x - 1$ in $\mathbb{Q}[x]$.

EXERCISE 3.196. **(a)** Let $p \in \mathbb{Z}$ be a positive prime and consider $\mathbb{Z}_{(p)} \equiv \{\frac{a}{b} \in \mathbb{Q} \mid (b,p) = 1\}$ (cf. Exercises 3.38 and 3.63). Show $\mathbb{Z}_{(p)}$ is a Euclidean domain.
(b) Show that any discrete valuation ring is a Euclidean domain. *Hint:* Use the power of the uniformizing parameter.

EXERCISE 3.197. Let p be a positive odd prime. Show that the following are equivalent:

(i) $p = n^2 + m^2$ for some $n, m \in \mathbb{Z}$.
(ii) $p \equiv 1 \pmod 4$.
(iii) The equation $x^2 = [-1]$ has a solution in $\mathbb{Z}_p$.

Hint: For the last two, write $U_p = \{g, g^2, \ldots, g^{p-1}\}$ (Corollary 3.31) and show $[-1] = g^{\frac{p-1}{2}}$; cf. the proof of Lemma 3.66.

EXERCISE 3.198 (Legendre Symbol). Let p be a positive odd prime.
(a) If $a \in \mathbb{Z}$, we say a is a *quadratic residue* $\bmod p$ if a is a perfect square $\bmod p$, i.e., if $[a] = [k]^2$ in $\mathbb{Z}_p$ for some $k \in \mathbb{Z}$. Show $\left|\{[k]^2 \mid [k] \in \mathbb{Z}_p^\times\}\right| = \frac{p-1}{2}$.
(b) Define the quadratic *Legendre symbol* by

$$\left(\frac{a}{p}\right) = \begin{cases} 0 & \text{if } p \mid a,\\ 1 & \text{if } a \text{ is a quadratic residue } \bmod p,\\ -1 & \text{if } a \text{ is not a quadratic residue } \bmod p.\end{cases}$$

Show $\left(\frac{a}{p}\right) = \left(\frac{a'}{p}\right)$ if $a \equiv a' \bmod p$.
(c) Show $\left(\frac{a}{p}\right)\left(\frac{b}{p}\right) = \left(\frac{ab}{p}\right)$.
(d) Calculate $\left(\frac{2}{5}\right)$, $\left(\frac{3}{7}\right)$, and $\left(\frac{4}{11}\right)$.

EXERCISE 3.199 (Euler's Criterion). Let p be an odd prime and let $a \in \mathbb{Z}$ with $p \nmid a$.

(a) *Euler's Criterion*: Show (cf. Exercise 3.198)

$$\left(\frac{a}{p}\right) \equiv a^{\frac{p-1}{2}} \mod p.$$

Hint: Factor $x^{p-1} - [1]$ as $(x^{\frac{p-1}{2}} - [1])(x^{\frac{p-1}{2}} + [1])$, use Fermat's Little Theorem, and show that the quadratic residues are the roots of the first term.

(b) *First Supplementary Law of Quadratic Reciprocity*: Show

$$\left(\frac{-1}{p}\right) = (-1)^{\frac{p-1}{2}}.$$

Hint: Use part (a) and the fact that $\left(\frac{-1}{p}\right) - (-1)^{\frac{p-1}{2}}$ is an integer between -2 and 2 or use Exercise 3.197.

EXERCISE 3.200 (Gauss's Lemma). Let p be an odd prime and let $a \in \mathbb{Z}$ with $p \nmid a$.

(a) *Gauss's Lemma*: Show (cf. Exercise 3.198)

$$\left(\frac{a}{p}\right) \equiv a^{\mu}$$

where $\mu = \left|\left\{[a], [2a], \ldots, \left[\frac{p-1}{2}a\right]\right\} \cap \left\{\left[\frac{p+1}{2}\right], \left[\frac{p+3}{2}\right], \ldots, [p-1]\right\}\right|$ in $\mathbb{Z}_p$. *Hint:* Let $z = [a][2a]\cdots\left[\frac{p-1}{2}a\right]$ and write z in two ways. For the first, factor out $a^{\frac{p-1}{2}}$. For the second, let $/x\backslash = [x]$ for $1 \le x \le \frac{p-1}{2}$ and let $/x\backslash = [-x]$ for $\frac{p+1}{2} \le x \le p-1$. Now apply $/\cdot\backslash$ to each term of z by factoring out $(-1)^n$ and show that the values of $/ka\backslash$ are distinct for $1 \le k \le \frac{p-1}{2}$. Compare the two factorizations and use Euler's Criterion (Exercise 3.199).

(b) *Second Supplementary Law of Quadratic Reciprocity*: Show

$$\left(\frac{2}{p}\right) = (-1)^{\frac{p^2-1}{8}}.$$

Hint: Use part (a) and consider $p = 4k+1$ and $p = 4k-1$ separately.

EXERCISE 3.201 (Quadratic Reciprocity). Let $p, q \in \mathbb{N}$ be distinct odd primes. This exercise proves the main *Law of Quadratic Reciprocity* (cf. Exercise 3.198):

$$\left(\frac{p}{q}\right)\left(\frac{q}{p}\right) = (-1)^{\frac{(p-1)(q-1)}{4}}.$$

(a) Write $\left(\frac{q}{p}\right) \equiv a^{\mu}$ with Gauss's Lemma as in Exercise 3.200. Show $\mu \bmod 2$ is the number of lattice points mod 2 in $[1, \frac{p-1}{2}] \times [1, \frac{q-1}{2}]$ lying beneath the line $py = qx$. *Hint:* Let $P = \{x \in \mathbb{Z} \mid 1 \le x \le \frac{p-1}{2}\}$ and write $\lfloor\cdot\rfloor$ for the integer part of a decimal. First show that qx can be written as either $p\left\lfloor\frac{qx}{p}\right\rfloor + x'$ or $p\left\lfloor\frac{qx}{p}\right\rfloor + p - x'$ for $x' \in P$. As x varies over all of P show x' varies over all of P. Finally, show $\sum_{x\in P} x \equiv \sum_x \left\lfloor\frac{qx}{p}\right\rfloor + \mu + \sum_{x\in P} x' \bmod 2$ to conclude $\mu \equiv \sum_{x\in P}\left\lfloor\frac{qx}{p}\right\rfloor \bmod 2$.

(b) Write $\left(\frac{p}{q}\right) \equiv a^{\nu}$ with Gauss's Lemma. Show $\nu \bmod 2$ is the number of lattice points mod 2 in $[1, \frac{p-1}{2}] \times [1, \frac{q-1}{2}]$ lying above the line $py = qx$.

(c) Show $\left(\frac{p}{q}\right)\left(\frac{q}{p}\right) = (-1)^{\frac{(p-1)(q-1)}{4}}$. *Hint:* Show there are no lattice points on the line $py = qx$.

(d) Evaluate $\left(\frac{12}{59}\right)$, $\left(\frac{17}{71}\right)$, $\left(\frac{12345}{331}\right)$, and $\left(\frac{299}{359}\right)$.

CHAPTER 4

Field Theory

Fields are some of the most useful types of rings. This chapter develops some of their important properties including the classification of finite fields and Galois theory.

1. Finite and Algebraic Extensions

1.1. Vector Spaces. The notion of a vector space is one of the most fundamental ideas in mathematics. Most students begin a serious study of vector spaces over $\mathbb{R}$ (or even over $\mathbb{C}$) in a linear algebra course. In this section we break free from the shackles of $\mathbb{R}$ and study vector spaces over an arbitrary field F. Almost all the basic ideas such as span, linear independence, and dimension trivially generalize to this fancier setting. In particular, the notion of dimension will be extremely important to our study of fields.

DEFINITION 4.1. Let F be a field. A *vector space over* F is an abelian group V (written additively) equipped with an operation called *scalar multiplication* that assigns to any $\lambda \in F$ and $v \in V$ an element called $\lambda v \in V$ satisfying the following properties. For $\lambda, \mu \in F$ and $v, w \in V$:
(1) $\lambda(\mu v) = (\lambda\mu)v$.
(2) $\lambda(v + w) = \lambda v + \lambda w$.
(3) $(\lambda + \mu)v = \lambda v + \mu v$.
(4) $1_F\, v = v$.

Elements of a vector space are often called *vectors.* Of course, any real or complex vector space from a linear algebra course is still a vector space over $\mathbb{R}$ or $\mathbb{C}$ under our expanded definition. For example, $\mathbb{R}^n$ and $\mathbb{C}^n$ are vector spaces over $\mathbb{R}$ and $\mathbb{C}$, respectively. More generally, if F is any field, then F^n is a vector space over F with vector addition given by standard addition in F^n,

$$(x_1, \ldots, x_n) + (y_1, \ldots, y_n) = (x_1 + y_1, \ldots, x_n + y_n),$$

and scalar multiplication given by

$$\lambda(x_1, \ldots, x_n) = (\lambda x_1, \ldots, \lambda x_n)$$

(Exercise 4.1).

As another example, $F[x]$ is a vector space over F with vector addition given by polynomial addition and with scalar multiplication given by multiplication by constants. As a final and extremely important example, if F and K are a fields with $F \subseteq K$, then K is a vector space over F with vector addition given by addition in K and scalar multiplication given by multiplication by F in K. For instance, $\mathbb{C}$ is a vector space over $\mathbb{R}$ and $\mathbb{Q}(\sqrt{2})$ is a vector space over $\mathbb{Q}$.

There are a number of basic notions from linear algebra over $\mathbb{R}$ that we need to carry over to more general vector spaces. The only real change is to essentially replace $\mathbb{R}$ by F everywhere. We start with the definition of span and its associated ideas.

DEFINITION 4.2. Let V be a vector space over a field F.

(1) Let $X \subseteq V$ be a nonempty collection of vectors. A *linear combination* of X is any vector of the form

$$\sum_{i=1}^{n} c_i v_i$$

for some $n \in \mathbb{N}$, $c_i \in F$, and $v_i \in X$.

(2) $X \subseteq V$ is said to *span* V over F if every vector in V is a linear combination of X, i.e., if every $v \in V$ can be written in the form

$$v = \sum_{i=1}^{n} c_i v_i$$

for some $n \in \mathbb{N}$, $c_i \in F$, and $v_i \in X$.

(3) If there exists a finite set spanning V, V is called *finite dimensional.* If V has no finite spanning set, V is called *infinite dimensional* and we write $\dim V = \infty$.

For example, in the vector space $\mathbb{R}[x]$ over $\mathbb{R}$, one linear combination of $X = \{x+1, x^2+1, x^3+1\}$ is the vector

$$5(x+1) + 6(x^2+1) + 7(x^3+1) = 7x^3 + 6x^2 + 5x + 18.$$

Since the arbitrary linear combination of X is of the form $ax^3+bx^2+cx+(a+b+c)$ for $a, b, c \in \mathbb{R}$, it is apparent that X does not span $\mathbb{R}[x]$. In fact by looking at degrees, it is easy to see that no finite set spans $\mathbb{R}[x]$ (Exercise 4.11) so that $\dim \mathbb{R}[x] = \infty$. On the other hand, obviously the set $\{1, x, x^2, x^3, \ldots\}$ spans $\mathbb{R}[x]$.

As another example, consider the vector space $\mathbb{C}$ over $\mathbb{R}$. The arbitrary linear combination of the set $\{1, i\}$ is $a + ib$ for $a, b \in \mathbb{R}$. Thus we see that $\{1, i\}$ spans $\mathbb{C}$ so that $\mathbb{C}$ is finite dimensional over $\mathbb{R}$. Of course the set $\{1\}$ does not span $\mathbb{C}$ since the set of linear combinations of $\{1\}$ is just $\mathbb{R}$. On the other extreme, the set $\{1, i, 2+2i\}$ still spans $\mathbb{C}$, though it is overkill. For instance we can write the vector $4+4i$ as a linear combination of $\{1, i, 2+2i\}$ in many ways: as $4 \cdot 1 + 4i + 0(2+2i)$, as $0 \cdot 1 + 0i + 2(2+2i)$, as $2 \cdot 1 + 2i + 1 \cdot (2+2i)$, and so forth. Eliminating such redundancy is the point of the next definition.

DEFINITION 4.3. Let V be a vector space over a field F.

(1) A nonempty set $X \subseteq V$ is said to be *linearly independent* if

$$\sum_{i=1}^{n} c_i v_i = 0 \text{ if only if } c_1 = c_2 = \cdots = c_n = 0$$

where $n \in \mathbb{N}$, $c_i \in F$, and $v_i \in X$ are distinct. Otherwise X is called *linearly dependent.*

(2) If $\mathcal{B}$ is a finite independent set spanning V, we say that $\mathcal{B}$ is a *basis* for V and that V has *dimension* $|\mathcal{B}|$, written $\dim_F V = |\mathcal{B}|$ or $\dim V = |\mathcal{B}|$.

The definition of linear independence is often viewed by linear algebra students as the poster child for confusing definitions. Part of the explanation lies with the fact that the notion of linear independence is one of the first abstract definitions a student encounters. Of course, by now, readers of this text are masters of confusing abstract definitions and will be able to amaze their friends and relations for hours. The other part of the explanation has to do with the fact that there is a bit of a disconnect between how mathematicians *think* about linear independence/dependence versus how they actually *use* the definition of linear independence/dependence.

To see this, recall a fact from linear algebra that remains true when we pass to an arbitrary field F: *a collection of at least two vectors is linearly dependent if and only if one of the vectors is a linear combination of the others* (Exercise 4.5). The connection between this statement and the definition can best be seen by an example. For instance, if v_1, v_2, v_3 are vectors in some vector space over $\mathbb{R}$, then the dependence relation

$$2v_1 + 4v_2 + 6v_3 = 0$$

is equivalent to writing, say, v_1 as a linear combination of the others,

$$v_1 = -2v_2 - 3v_3.$$

As a result, when mathematicians read that a set X is linearly independent, they often mentally reinterpret the statement as saying that no vector in X is a linear combination of the remaining vectors. This interpretation gives a much more concrete picture of the meaning of linear independence. Even so, mathematicians usually use the more abstract (though equivalent) definition of linear independence given in Definition 4.3 whenever they write proofs because of its ease of use and efficiency.

At last, turning to an example, consider the vector space $\mathbb{C}$ over $\mathbb{R}$. First we show that $\{1, i\}$ is linearly independent. As is typical in these sorts of proofs, suppose there exist $a, b \in \mathbb{R}$ so that $a \cdot 1 + bi = 0$. To establish linear independence, we must show that the only time this happens is when $a = b = 0$. Of course this is obvious since a complex number is zero if and only if its real and imaginary parts are zero. Thus $\{1, i\}$ is linearly independent. Since we have already seen that $\{1, i\}$ spans $\mathbb{C}$ over $\mathbb{R}$, it follows that $\{1, i\}$ is a basis and $\dim_{\mathbb{R}} \mathbb{C} = 2$.

As a slightly more involved example in $\mathbb{C}$ over $\mathbb{R}$, consider $\{z, w\}$ where $z = 1+i$ and $w = 2-i$. To see this set is linearly independent, again suppose that $az+bw = 0$ for some $a, b \in \mathbb{R}$. We must show that $a = b = 0$. This just involves some arithmetic:

$$\begin{aligned} az + bw &= 0, \\ a\,(1+i) + b\,(2-i) &= 0, \\ (a+2b) + (a-b)\,i &= 0. \end{aligned}$$

Thus $az + bw = 0$ implies $a + 2b = 0$ and $a - b = 0$. Subtracting these equations gives $3b = 0$ so that $b = 0$. Substituting $b = 0$ back in to either equation then gives $a = 0$ as well. Thus $\{z, w\}$ is an independent set. Since it is straightforward to see $\{z, w\}$ also spans $\mathbb{C}$, it follows that $\{z, w\}$ is another basis for $\mathbb{C}$ over $\mathbb{R}$.

The alert reader will note that we need to check that the definition of $\dim V$ in part (2) of Definition 4.3 is well defined. For this, it is necessary to verify that any two bases for V have the same cardinality. Even more fundamental, we would like to know that a basis exists for every finite dimensional vector space.

THEOREM 4.4. *Let V be a nonzero finite dimensional vector space over a field F.*

(1) *There is a basis for V.*

(2) *Any two bases for V have the same number of elements. In particular, $\dim V$ is well defined.*

PROOF. Start with part (1). Since V is finite dimensional, there exist distinct $v_1, \ldots, v_n \in V$ so that $\text{span}\{v_1, \ldots, v_n\} = V$. If $\{v_1, \ldots, v_n\}$ is independent, we are done. Otherwise, one of the vectors is a linear combination of the others, say v_n after relabeling. Then $\{v_1, \ldots, v_{n-1}\}$ clearly still spans V. Now we repeat the argument. If we eventually get down to an independent set, we are done. If we never arrive at an independent set, then we would end up with $V = \text{span}\{v_1\}$ with $\{v_1\}$ dependent. Since the first condition requires $v_1 \neq 0$ and the second condition requires $v_1 = 0$, we have a contradiction.

Turn to part (2). Let $\{v_1, \ldots, v_n\}$ be a spanning set for V and let $\{w_1, \ldots, w_m\}$ be an independent set. It suffices to show that $n \geq m$. To see that this is sufficient, let $\mathcal{B}_i$ be two bases for V. Since $\mathcal{B}_1$ spans and $\mathcal{B}_2$ is independent, we get $|\mathcal{B}_1| \geq |\mathcal{B}_2|$. Switching the roles of $\mathcal{B}_1$ and $\mathcal{B}_2$ finishes the proof.

Now we show $n \geq m$. Write $w_1 = c_1v_1 + \cdots + c_nv_n$ for some $c_i \in F$. Since $w_1 \neq 0$, some $c_i \neq 0$. After relabeling, assume $c_1 \neq 0$. Multiplying by c_1^{-1} and moving the rest of the terms to the other side, it follows that $v_1 \in \text{span}\{w_1, v_2, \ldots, v_n\}$ so that $\{w_1, v_2, \ldots, v_n\}$ is also a spanning set for V. Next write $w_2 = d_1w_1 + d_2v_2 + \cdots + d_nv_n$ for some $d_i \in F$. Since $w_2 \neq 0$, some $d_i \neq 0$. In fact, since $\{w_1, w_2\}$ are independent, we must have some $d_i \neq 0$ for $i \geq 2$. After relabeling, assume $d_2 \neq 0$. As before, we get $\{w_1, w_2, v_3, \ldots, v_n\}$ is a spanning set for V. If $n < m$, we could repeat this procedure and eventually get a spanning set of the form $\{w_1, \ldots, w_n\}$. But then w_{n+1} would be a linear combination of $\{w_1, \ldots, w_n\}$, which is a contradiction to independence. □

As examples, we have $\dim F^n = n$ since clearly $\{e_1, \ldots, e_n\}$ is a basis where e_i is the ith *standard basis vector*, $e_i = (0, \ldots, 0, 1, 0, \ldots, 0)$ with the 1 appearing in the ith spot. As another example, we have seen $\dim_{\mathbb{R}} \mathbb{C} = 2$ since $\{1, i\}$ is a basis for $\mathbb{C}$ over $\mathbb{R}$. As a twist of this example, observe that $\dim_{\mathbb{C}} \mathbb{C} = 1$ since $\{1\}$ is a basis for $\mathbb{C}$ over $\mathbb{C}$. As a final example, $\mathbb{Q}(\sqrt{2})$ is a vector space over $\mathbb{Q}$ with basis $\{1, \sqrt{2}\}$ so that $\dim_{\mathbb{Q}} \mathbb{Q}(\sqrt{2}) = 2$ (Exercise 4.7).

1.2. Finite and Algebraic Extensions.

1.2.1. *Definitions.* We now begin the application of linear algebra to fields.

DEFINITION 4.5. Let $F \subseteq K$ be fields.

(1) F is called a *subfield* of K and K is called an *extension field* of F.

(2) Viewing K as a vector space over F, we say K is a *finite extension* if $\dim_F K < \infty$ and we write

$$[K : F] = \dim_F K,$$

called the *degree* of the extension. Otherwise we say K is an *infinite extension*.

As an example of subfields/extension fields, consider $\mathbb{R} \subseteq \mathbb{C}$ or $\mathbb{Q} \subseteq \mathbb{Q}(\sqrt{2})$. We have already noted that $\dim_{\mathbb{R}} \mathbb{C} = 2$ and $\dim_{\mathbb{Q}} \mathbb{Q}(\sqrt{2}) = 2$. Thus $[\mathbb{C} : \mathbb{R}] = 2$ and $[\mathbb{Q}(\sqrt{2}) : \mathbb{Q}] = 2$ so that both are finite extensions of degree 2. Almost all of our remaining work will be devoted to studying finite extension fields. However, for the sake of completeness, we note that $\mathbb{Q} \subseteq \mathbb{R}$ is an example of an infinite extension. A fast proof of this result makes use of countability arguments (Exercise 4.11).

Next we turn to a common procedure for constructing extension fields. It is helpful to recall Definition 3.6 and Theorem 3.7 for generators of rings.

DEFINITION 4.6. Let $F \subseteq K$ be fields with $S \subseteq K$.

(1) Write

$$F[S]$$

for the ring generated by F and S in K, namely $F[S] = \langle F \cup S \rangle$.

(2) Write

$$F(S)$$

for the smallest subfield of K containing F and S, called the *field obtained by adjoining* S to F. In particular, $F(S)$ is the intersection of all subfields of K containing F and S (Exercise 4.14).

(3) The extension $F(S)$ of F is called *simple* if $|S| = 1$ and it is called *finitely generated* if $|S| < \infty$.

Concentrating on the case where $|S| = 1$, recall that Theorem 3.7 shows that

$$F[\alpha] = \left\{ \sum_{i=0}^{n} c_i \alpha^i \mid n \in \mathbb{Z}_{\geq 0},\, c_i \in F \right\}$$

for any $\alpha \in K$. If it turns out that $\alpha^i \in F$ for some i, then many of the above terms are unnecessary. For example, since $\left(\sqrt{2}\right)^2 \in \mathbb{Q}$, there is no point in writing complicated redundant expressions such as $a_0 + a_1\sqrt{2} + a_2\left(\sqrt{2}\right)^2 + \cdots$, $a_i \in \mathbb{Q}$, when just $a_0 + a_1\sqrt{2}$ will suffice. Thus $\mathbb{Q}[\sqrt{2}] = \{a_0 + a_1\sqrt{2} \mid a_0, a_1 \in \mathbb{Q}\}$.

Of course by construction $F[\alpha] \subseteq F(\alpha)$, though $F(\alpha)$ can be larger since it also needs inverses. In fact, it is easy to see

$$F(\alpha) = \left\{ \left(\sum_{i=0}^{n} c_i \alpha^i\right) \left(\sum_{i=0}^{m} d_i \alpha^i\right)^{-1} \mid n, m \in \mathbb{Z}_{\geq 0},\, c_i, d_i \in F,\, \sum_{i=0}^{m} d_i \alpha^i \neq 0 \right\}$$

(Exercise 4.15). While this looks pretty unpleasant, we will often find ourselves in a simpler situation. For example, when α is the root of an irreducible $f \in F[x]$ of degree n, we will see below (Theorem 4.12) that we actually have $F(\alpha) = F[\alpha]$ and that, moreover, $F[\alpha]$ is spanned by $\{1, \alpha, \ldots, \alpha^{n-1}\}$ over F.

We can see this concretely in the case of $F = \mathbb{Q}$, $K = \mathbb{R}$, and $\alpha = \sqrt{2}$ (use $x^2 - 2$). Since it is easy to directly check that $\mathbb{Q}[\sqrt{2}]$ is a field (Exercise 3.13), it follows from the definition that $\mathbb{Q}(\sqrt{2}) \subseteq \mathbb{Q}[\sqrt{2}]$. Since we always have $\mathbb{Q}[\sqrt{2}] \subseteq \mathbb{Q}(\sqrt{2})$, we get $\mathbb{Q}(\sqrt{2}) = \mathbb{Q}[\sqrt{2}]$ as desired. Furthermore, it turns out that every element of $\mathbb{Q}[\sqrt{2}]$ is a root of some polynomial with coefficients in $\mathbb{Q}$. For instance, $a + b\sqrt{2}$ is a root of the polynomial

$$(x - a - b\sqrt{2})(x - a + b\sqrt{2}) = x^2 - 2ax + (a^2 - 2b^2) \in \mathbb{Q}[x].$$

These observations lead to the following crucial definition.

DEFINITION 4.7. Let $F \subseteq K$ be fields.

(1) An element $u \in K$ is called *algebraic over* F if there is a nonzero $f \in F[x]$ so that $f(u) = 0$. Otherwise u is called *transcendental*.

(2) K is called an *algebraic extension* if every $u \in K$ is algebraic over F.

In the preceding discussion, we have seen that $\sqrt{2}$ is algebraic over $\mathbb{Q}$ and that $\mathbb{Q}[\sqrt{2}]$ is an algebraic extension. We will see below (Theorem 4.12) that attaching an algebraic element always produces an algebraic extension.

Of course an algebraic extension is not the only possibility. By countability arguments, it turns out that almost all elements of $\mathbb{R}$ are actually transcendental over $\mathbb{Q}$ (Exercise 4.30). For example, the numbers π and e are transcendental, though the proofs are not easy.

1.2.2. *Theorems.* With that plethora of definitions out of the way, we start our study of extensions with a few small, though useful, remarks.

LEMMA 4.8. *Let $F \subseteq K$ be fields with $\alpha, \beta \in K$.*

(1) $[K : F] = 1$ *if and only if* $K = F$.

(2) $\alpha \in F$ *if and only if* $F(\alpha) = F$ *if and only if* $[F(\alpha) : F] = 1$.

(3) $(F(\alpha))(\beta) = F(\alpha, \beta)$.

PROOF. For part (1), observe that $[K : F] = 1$ if and only if $\dim_F K = 1$. If $\dim_F K = 1$, choose $\alpha \in K$ so that $\{\alpha\}$ is a basis for K over F. In particular, there is $c \in F$ so that $1 = c\alpha$ so that $\alpha = c^{-1} \in F$. From this it follows that $K = F$. On the other hand, if $K = F$, then clearly $\{1\}$ is a basis for K over F so that $\dim_F K = 1$.

For part (2), first suppose $\alpha \in F$. Then clearly F is the smallest subfield of K containing F and α so that $F(\alpha) = F$. The converse if obvious. Since we already know $F(\alpha) = F$ if and only if $[F(\alpha) : F] = 1$, we are done.

Turn to part (3). Since $(F(\alpha))(\beta)$ is a field containing F, α, and β, $F(\alpha, \beta) \subseteq (F(\alpha))(\beta)$ by definition. For the reverse inclusion, observe that $F(\alpha, \beta)$ is a field containing $F(\alpha)$ and β so that $(F(\alpha))(\beta) \subseteq F(\alpha, \beta)$. □

Our first real result gives the dimension of an extension of an extension and will be used repeatedly. Just as Lagrange's Theorem gives a strong combinatorial constraint on the order of a subgroup of a finite group, the following theorem places a strong combinatorial constraint on the degree of an extension of a subfield.

THEOREM 4.9 (Multiplication of Degrees). *Let $F \subseteq E \subseteq K$ be fields. Then*

$$[K : F] = [K : E]\,[E : F].$$

PROOF. Assume first that $[K : E] = m$ and $[E : F] = n$ are finite. Choose bases $\{u_1, \ldots, u_m\}$ for K over E and $\{v_1, \ldots, v_n\}$ for E over F. To complete the proof, we will show $\mathcal{B} = \{u_i v_j \mid 1 \leq i \leq m \text{ and } 1 \leq j \leq n\}$ is a basis for K over F.

We begin by showing $\mathcal{B}$ spans K over F. Suppose $k \in K$. By definition, we can write $k = \sum_i e_i u_i$ for some $e_i \in E$. Similarly, we can write $e_i = \sum_j f_{i,j} v_j$ for some $f_j \in F$. Thus $k = \sum_{i,j} f_{i,j}\, u_i v_j$ as desired.

Next we show $\mathcal{B}$ is independent. If $0 = \sum_{i,j} f_{i,j}\, u_i v_j$ for some $f_{i,j} \in F$, then $0 = \sum_i \left(\sum_j f_{i,j} v_j \right) u_i$. Since $\{u_i\}$ is independent, $0 = \sum_j f_{i,j} v_j$. Since $\{v_j\}$ is independent, $f_{i,j} = 0$ as desired.

Finally, if either m or n is ∞, the above argument shows K has an infinite number of independent vectors over F. From the proof of Theorem 4.4, we see that K has no finite basis over K so that $\dim_F K = \infty$ as well. □

We begin our study of finite extensions by making explicit the structure of simple algebraic extensions. We need a lemma first.

LEMMA 4.10. *Let $F \subseteq K$ be fields and let $\alpha \in K$ be algebraic over F. There exists a unique monic polynomial $f \in F[x]$ so that $(f) = \{g \in F[x] \mid g(\alpha) = 0\}$. Moreover, f is irreducible over F.*

PROOF. Let I be the nonzero ideal $\{g \in F[x] \mid g(\alpha) = 0\}$ in $F[x]$. Since $F[x]$ is a PID, there exists $f \in F[x]$ so that $(f) = I$. By multiplying by the inverse of the leading coefficient, we may assume f is monic. Clearly the only other generators for I are associates of f (Exercise 3.173). In particular, looking at leading coefficients shows that the only monic associate is f itself.

Turning to the question of irreducibility, argue by contradiction. If f were not irreducible, we could write $f = gh$ for some nonzero $g, h \in F[x]$ of lower degree. Since $f(\alpha) = 0$, this would imply that, say, $g(\alpha) = 0$ so that $g \in I$. However this contradicts the fact that nonzero elements of I have degree at least $\deg f$ since $I = (f)$. Thus f is irreducible as desired. □

DEFINITION 4.11. Let $F \subseteq K$ be fields and let $\alpha \in K$ be algebraic. Let f be the unique irreducible monic polynomial $f \in F[x]$ so that $(f) = \{g \in F[x] \mid g(\alpha) = 0\}$. We say f is the *minimal polynomial* of α over F and we say the *degree* of α over F is $\deg f$.

Returning again to the case of $\mathbb{Q} \subseteq \mathbb{R}$, we see that the minimal polynomial of $\sqrt{2}$ over $\mathbb{Q}$ is $x^2 - 2$ so that the degree of $\sqrt{2}$ over $\mathbb{Q}$ is 2. Similarly in the case of $\mathbb{R} \subseteq \mathbb{C}$, we see that the minimal polynomial of i is $x^2 + 1$ so that the degree of i over $\mathbb{R}$ is also 2.

We are now in a position to express the structure of simple algebraic extensions in terms of quotients of polynomial rings.

THEOREM 4.12 (Simple Algebraic Extensions). *Let $F \subseteq K$ be fields and let $\alpha \in K$ be algebraic with minimal polynomial f of degree n.*

(1) $F(\alpha) = F[\alpha] \cong F[x]/(f)$.

(2) $\{1, \alpha, \ldots, \alpha^{n-1}\}$ *is a basis for $F(\alpha)$ over F. In particular, $[F(\alpha) : F] = n$.*

(3) $n \mid [K : F]$.

PROOF. For part (1), consider the surjective ring homomorphism $\varphi_\alpha : F[x] \to F[\alpha]$ by $\varphi_\alpha(g) = g(\alpha)$, $g \in F[x]$ (Theorem 3.29). By definition, $\ker \varphi_\alpha = (f)$ so that $F[\alpha] \cong F[x]/(f)$. Since f is irreducible and therefore prime (Theorem 3.51), (f) is maximal (Corollary 3.61) so that $F[x]/(f)$ is a field (Theorem 3.22). In particular, $F[\alpha]$ is a field and so $F(\alpha) = F[\alpha]$.

For part (2), the Division Algorithm allows us to write any $g \in F[x]$ as $g = fq+r$ with $q, r \in F[x]$ and $\deg r < \deg f$. Applying φ_α, it follows that $F[\alpha]$ is spanned by $\{1, \alpha, \ldots, \alpha^{n-1}\}$ over F. It remains to see that $\{1, \alpha, \ldots, \alpha^{n-1}\}$ is independent. For this, suppose $\sum_{i=0}^{n-1} c_i \alpha^i = 0$ for some $c_i \in F$. Then $h = \sum_{i=0}^{n-1} c_i x^i \in (f)$. Since $\deg h < \deg f$, we see that $h = 0$ so that each $c_i = 0$ as desired. Finally, part (3) follows from Theorem 4.9. □

The next theorem classifies all finite extensions of F. In particular, it shows they are algebraic. As a corollary, it also proves the nonobvious result that attaching an algebraic element to a field results in an algebraic extension. For example, $\sqrt[5]{2}$ is algebraic over $\mathbb{Q}$ (use $x^5 - 2$). Therefore the next theorem says that $\mathbb{Q}\left(\sqrt[5]{2}\right)$ is an algebraic extension. For example, since $\alpha = \frac{9+\sqrt[5]{2}+7\sqrt[5]{8}}{1+11\sqrt[5]{16}} \in \mathbb{Q}(\sqrt[5]{2})$, it follows that α is algebraic. However, it takes a bit of effort to find a polynomial over $\mathbb{Q}$ with α as a root.

THEOREM 4.13 (Finite Extensions). *Let $F \subseteq K$ be fields.*

(1) *If K is a finite extension, then it is algebraic.*

(2) *K is a finite extension of F if and only if K is finitely generated by algebraic elements, i.e., if and only if $K = F[\alpha_1, \ldots, \alpha_n]$ for algebraic $\alpha_i \in K$ over F.*

PROOF. For part (1), suppose K is a finite extension and let $\alpha \in K$. Applying Theorem 4.9 to $F \subseteq F(\alpha) \subseteq K$ shows

$$[K : F(\alpha)][F(\alpha) : F] = [K : F] < \infty.$$

In particular, $n = \dim_F F(\alpha) = [F(\alpha) : F] < \infty$ so that any $n+1$ elements of $F(\alpha)$ are dependent over F. Thus there exists $c_0, \ldots, c_n \in F$, not all zero, so that $\sum_{i=0}^n c_i \alpha^i = 0$. If we let $f = \sum_{i=0}^n c_i x^i \in F[x]$, clearly $f(\alpha) = 0$ with $f \neq 0$ so that α is algebraic over F. Hence K is an algebraic extension of F.

Turning to part (2), first suppose K is a finite extension. If $K = F$, there is nothing to do. Otherwise, there exists $\alpha_1 \in K \backslash F$. Notice that since $\alpha_1 \notin F$, $[F(\alpha_1) : F] \geq 2$ so that

$$[K : F] = [K : F(\alpha_1)][F(\alpha_1) : F] \geq 2[K : F(\alpha_1)] \geq 2.$$

Of course, if $K = F(\alpha_1)$, we are done. Otherwise there exists $\alpha_2 \in K \backslash F(\alpha_1)$. Again notice that since $\alpha_2 \notin F(\alpha_1)$, $[F(\alpha_1, \alpha_2) : F(\alpha_1)] \geq 2$ so that

$$\begin{aligned}[K : F] &\geq 2[K : F(\alpha_1)] = 2[K : F(\alpha_1, \alpha_2)][F(\alpha_1, \alpha_2) : F(\alpha_1)] \\ &\geq 2^2[K : F(\alpha_1, \alpha_2)] \geq 2^2.\end{aligned}$$

Of course, if $K = F(\alpha_1, \alpha_2)$, we are done. Otherwise we repeat the process. Since $[K : F] < \infty$, it is clear that the procedure terminates in a finite number of steps and gives us $K = F(\alpha_1, \ldots, \alpha_n)$ for some $\alpha_i \in K$. Each α_i is algebraic by part (1).

Conversely, suppose $K = F(\alpha_1, \ldots, \alpha_n)$ for some algebraic element $\alpha_i \in K$ over F. By Theorem 4.9 and induction, it suffices to show $[F(\alpha_1, \ldots, \alpha_i) : F(\alpha_1, \ldots, \alpha_{i-1})]$ is finite. But since α_i is algebraic over F, it is certainly algebraic over $F(\alpha_1, \ldots, \alpha_{i-1})$. Thus Theorem 4.12 shows that the extension is finite. □

The converse to Theorem 4.13, part (1), can be false—there exist infinite algebraic extensions (Exercise 4.30).

COROLLARY 4.14. *Let $F \subseteq K$ be fields. If $\alpha, \beta \in K$ are algebraic over F, then so are $\alpha \pm \beta$, $\alpha\beta$, and $\alpha\beta^{-1}$ (for $\beta \neq 0$).*

1.3. Exercises 4.1–4.34.

1.3.1. *Vector Spaces.*

EXERCISE 4.1. **(a)** If F is a field, show F^n is a vector space over F with vector addition given by the usual addition in the ring F^n and with scalar multiplication given by $\lambda(x_1, \ldots, x_n) = (\lambda x_1, \ldots, \lambda x_n)$.

(b) Show $F[x]$ is a vector space over F with vector addition given by polynomial addition and with scalar multiplication given by multiplication by constants.

(c) Let F be a field and let $F[x]_n = \{f \in F[x] \mid \deg f \leq n\}$. Show $F[x]_n$ is a vector space (though not a subring of $F[x]$ when $n \in \mathbb{N}$).

(d) If $F \subseteq K$ are fields, show K is a vector space over F with vector addition given by addition in K and with scalar multiplication given by multiplication of F in K.

(e) If F is a field of characteristic p, show F is a vector space over $\mathbb{Z}_p$ (cf. Exercise 3.81).

(f) Let F be a field. Show $M_{n,m}(F)$ is a vector space with vector addition given by matrix addition and with scalar multiplication given by scalar multiplication.

EXERCISE 4.2. Let V be a vector space over a field F. Show:

(a) $0_F\, v = 0_V$ for $v \in V$.

(b) $(-\lambda)v = \lambda(-v) = -(\lambda v)$ for $\lambda \in F$.

EXERCISE 4.3. **(a)** Show $\{1, \sqrt{2}\}$ spans $\mathbb{Q}[\sqrt{2}]$ over $\mathbb{Q}$.

(b) Show $\{1+\sqrt{2}, 1-\sqrt{2}\}$ spans $\mathbb{Q}[\sqrt{2}]$ over $\mathbb{Q}$.

(c) Show $\{1, x, x^2, \ldots\}$ spans $F[x]$ over F where F is a field.

(d) Show $\{1, x, \ldots, x^n\}$ spans $F[x]_n$, the polynomials of degree at most n.

(e) Show $\{E_{i,j} \mid 1 \leq i \leq n,\ 1 \leq j \leq m\}$ spans $M_{n,m}(F)$ over F where $E_{i,j}$ is the matrix with a 1 in the i, j^{th} position and 0's elsewhere.

(f) Show $\{([1], [0], [0]), ([1], [2], [0]), ([1], [2], [3])\}$ spans $\mathbb{Z}_5^3$ over $\mathbb{Z}_5$.

EXERCISE 4.4. **(a)** Show $\{1, \sqrt{2}\}$ is linearly independent over $\mathbb{Q}$. *Hint:* $\sqrt{2}$ is not rational.

(b) Show $\{1+\sqrt{2}, 1-\sqrt{2}\}$ is linearly independent over $\mathbb{Q}$.

(c) Show $\{1, x, x^2, \ldots\}$ is linearly independent in $F[x]$ where F is a field.

(d) Show $\{E_{i,j} \mid 1 \leq i \leq n,\ 1 \leq j \leq m\}$ is linearly independent in $M_{n,m}(F)$.

(e) Show $\{([1], [0], [0]), ([1], [2], [0]), ([1], [2], [3])\}$ is linearly independent in $\mathbb{Z}_5^3$.

EXERCISE 4.5 (Linear Dependence). Let V be a vector space over a field F and let $X \subseteq V$ be a collection of vectors with $|X| \geq 2$. Show X is linearly dependent if and only if there exist distinct $v_0, v_1, \ldots, v_n \in X$ so that v_0 is a linear combination of $\{v_1, \ldots, v_n\}$, some $n \in \mathbb{N}$.

EXERCISE 4.6 (Unique Linear Combinations). Let V be a vector space over a field F and let $\{v_\alpha\}$ be a basis for V. Show that each element of V can be written as a *unique* linear combination of $\{v_\alpha\}$.

EXERCISE 4.7. By finding a basis in each case, show:

(a) Show $\dim_{\mathbb{Q}} \mathbb{Q}[\sqrt{2}] = 2$.

(b) Show $\dim_F F[x]_n = n+1$ where F is a field and $F[x]_n$ is the set of polynomials of degree at most n.

(c) Show $\dim_F M_{n,m}(F) = nm$.

EXERCISE 4.8. **(a)** Let F be a field. Show $\{1, (1+x), (1+x)^2, \ldots\}$ spans $F[x]$.

(b) Show $\{1, (1+x), (1+x)^2, \ldots, (1+x)^n\}$ is linearly independent for each $n \in \mathbb{Z}_{\geq 0}$.

(c) For any $f \in F[x]$ with $\deg f = n$, show there exist unique $c_0, \ldots, c_n \in F$ so that $f = \sum_i c_i(1+x)^i$.

EXERCISE 4.9. If $\{v_1, \ldots, v_n\}$ is a basis of a vector space V over F, show that $\{c_1v_1,\ c_2v_2 + b_{2,1}v_1,\ c_3v_3 + b_{3,2}v_2 + b_{3,1}v_1, \ldots\}$ is also a basis for any $c_i, b_{i,j} \in F$ with $c_i \neq 0$.

EXERCISE 4.10. Let V be a vector space over a field F.

(a) If $\{v_1, \ldots, v_n\}$ spans V and $v \in V$, show $\{v_1, \ldots, v_n, v\}$ is dependent.

(b) If $\{v_1, \ldots, v_n\}$ is linearly independent in V, show $\{v_1, \ldots, v_{n-1}\}$ does not span V.

EXERCISE 4.11. **(a)** Let F be a field. Show $\dim F[x] = \infty$.

(b) Show $\dim_{\mathbb{Q}} \mathbb{R} = \infty$. *Hint:* Argue by contradiction and use the fact that $\mathbb{Q}$ is countable, but $\mathbb{R}$ is not.

EXERCISE 4.12 (Linear Independence of Characters). Let G be a group and let F be a field. An F-valued *character* of G is a homomorphism $\chi : G \to F^\times$; i.e., $\chi(g) \in F\backslash\{0\}$ and $\chi(gg') = \chi(g)\chi(g')$ for all $g, g' \in G$. This exercise shows that if $\chi_1, \ldots, \chi_n$ are distinct characters on G, then they are linearly independent as functions on G. This result is sometimes called *Dedekind's Lemma.*

(a) Prove the result when $n = 1$.

(b) Assume $n \geq 2$ and, arguing by contradiction, suppose $\{\chi_1, \ldots, \chi_n\}$ is linearly dependent. Among all the linear dependence relations, show it is possible to choose one with a minimal number of nonzero coefficients, m with $2 \leq m \leq n$. After possibly relabeling, write $\sum_{j=1}^m a_j \chi_j = 0$ for $a_j \in F^\times$.

(c) Show there is a $g_0 \in G$ so that $\chi_1(g_0) \neq \chi_m(g_0)$.

(d) Show $\sum_{j=1}^m a_j \chi_j(g_0 g) = 0$ and $\sum_{j=1}^m a_j \chi_m(g_0) \chi_j(g) = 0$. Then subtract one equation from the other to get a contradiction to the choice of minimal m.

EXERCISE 4.13 (Classification of Finite Dimensional Vector Spaces). A *linear transformation* between vector spaces V_1 and V_2 over a field F is a map $T : V_1 \to V_2$ so that $T(\lambda v_1 + v_2) = \lambda T(v_1) + T(v_2)$ for all $v_i \in V_1$ and $\lambda \in F$. We say V_1 and V_2 are *isomorphic*, $V_1 \cong V_2$, if there exists a bijective linear transformation from V_1 to V_2. If V is a vector space of dimension n over F, show

$$V \cong F^n.$$

Hint: Pick a basis for V.

1.3.2. *Finite and Algebraic Extensions.*

EXERCISE 4.14. **(a)** Let F_α, $\alpha \in \mathcal{A}$, be subfields of K. Show $\bigcap_\alpha F_\alpha$ is a field.

(b) Let $F \subseteq K$ be fields and let S be a subset of K. Set $\mathcal{G} = \{$subfields $E \subseteq K \mid F \subseteq E$ and $S \subseteq E\}$. Show $\bigcap_{E \in \mathcal{G}} E$ is the smallest subfield of K containing F and S.

EXERCISE 4.15. Let $F \subseteq K$ be fields with $\alpha \in K$. Show

$$F(\alpha) = \left\{ \left(\sum_{i=0}^{n} c_i \alpha^i \right) \left(\sum_{i=0}^{m} d_i \alpha^i \right)^{-1} \mid n, m \in \mathbb{Z}_{\geq 0},\, c_i, d_i \in F,\, \sum_{i=0}^{m} d_i \alpha^i \neq 0 \right\}.$$

EXERCISE 4.16. Show that the following elements are algebraic over $\mathbb{Q}$:

(a) $\sqrt{3}$,

(b) $\sqrt[3]{2}$,

(c) $n^{\frac{k}{m}}$, $n, m \in \mathbb{N}$ and $k \in \mathbb{Z}$,

(d) $e^{\frac{2\pi k i}{n}}$, $k \in \mathbb{Z}$ and $n \in \mathbb{N}$,

(e) $\left(\frac{6}{11}\sqrt{3} + 71\sqrt[3]{2}\right)^{213} e^{\frac{i\pi}{3}} + \frac{27}{\sqrt{3} - \sqrt[3]{2} + e^{-\frac{i\pi}{3}}}$. *Hint:* Theorem 4.13.

EXERCISE 4.17. Find the minimal polynomial over $\mathbb{Q}$ for the following:
(a) $\sqrt{5}$,
(b) $\sqrt{1+\sqrt{5}}$,
(c) $\sqrt{5}+i$.

EXERCISE 4.18. **(a)** Show $(\sqrt{2}+\sqrt{3})^3 = 11\sqrt{2}+9\sqrt{3}$.
(b) Show $\sqrt{2}, \sqrt{3} \in \mathbb{Q}\left[\sqrt{2}+\sqrt{3}\right]$.
(c) Show $\mathbb{Q}\left[\sqrt{2}+\sqrt{3}\right] = \mathbb{Q}[\sqrt{2}, \sqrt{3}]$.

EXERCISE 4.19. **(a)** Show $\left[\mathbb{Q}[\sqrt{3}] : \mathbb{Q}\right] = 2$.
(b) Show $\left[\mathbb{Q}[\sqrt[3]{2}] : \mathbb{Q}\right] = 3$.
(c) Show $\left[\mathbb{Q}[\sqrt{3}, \sqrt[3]{2}] : \mathbb{Q}\right] = 6$. *Hint:* First show $6 \mid \left[\mathbb{Q}[\sqrt{3}, \sqrt[3]{2}] : \mathbb{Q}\right]$ and then show $\left[\mathbb{Q}[\sqrt{3}, \sqrt[3]{2}] : \mathbb{Q}[\sqrt[3]{2}]\right] \leq 2$.
(d) Show $\mathbb{Q}[\sqrt[3]{2}] = \mathbb{Q}[\sqrt[3]{4}]$. *Hint:* Show $\left[\mathbb{Q}[\sqrt[3]{4}] : \mathbb{Q}\right] = 3$ and that $\mathbb{Q}[\sqrt[3]{2}] \supseteq \mathbb{Q}[\sqrt[3]{4}]$.

EXERCISE 4.20. **(a)** Show $\left[\mathbb{Q}[e^{\frac{i\pi}{3}}] : \mathbb{Q}\right] = 2$.
(b) Show $[\mathbb{Q}[i] : \mathbb{Q}] = \left[\mathbb{Q}[\sqrt{3}] : \mathbb{Q}\right] = 2$.
(c) Show $\mathbb{Q}[e^{\frac{i\pi}{3}}, i] = \mathbb{Q}[e^{\frac{i\pi}{3}}, \sqrt{3}] = \mathbb{Q}[\sqrt{3}, i]$. *Hint:* $e^{\frac{i\pi}{3}} = \frac{1}{2} + \frac{\sqrt{3}}{2}i$.
(d) Show $\left[\mathbb{Q}[e^{\frac{i\pi}{3}}, i] : \mathbb{Q}\right] = 4$. *Hint:* Since $i \notin \mathbb{Q}[\sqrt{3}]$, show $\left[\mathbb{Q}[\sqrt{3}, i] : \mathbb{Q}[\sqrt{3}]\right] = 2$.

EXERCISE 4.21. If $F \subseteq K$ are fields with $\alpha \in K$, show α is algebraic if and only if $[F(\alpha) : F] < \infty$.

EXERCISE 4.22. If $F \subseteq E \subseteq K$ are fields with $[K : F]$ prime, show $E = F$ or $E = K$.

EXERCISE 4.23. Suppose $F \subseteq E \subseteq K$ are fields and $\alpha \in K$ is algebraic over F.
(a) Show $[E(\alpha) : E] \leq [F(\alpha) : F]$. *Hint:* Minimal polynomial.
(b) Show $[E(\alpha) : F(\alpha)] \leq [E : F]$. *Hint:* Show that a basis for E over F still spans $E(\alpha)$ over $F(\alpha)$ or look at the chains $F \subseteq F(\alpha) \subseteq E(\alpha)$ and $F \subseteq E \subseteq E(\alpha)$.

EXERCISE 4.24. Suppose $F \subseteq K$ are fields with $\alpha, \beta \in K$ algebraic over F of degree n and m, respectively.
(a) Show $[F(\alpha, \beta) : F] \leq nm$. *Hint:* Look at $F \subseteq F(\alpha) \subseteq F(\alpha, \beta)$; cf. Exercise 4.23.
(b) If $(n, m) = 1$, show $[F(\alpha, \beta) : F] = nm$. *Hint:* Show $n, m \mid [F(\alpha, \beta) : F]$.
(c) Show by example that part (b) can be false when $(n, m) \neq 1$.

EXERCISE 4.25 (Algebraic Extensions of Algebraic Extensions Are Algebraic). If $F \subseteq E \subseteq K$ are fields with K algebraic over E and E algebraic over F, show K is algebraic over F. *Hint:* For $\alpha \in K$, write its minimal polynomial over E as $f = \sum_{i=1}^{n} c_i x^i$ for $c_i \in E$. Now look at $F \subseteq F[c_1, \ldots, c_n] \subseteq F[c_1, \ldots, c_n, \alpha]$.

EXERCISE 4.26. If K is a finite extension of F of odd degree and if $\alpha \in K$, show $F(\alpha) = F(\alpha^2)$. *Hint:* Theorem 4.9.

EXERCISE 4.27. Let F be a field with $\mathbb{Q} \subseteq F \subseteq \mathbb{C}$ and $[F : \mathbb{Q}] = 2$. Show there exists $\alpha \in \mathbb{C} \backslash \mathbb{Q}$ with $\alpha^2 \in \mathbb{Q}$ so that $F = \mathbb{Q}[\alpha]$. *Hint:* Start with $\alpha \in F \backslash \mathbb{Q}$, complete the square of the minimal polynomial, and then shift α by the corresponding rational number.

Exercise 4.28. Suppose K is an algebraic extension of F and that R is a ring satisfying $F \subseteq R \subseteq K$. Show R is a field. *Hint:* To get an inverse for $\alpha \in R \backslash F$, start by subtracting the constant term (can it be zero?) from the minimal polynomial and factoring out x.

Exercise 4.29 (Algebraic Closure). **(a)** Let $F \subseteq K$ be fields. If $\alpha, \beta \in K$ are algebraic over F, show there is an algebraic subfield $E \subseteq K$ containing F so that $\alpha, \beta \in E$.

(b) Let $\mathcal{A} = \{\alpha \in K \mid \alpha \text{ is algebraic over } F\}$. Show $\mathcal{A}$ is an algebraic subfield of K containing F, called the *algebraic closure* of F in K.

Exercise 4.30 (Algebraic Numbers). **(a)** Let $\mathbb{A} = \{a \in \mathbb{C} \mid a \text{ is algebraic over } \mathbb{Q}\}$, called the field of *algebraic numbers*. Show $\mathbb{A}$ is an algebraic extension of $\mathbb{Q}$ (cf. Exercise 4.29).

(b) Show $[\mathbb{A} : \mathbb{Q}] = \infty$. *Hint:* $\mathbb{Q} \subseteq \mathbb{Q}\left[\sqrt[n]{2}\right] \subseteq \mathbb{A}$.

(c) Show that the set of algebraic numbers is countable and that the set of transcendental numbers is uncountable. *Hint:* How many polynomials are there over $\mathbb{Q}$?

1.3.3. *Transcendentals.*

Exercise 4.31. Suppose $F \subseteq K$ are fields with $\tau \in K$ transcendental over F.

(a) If $\alpha \in K \backslash F$ is algebraic over F, show $\alpha \notin F(\tau)$. *Hint:* Let $p \in F[x]$ be the minimal polynomial of α. Argue by contradiction and suppose $\alpha = f(\tau)g(\tau)^{-1}$ for some $f, g \in F[x]$ with $g(\tau) \neq 0$. Now clear denominators in $p\left(\frac{f(\tau)}{g(\tau)}\right) = 0$.

(b) Conclude that any $\beta \in F(\tau) \backslash F$ is transcendental over F.

Exercise 4.32. Suppose $F \subseteq K$ are fields with $\tau \in K$ transcendental over F. Show $F(\tau) \cong F(x)$, the field of rational functions over F.

Exercise 4.33 (Algebraically Independent). Suppose $F \subseteq K$ are fields with $\alpha_1, \ldots, \alpha_n \in K$. We say $\{\alpha_1, \ldots, \alpha_n\}$ is *algebraically dependent* if there is a nonzero $f \in F[x_1, \ldots, x_n]$ so that $f(\alpha_1, \ldots, \alpha_n) = 0$. Otherwise, they are called *algebraically independent* over F. If $\{\alpha_1, \ldots, \alpha_n\}$ are algebraically independent, show $F(\alpha_1, \ldots, \alpha_n) \cong F(x_1, \ldots, x_n)$, the field of rational functions in n-variables.

Exercise 4.34. Suppose $F \subseteq K$ are fields. We say $\{\alpha_1, \ldots, \alpha_n\} \subseteq K$ is a *finite transcendence basis* if $\{\alpha_1, \ldots, \alpha_n\}$ is algebraically independent over F (Exercise 4.33) and if K is algebraic over $F(\alpha_1, \ldots, \alpha_n)$. This exercise shows that two transcendence bases have the same number of elements. This common number is called the *transcendence degree* and is written $\operatorname{tr}\deg_F K$.

(a) Suppose that $\{\alpha_1, \ldots, \alpha_m\}$ and $\{\beta_1, \ldots, \beta_n\} \subseteq K$ with K algebraic over $F(\beta_1, \ldots, \beta_n)$ and with $\{\alpha_1, \ldots, \alpha_m\}$ algebraically independent. Show $m \leq n$ and conclude that the definition of transcendence degree is well defined. *Hint:* See the ideas of Theorem 4.4.

(b) Show that $\{\alpha_1, \ldots, \alpha_m\}$ can be completed to a transcendence basis over F.

(c) If $F \subseteq K \subseteq L$, show $\operatorname{tr}\deg_F L = \operatorname{tr}\deg_K L + \operatorname{tr}\deg_F K$.

2. Splitting Fields

2.1. Splitting Fields. Consider the irreducible polynomial $f = x^2 + 1 \in \mathbb{R}[x]$. Clearly f has no roots over $\mathbb{R}$. However, we are well familiar with the fact that it is possible to factor f by passing to a larger field, namely $\mathbb{R}[i] = \mathbb{C}$. It turns out that this sort of construction can always be generalized.

DEFINITION 4.15. Let $F \subseteq K$ be fields and let $f \in F[x]$ be nonconstant with $\deg f = n$. We say K is a *splitting field* for f over F if:

(1) There exist nonzero $c_n \in F$ and $\alpha_1, \ldots, \alpha_n \in K$ so that $f = c_n(x - \alpha_1)\cdots(x - \alpha_n)$.

(2) $K = F[\alpha_1, \ldots, \alpha_n]$.

More generally, when condition (1) holds in Definition 4.15, we say that f *splits* over K. Condition (2) is then a type of minimality condition ensuring that K is no bigger than strictly necessary in order for f to completely factor (cf. Exercise 4.37). As an example, $\mathbb{C}$ is a splitting field for $x^2 + 1 \in \mathbb{R}[x]$ over $\mathbb{R}$ since $x^2 + 1$ completely factors over $\mathbb{C}$ as $(x - i)(x + i)$ and since $\mathbb{C} = \mathbb{R}[i] = \mathbb{R}[i, -i]$. Switching the fields a bit, $\mathbb{Q}[i]$ is a splitting field for $x^2 + 1 \in \mathbb{Q}[x]$ over $\mathbb{Q}$ for the same reasons. In a similar spirit, $\mathbb{Q}[\sqrt{2}]$ is a splitting field for $x^2 - 2 \in \mathbb{Q}[x]$ over $\mathbb{Q}$. However, $\mathbb{R}$ is not a splitting field for $x^2 - 2 \in \mathbb{Q}[x]$ over $\mathbb{Q}$. Even though $x^2 - 2$ splits over $\mathbb{R}$, $\mathbb{R}$ is much bigger than just $\mathbb{Q}[\sqrt{2}]$.

Our first result is that splitting fields always exist.

THEOREM 4.16. *Let F be a field and let $f \in F[x]$ be nonconstant.*

(1) *There exists an extension field E of F so that $E = F[\alpha]$ with $\alpha \in E$ satisfying $f(\alpha) = 0$.*

(2) *Splitting fields for f over F exist.*

PROOF. For part (1), use the fact that $F[x]$ is a UFD to find an irreducible factor $h \in F[x]$ of f. The big trick is to look at $E = F[x]/(h)$. Since $F[x]$ is a Euclidean domain and therefore a PID, E is a field (Theorem 3.51, Corollary 3.61, and Theorem 3.22). Clearly the inclusion $F \subseteq F[x]$ descends to an inclusion $F \subseteq E$ by identifying $c \in F$ with $c + (h) \in E$.

Set $\alpha = x + (h) \in E$. Note that $\alpha^k = x^k + (h)$ and $c\alpha = cx + (h)$ by definition of the quotient structure. Thus $g(\alpha) = g(x) + (h)$ for any $g \in F[x]$. In particular, we set everything up so that $h(\alpha) = h(x) + (h) = (h) = 0$ in E. Since $h \mid f$, we also have $f(\alpha) = 0$.

It only remains to check that $E = F[\alpha]$. For this, write $m = \deg h$. Then the Division Algorithm shows that any element of $F[x]$ is equivalent to a polynomial of degree at most $m - 1$ modulo multiples of h. Using the identity $E = F[x]/(h)$ and passing to the quotient, it follows that $\{1, \alpha, \ldots, \alpha^{m-1}\}$ spans E over F. In particular, $E = F[\alpha]$ as desired.

For part (2), write E_1 for E and α_1 for α. Since f has a root $\alpha_1 \in E_1 = F[\alpha_1]$, we can write $f = (x - \alpha_1)f_1$ for $f_1 \in E_1[x]$. If f_1 is nonconstant, part (1) again shows there is an extension field E_2 of $F[\alpha_1]$ with $\alpha_2 \in E_2$ so that $E_2 = (F[\alpha_1])[\alpha_2] = F[\alpha_1, \alpha_2]$ and $f_1(\alpha_2) = 0$. Thus we can write $f_1 = (x - \alpha_2)f_2$ for $f_2 \in E_2[x]$ and so $f = (x - \alpha_1)(x - \alpha_2)f_2$. Repeating this a total of $n = \deg f$ times, we end up with a field E_n and elements $\alpha_1, \ldots, \alpha_n \in E_n$ so that $E_n = F[\alpha_1, \ldots, \alpha_n]$ and $f = (x - \alpha_1)\cdots(x - \alpha_n)f_n$ with $\deg f_n = 0$ in $E_n[x]$. Looking at the leading coefficient shows that $f_n \in F$ as desired. $\square$

In fact, it turns out that splitting fields are essentially unique. To state this precisely, we need the following definition.

DEFINITION 4.17. Let F be a field and suppose E and K are extension fields of F.

(1) An *F-isomorphism* from E to K is an isomorphism $\varphi : E \to K$ satisfying $\varphi(c) = c$ for all $c \in F$. In other words, there is a commutative diagram (where the upward arrows are inclusions):

$$\begin{array}{ccc} E & \xrightarrow{\varphi} & K \\ \nwarrow & & \nearrow \\ & F & \end{array}$$

If $E = K$, then φ is called an *F-automorphism* of E.

(2) If an F-isomorphism from E to K exists, E and K are called *isomorphic extensions* of F.

We next show that any two splitting fields of a polynomial are isomorphic extensions. For use later, we actually prove a more general result. To set things up, suppose $\varphi : F_1 \to F_2$ is an isomorphism of fields. Then it is simple to check that φ extends to an isomorphism of rings $\varphi : F_1[x] \to F_2[x]$ by defining

$$\varphi\left(\sum_k c_k x^k\right) = \sum_k \varphi(c_k)\, x^k$$

for $c_k \in F_1$ (Exercise 3.92).

THEOREM 4.18. *Let $F_i \subseteq E_i$ be fields, let $\varphi : F_1 \to F_2$ be an isomorphism of fields, and let $h, f \in F_1[x]$ be nonconstant. Extend φ to an isomorphism $\varphi : F_1[x] \to F_2[x]$.*

(1) *Suppose that h is irreducible in $F_1[x]$, that $\alpha \in E_1$ is a root of h, and that $\beta \in E_2$ is a root of $\varphi(h)$. There exists an isomorphism $\Psi : F_1[\alpha] \to F_2[\beta]$ so that $\Psi(\alpha) = \beta$ and $\Psi(c) = \varphi(c)$ for all $c \in F$. In other words, there is a commutative diagram extending φ (where the up arrows are inclusions) and taking α to β:*

$$\begin{array}{ccc} E_1 & & E_2 \\ \uparrow & & \uparrow \\ F_1[\alpha] & \overset{\Psi}{\dashrightarrow} & F_2[\beta] \\ \uparrow & & \uparrow \\ F_1 & \underset{\varphi}{\longrightarrow} & F_2 \end{array}$$

(2) *Suppose E_1 is a splitting field for f over F_1 and E_2 is a splitting field for $\varphi(f)$ over F_2. There exists an isomorphism $\Phi : E_1 \to E_2$ so that $\Phi(c) = \varphi(c)$ for all $c \in F$. In other words, there is a commutative diagram extending φ (where the up arrows are inclusions):*

$$\begin{array}{ccc} E_1 & \overset{\Phi}{\dashrightarrow} & E_2 \\ \uparrow & & \uparrow \\ F_1 & \underset{\varphi}{\longrightarrow} & F_2 \end{array}$$

(3) *Any two splitting fields of a polynomial are isomorphic extensions.*

PROOF. Part (3) clearly follows from part (2) by using $F_1 = F_2$ and $\varphi = \mathrm{Id}$. In turn, part (2) will follow from part (1). To see this, induct on $n = \deg f$. If $n = 1$, there is nothing to do since $E_i = F_i$. Otherwise, let h be an irreducible factor of f. Since E_1 is a splitting field, there exists $\alpha_1, \ldots, \alpha_n \in E_1$ so that $E_1 = F_1[\alpha_1, \ldots, \alpha_n]$ and α_i is a root of f. Since $E_1[x]$ is a UFD, we can reindex and assume α_1 is a root of h. Similarly, there exists $\beta_1, \ldots, \beta_n \in E_2$ so that $E_2 = F_2[\beta_1, \ldots, \beta_n]$ and β_i is a root of $\varphi(f)$. As $\varphi(h)$ is a factor of $\varphi(f)$, we can reindex and assume β_1 is a root of $\varphi(h)$. Part (1) shows there exists an isomorphism $\Psi : F_1[\alpha_1] \to F_2[\beta_1]$ satisfying $\Psi|_F = \varphi$.

Now write $f = (x - \alpha_1) f_1$ for $f_1 \in F_1[\alpha_1]$. Notice E_1 is clearly a splitting field for f_1 over $F_1[\alpha_1]$ and E_2 is a splitting field for $\varphi(f_1)$ over $F_2[\beta_1]$. Thus we can apply our inductive hypothesis in the setting of f_1 with base field $F_1[\alpha_1]$ to get an isomorphism $\Phi : E_1 \to E_2$ so that $\Phi|_{F_1[\alpha_1]} = \Psi$. In particular, $\Phi|_{F_1} = \Psi|_{F_1} = \varphi$ as desired.

At last, turn to part (1). Observe that irreducibility implies h is the minimal polynomial for α over F up to a nonzero scalar multiple. In particular, we can apply Theorem 4.12 to see that $F_1[\alpha] \cong F_1[x]/(h)$. Similarly, $\varphi(h)$ is irreducible (Exercise 3.92) so that $F_2[\beta] \cong F_2[x]/(\varphi(h))$. Recall that the isomorphisms are induced by the evaluation maps so that $\varphi_\alpha : F_1[x]/(h) \to F_1[\alpha]$ is given by $\varphi_\alpha(f_1 + (h)) = f_1(\alpha)$ and $\varphi_\beta : F_2[x]/(\varphi(h)) \to F_2[\beta]$ is given by $\varphi_\alpha(f_2 + (\varphi(h))) = f_2(\beta)$ for $f_i \in F_i[x]$. For the final piece of the puzzle, observe that the isomorphism $\varphi : F_1[x] \to F_2[x]$ clearly descends to an isomorphism $\overline{\varphi} : F_1[x]/(h) \to F_2[x]/(\varphi(h))$ given by $\overline{\varphi}(f_1 + (h)) = \varphi(f_1) + (\varphi(h))$.

In diagram form, we thus have the maps

$$\begin{array}{ccc}
F_1[x]/(h) & \xrightarrow[\cong]{\overline{\varphi}} & F_2[x]/(\varphi(h)) \\
\varphi_\alpha \ \downarrow \ \cong & & \cong \ \downarrow \ \varphi_\beta \\
F_1[\alpha] & \overset{\text{hoped for } \Psi}{\dashrightarrow} & F_2[\beta] \\
\uparrow & & \uparrow \\
F_1 & \xrightarrow[\varphi]{} & F_2
\end{array}$$

To finish the proof, let $\Psi = \varphi_\beta \circ \overline{\varphi} \circ \varphi_\alpha^{-1}$. By construction, Ψ is an isomorphism. Finally we calculate that

$$\Psi(\alpha) = \left(\varphi_\beta \circ \overline{\varphi} \circ \varphi_\alpha^{-1}\right)(\alpha) = \left(\varphi_\beta \circ \overline{\varphi}\right)(x + (h)) = \left(\varphi_\beta\right)(x + (\varphi(h))) = \beta$$

and, for $c \in F_1$,

$$\Psi(c) = \left(\varphi_\beta \circ \overline{\varphi} \circ \varphi_\alpha^{-1}\right)(c) = \left(\varphi_\beta \circ \overline{\varphi}\right)(c + (h)) = \varphi_\beta(\varphi(c) + (\varphi(h))) = \varphi(c)$$

as desired. □

COROLLARY 4.19. *Let F be a field, let $f \in F[x]$ with $n = \deg f \geq 1$, and let K be a splitting field for f over F. Then $[K : F] \leq n!$.*

PROOF. Since all splitting fields are isomorphic extensions, they all have the same degree over F. Thus it suffices to look at the splitting field constructed in Theorem 4.16. Continuing the notation found there, set $E_0 = F$ and $f = f_0$. For $1 \leq i \leq n$, observe that the minimal polynomial for α_i over E_{i-1} divides f_{i-1} and that $\deg f_{i-1} = n + 1 - i$. Thus $[E_i : E_{i-1}] \leq n + 1 - i$. Theorem 4.9 finishes the proof by looking at the chain of fields $F = E_0 \subseteq E_1 \subseteq \cdots \subseteq E_n = K$. □

Of course the upper bound of $n!$ in Corollary 4.19 can sometimes be overkill. For instance, if $p \in \mathbb{N}$ is prime, then the polynomial $x^p - 1 \in \mathbb{Q}[x]$ factors as

$$x^p - 1 = (x - 1)(x^{p-1} + x^{p-2} + \cdots + x + 1)$$

with the cyclotomic polynomial $\Phi_p = x^{p-1} + x^{p-2} + \cdots + x + 1$ irreducible over $\mathbb{Q}$ (Exercise 3.145). However, the roots of Φ_p are $e^{2\pi i k/p}$ where $1 \leq k < p$. Since

$\mathbb{Q}[e^{2\pi i/p}]$ clearly contains all the roots, it follows that $K = \mathbb{Q}[e^{2\pi i/p}]$ is the splitting field for Φ_p (and $x^p - 1$). Thus $[K : \mathbb{Q}] = p$ which is much less than $p!$ (when $p > 2$).

Splitting fields turn out to have an important structural property that can be used to characterize them. Justification for the use of the word *normal* below will be seen in Corollary 4.45.

DEFINITION 4.20. Let E be an algebraic extension of F. We say E is a *normal extension* of F if any irreducible $f \in F[x]$ having a root in E actually splits over E.

For example, $\mathbb{R} \subseteq \mathbb{C}$ is a normal extension since, in fact, every real polynomial factors completely over $\mathbb{C}$ by the Fundamental Theorem of Algebra. However, $\mathbb{Q} \subseteq \mathbb{Q}[3^{1/3}]$ is not a normal extension since the irreducible polynomial $f = x^3 - 3 \in \mathbb{Q}[x]$ has a root in $\mathbb{Q}[3^{1/3}]$ but does not split. This can be seen by observing that the other two roots of f in $\mathbb{C}$, namely $3^{1/3}e^{\pm 2\pi i/3}$, are not real.

THEOREM 4.21. *Let $F \subseteq K$ be fields. Then K is the splitting field of some $f \in F[x]$ if and only if K is a finite, normal extension of F.*

PROOF. First suppose K is the splitting field of f over F. By definition, K is a finite extension. If $g \in F[x]$ is irreducible and $g(\beta) = 0$ for some $\beta \in K$, it remains to see that g splits over K. Let E be a splitting field for g over K. If $\gamma \in E$ is a root of g, it suffices to show $\gamma \in K$.

Toward this end, consider the chain of fields $F \subseteq F(\beta) \subseteq K$ and $F \subseteq F(\gamma) \subseteq K(\gamma)$. By Theorem 4.18, part (1), using $\varphi = \mathrm{Id}$, there is an F-isomorphism $\Psi : F(\beta) \to F(\gamma)$. Moving to the next link, observe that the last field of each chain is obtained from the middle field by attaching the roots of f. Thus K is still a splitting field of f over $F(\beta)$ and $K(\gamma)$ is a splitting field of f over $F(\gamma)$ (Exercise 4.48). By Theorem 4.18, part (2), with $\varphi = \Psi$, Ψ extends to an F-isomorphism (of the same name) $\Psi : K \to K(\gamma)$. Since isomorphisms preserve bases and therefore dimensions (Exercise 4.49), $[K : F] = [K(\gamma) : F]$. Thus looking at the chain $F \subseteq K \subseteq K(\gamma)$, we see that

$$[K(\gamma) : K] = \frac{[K(\gamma) : F]}{[K : F]} = 1$$

so that $K(\gamma) = K$ and $\gamma \in K$ as desired.

For the other direction, assume K is finite and normal. Use Theorem 4.13 to write $K = [\alpha_1, \dots, \alpha_n]$ with $\alpha_i \in K$ algebraic over F. Let $f_i \in F[x]$ be the minimal polynomial for α_i and let $f = \prod_i f_i$. Since K is normal, it contains all the roots of f. By construction, it is therefore the splitting field of f as desired. □

2.2. Algebraic Closures. Now that we know there always exists an extension field in which a particular polynomial factors completely, a natural question is whether there exists an extension field in which *every* polynomial factors completely. In one case we know the answer. The Fundamental Theorem of Algebra says that every polynomial over $\mathbb{R}$ (in fact, over $\mathbb{C}$) factors completely over $\mathbb{C}$.

It turns out that this result is true in general; however the proof requires the use of a mathematical axiom called Zorn's Lemma. This axiom is rather strange looking, but it is equivalent to the famous Axiom of Choice and most modern mathematicians use it freely. The Axiom of Choice appears to be a completely innocuous assumption; however it has quite surprising consequences. The relation between Zorn's Lemma and the Axiom of Choice is discussed in Exercise 4.62. For now, we simply state the definitions needed for Zorn's Lemma.

DEFINITION 4.22. A *partially ordered set* (or a *poset*) is a nonempty set S together with a relation $\leq$ on S satisfying the following:

(1) *(Reflexive)* $a \leq a$ for all $a \in S$.
(2) *(Antisymmetric)* If $a \leq b$ and $b \leq a$, then $a = b$.
(3) *(Transitive)* If $a \leq b$ and $b \leq c$, then $a \leq c$.

As an example, $\mathbb{R}$ is a partially ordered set with the standard inequality, $\leq$. In this case for any $a, b \in \mathbb{R}$, either $a \leq b$ or $b \leq a$ (which makes $\mathbb{R}$ into a *totally ordered set*). However, in general, we do not assume the relation in a partially ordered set allows us to compare every two elements. For example, let X be a nonempty set and consider the power set $S = \mathcal{P}(X)$ with the partial order $\leq$ given by inclusion. In other words for $A, B \in S$, we say $A \leq B$ if $A \subseteq B$. The necessary hypotheses of a partially ordered set are clearly satisfied (Exercise 4.54). However, if $|X| \geq 2$, there are many subsets $A, B \in S$ satisfying $A \nsubseteq B$ and $B \nsubseteq A$. In that case, we cannot write $A \leq B$ and we cannot write $B \leq A$. When this happens, we say A and B are *not comparable.*

DEFINITION 4.23. Let $(S, \leq)$ be a partially ordered set.

(1) A nonempty subset $T \subseteq S$ is *linearly ordered* (or a *chain*) if for each $a, b \in T$, either $a \leq b$ or $b \leq a$.

(2) A nonempty subset $T \subseteq S$ has an *upper bound* in S if there exists $u \in S$ so that $t \leq u$ for all $t \in T$.

(3) An element $m \in S$ is called *maximal* if the only element $c \in S$ satisfying $m \leq c$ is $c = m$.

Every subset of $\mathbb{R}$ is linearly ordered since $\mathbb{R}$ is totally ordered. However, not all subsets of $\mathbb{R}$ have an upper bound. For instance, $\mathbb{Z} \subseteq \mathbb{R}$ does not. The same reasoning shows that $\mathbb{R}$ has no maximal elements. The situation is completely different for $\mathcal{P}(X)$ with the partial ordering given by inclusion (as above). In this setting, every subset $\mathcal{P}(X)$ has an upper bound. For instance, X itself works. Moreover, X also counts as a maximal element of $\mathcal{P}(X)$.

With this notation, we are able to state Zorn's Lemma, which we take as an axiom.

AXIOM 3 (Zorn's Lemma). Let S be a partially ordered set in which each linearly ordered subset has an upper bound. Then S has a maximal element.

The following is a typical application of Zorn's Lemma.

THEOREM 4.24. *Let I be an ideal of a commutative ring R with identity. If $I \neq R$, there exists a maximal ideal M of R containing I.*

PROOF. Let $\mathcal{P} = \{J \mid I \subseteq J,\ J \neq R, \text{ and } J \text{ is an ideal in } R\}$. Notice $\mathcal{P}$ is nonempty since $I \in \mathcal{P}$. Partially order $\mathcal{P}$ by inclusion. If $T = \{J_\alpha\}$ is a linearly ordered subset of $\mathcal{P}$, we claim that $K = \bigcup_\alpha J_\alpha$ is an upper bound for T. If $K \in \mathcal{P}$, clearly $J_\alpha \leq K$ for all α as needed. Thus it is only necessary to check that $K \in \mathcal{P}$. First note that $I \subseteq K$ since $I \subseteq J_\alpha$. Next we show that K is an ideal. To first see that K is a subring, suppose $k_i \in K$ so that $k_i \in J_{\alpha_i}$. Since T is linearly ordered, we may reindex and assume $J_{\alpha_1} \subseteq J_{\alpha_2}$. Since J_{α_2} is a subring,

$k_1 - k_2, k_1k_2 \in J_{\alpha_2} \subseteq K$ as needed. To check that K is an ideal, suppose $r \in R$. Since J_{α_1} is an ideal, $rk_1 \in J_{\alpha_1} \subseteq K$ as desired. Finally, we need to see that $K \neq R$. Arguing by contradiction, suppose $1 \in K$. Then $1 \in J_\alpha$ for some α. Since J_α is an ideal, this forces $R = J_\alpha$, which is a contradiction to the fact that $J_\alpha \in \mathcal{P}$. Thus $1 \notin K$ and $K \neq R$.

Zorn's Lemma therefore says there exists a maximal element M in $\mathcal{P}$. By definition, M is an ideal containing I with $M \neq R$. If M were not a maximal ideal in R, there would exist an ideal L with $M \subsetneq L$ and $L \neq R$. However, this would imply $L \in \mathcal{P}$ with $M \leq L$ and $L \neq M$, which is a contradiction to the fact that M is a maximal element in $\mathcal{P}$. As a result, M is a maximal ideal in R as desired. □

Note that if the hypotheses of Theorem 4.24 are relaxed, it can easily happen that maximal ideals do not exist (cf. Exercise 4.58). We turn now to the question of finding an extension field in which all polynomials completely factor.

DEFINITION 4.25. **(1)** Let F be a field. An extension field $\overline{F}$ is called an *algebraic closure* of F if $\overline{F}$ is an algebraic extension in which every nonconstant $f \in F[x]$ can be written as $f = c_n(x - \alpha_1) \cdots (x - \alpha_n)$ for a nonzero $c_n \in F$ and $\alpha_1, \ldots, \alpha_n \in \overline{F}$.

(2) A field K is called *algebraically closed* if K is an algebraic closure of itself.

For example, $\mathbb{C}$ is an algebraic closure of $\mathbb{R}$ and $\mathbb{C}$ is algebraically closed. In general, algebraic closures exist, though the proof requires some care. One hard part is to avoid writing down a class of objects that is just too large to actually be a set, such as the "set of all sets" (Exercise 4.61). In this case, it is tempting to try to apply Zorn's Lemma to the "set of all algebraic extension fields of a field". However, this "set" is not really a set either. The following trick of Artin allows us to bypass some of these technicalities.

THEOREM 4.26. *Let F be a field. An algebraic closure exists.*

PROOF. First we show there exists an algebraic extension F_1 of F in which each nonconstant $f \in F[x]$ has a root. Towards this end, introduce a formal indeterminate Y_f for each nonconstant $f \in F[x]$ and let $S = \{Y_f \mid f \in F[x] \backslash F\}$. Form the polynomial ring $F[\{Y_f\}]$ of polynomials over F in the variables $\{Y_f\}$. Let I be the ideal of $F[\{Y_f\}]$ generated by $\{f(Y_f) \mid f \in F[x] \backslash F\}$. We claim that $I \neq F[\{Y_f\}]$. If not, then we can write

$$1 = \sum_{k=1}^{n} g_k f_k(Y_{f_k}) \tag{4.27}$$

for some $f_k \in F[x] \backslash F$ and $g_k \in F[\{Y_f\}]$. Since polynomials are finite sums, we can make a list of all the variables used and assume (possibly making n larger) that $g_k = g_k(Y_{f_1}, \ldots, Y_{f_n})$. Using Theorem 4.16 a finite number of times, there is an extension field E of F so that f_k has a zero $\alpha_k \in E$, $1 \leq k \leq n$. Using the evaluation map sending $Y_{f_k} \to \alpha_k$, (4.27) would show $1 = 0$, which is a contradiction. Thus $I \neq F[\{Y_f\}]$.

Theorem 4.24 therefore shows there exists a maximal ideal M in $F[\{Y_f\}]$ containing I. Thus $F_1 = F[\{Y_f\}]/M$ is a field. The map $F \to F_1$ given by $c \to c+M$ is clearly an injective homomorphism. By identifying F with its image, we may view F_1 as an extension field of F. By construction, $f \in F[x] \backslash F$ has a root in F_1, namely $Y_f + M$. It remains only to see that F_1 is algebraic. For this, note that any element of F_1 is contained in some finite extension of the form $F[Y_{f_1}+M,\ldots,Y_{f_k}+M]$ and so is algebraic.

Now we show that an algebraic closure exists. Form a chain of algebraic extensions

$$F = F_0 \subseteq F_1 \subseteq F_2 \subseteq F_3 \subseteq \cdots$$

so that each nonconstant polynomial in $F_k[x]$ has a root in F_{k+1}. Let $\overline{F} = \bigcup_{k=0}^{\infty} F_k$. Since any two elements of $\overline{F}$ are contained in some large F_k, it follows trivially that $\overline{F}$ is a field containing F. If $f \in \overline{F}[x]$ (though, in the end, we only care about starting with $f \in F[x]$), the coefficients of f are also in some large F_k so that f has a root in $F_{k+1} \subseteq \overline{F}$. In particular, it follows from induction that f can be written as $f = c_n(x-\alpha_1)\cdots(x-\alpha_n)$ for a nonzero $c_n \in \overline{F}$ and $\alpha_1,\ldots,\alpha_n \in \overline{F}$. If $f \in F[x]$, then of course $c_n \in F$. It remains only to see that $\overline{F}$ is algebraic over F. But this follows since any element of $\overline{F}$ is in some F_k which is algebraic over F. □

COROLLARY 4.28. *Let F be a field and let $\overline{F}$ be an algebraic closure of F. Then $\overline{F}$ is algebraically closed.*

PROOF. Let $\overline{f} \in \overline{F}[x]$ be nonconstant. We know there exists an extension field E of $\overline{F}$ so that $E = \overline{F}[\alpha]$ with $\alpha \in E$ and $\overline{f}(\alpha) = 0$. Then E is algebraic over F (Exercise 4.25). In particular, α is a root of a polynomial in $F[x]$. However, all polynomials in $F[x]$ have roots in $\overline{F}$ so that $\alpha \in \overline{F}$ as desired. Induction finishes the proof. □

It turns out that algebraic closures are unique up to an F-isomorphism, but we leave the proof to Exercise 4.59. However, we do prove the following result which is very useful for subfields of $\mathbb{C}$.

THEOREM 4.29. *Let $F \subseteq K$ be fields with K algebraically closed. Then $\overline{F} = \{\alpha \in K \mid \alpha \text{ is algebraic over } F\}$ is an algebraic closure for F.*

PROOF. $\overline{F}$ is easily seen to be an algebraic extension of K (Exercise 4.29). On the other hand, $\overline{F}$ clearly has enough roots to factor each nonconstant polynomial over F since K is algebraically closed. □

2.3. Exercises 4.35–4.62.

2.3.1. *Splitting Fields.*

EXERCISE 4.35. Find a splitting field $K \subseteq \mathbb{C}$ for the following polynomials over $\mathbb{Q}$ and find $[K : \mathbb{Q}]$:

(a) $x^2 + 2x - 3$,
(b) $x^3 - 2$,
(c) $x^4 - 5x^2 + 6$,
(d) $x^4 - 4x^2 - 5$,
(e) $x^4 + 1$,
(f) $x^5 - 3$.

EXERCISE 4.36. Show that $x^2 - 3$ and $x^2 + 2x - 2$ have the same splitting field in $\mathbb{C}$ over $\mathbb{Q}$.

EXERCISE 4.37. Let $F \subseteq K$ be fields and let $f \in F[x]$. Show K is a splitting field for f if and only if f splits over K but not over any proper subfield of K containing F.

EXERCISE 4.38. Let $F \subseteq K$ be fields. Show K is a splitting field for $f \in F[x]$ if and only if K is a splitting field for the product of the distinct (nonassociate) irreducible factors of f.

EXERCISE 4.39. For $f \in \mathbb{R}[x]$ nonconstant, show the splitting field for f is either $\mathbb{R}$ or $\mathbb{C}$.

EXERCISE 4.40. Let F be a field and let $f \in F[x]$ with $\deg f = 2$. Show that the splitting field for f is a simple extension of F. *Hint:* Express one root in terms of the other root and coefficients of f.

EXERCISE 4.41. Let $p \in \mathbb{N}$ be a prime. Find a splitting field K for $x^p - 2$ over $\mathbb{Q}$ and show $[K : \mathbb{Q}] = p(p-1)$. *Hint:* Attach first $2^{\frac{1}{p}}$ and then consider the cyclotomic polynomial (cf. Exercises 3.145 and 4.53).

EXERCISE 4.42. Let F be a field, let $f \in F[x]$, and let K be a splitting field for f over F with $[K : F]$ prime. Show $K = F[\alpha]$ for any $\alpha \notin F$ that is a root of f.

EXERCISE 4.43. Let $f \in \mathbb{Q}[x]$ have $\deg f = 3$ and let K be a splitting field for f over $\mathbb{Q}$.

(a) Show $[K : \mathbb{Q}] \in \{1, 2, 3, 6\}$.

(b) If f is irreducible, show $[K : \mathbb{Q}] \in \{3, 6\}$.

EXERCISE 4.44. Construct a splitting field for $x^2 + 1 \in \mathbb{Z}_3[x]$.

EXERCISE 4.45. Let $F \subseteq K$ be fields with $f, g \in F[x]$.

(a) Suppose that K contains a common root of f and g. Show f and g have a common divisor in $F[x]$ of degree at least 1. *Hint:* Try the minimal polynomial and Exercise 3.97.

(b) Conclude that f and g are relatively prime if and only if they have no common roots in any extension K of F.

EXERCISE 4.46. **(a)** Find all $\mathbb{R}$-automorphisms of $\mathbb{C}$. *Hint:* If φ is such an automorphism, what are the possible values of $\varphi(i)$?

(b) Find all $\mathbb{Q}$-automorphisms of $\mathbb{Q}[\sqrt{2}]$.

(c) Find all $\mathbb{Q}$-automorphisms of $\mathbb{R}$. *Hint:* If φ is such an automorphism, show that $\varphi(\pm x^2) \in \pm\mathbb{R}^+$ for $x \in \mathbb{R}^\times$, then show φ preserves $(0, \infty)$ and $(-\infty, 0)$, and then show φ preserves intervals with rational endpoints.

EXERCISE 4.47. Let $F \subseteq K \subseteq E$ be fields with K a splitting field for $f \in F[x]$. If φ is an F-automorphism of E, show $\varphi(K) = K$. *Hint:* Show roots of f get mapped to roots of f.

EXERCISE 4.48. Let $F \subseteq E \subseteq K$ be fields with K a splitting field for $f \in F[x]$. Show K is also a splitting field for f over E.

EXERCISE 4.49. Let $F_i \subseteq E_i$ be fields and let $\varphi : E_1 \to E_2$ be an isomorphism that restricts to an isomorphism $F_1 \to F_2$.

(a) Show φ maps a basis of E_1 over F_1 to a basis of E_2 over F_2.

(b) Show $[E_1 : F_1] = [E_2 : F_2]$.

EXERCISE 4.50. **(a)** Show $\mathbb{Q}[2^{\frac{1}{2}}]$ is a normal extension of $\mathbb{Q}$.

(b) Show $\mathbb{Q}[2^{\frac{1}{3}}]$ is not a normal extension of $\mathbb{Q}$.

EXERCISE 4.51. Let $F \subseteq K$ be fields with $[K : F] = 2$. Show K is normal. *Hint:* Division Algorithm or Exercise 4.40.

EXERCISE 4.52. Show $\mathbb{Q}[2^{\frac{1}{4}}]$ is not a splitting field of any polynomial over $\mathbb{Q}$.

EXERCISE 4.53. Let F be a field and let $a \in F^{\times}$. If K is a splitting field for $x^n - a \in F[x]$ over F, show K is also a splitting field for $(x^n - a)(x^n - 1) \in F[x]$. *Hint:* Let u be a root of $x^n - 1 \in K[x]$ in some splitting field. If $\{\alpha_j\}$ are the roots of $x^n - a$, show $u\alpha_1 = \alpha_j$ for some j.

2.3.2. *Zorn and Algebraic Closures.*

EXERCISE 4.54. Let X be a nonempty set and consider the power set $S = \mathcal{P}(X)$.

(a) For $A, B \in S$, define $A \leq B$ if $A \subseteq B$. Show $\leq$ gives a partial order on X. It is called the *inclusion* partial ordering.

(b) For $A, B \in S$, define $A \leq' B$ if $A \supseteq B$. Show $\leq'$ gives a partial order on X. It is called the *reverse inclusion* partial ordering.

EXERCISE 4.55. **(a)** For $n, m \in \mathbb{N}$, define $n \leq' m$ if $n \mid m$. Show $\leq'$ gives a partial order on $\mathbb{N}$ that is not totally ordered and has no maximal elements.

(b) For $n, m \in \mathbb{N}$, define $n \leq'' m$ if $m \mid n$. Show $\leq''$ gives a partial order on $\mathbb{N}$ that is not totally ordered but has a maximal element.

EXERCISE 4.56. Let $S = \{\text{subgroups } H \subseteq (\mathbb{Q}, +) \mid H \neq \mathbb{Q}\}$ be partially ordered by inclusion. For $k \in \mathbb{N}$, let $A_k = \left\langle \frac{1}{k!} \right\rangle$ and $B_k = \left\langle \frac{1}{2^k} \right\rangle$. Define $\mathcal{A} \subseteq S$ and $\mathcal{B} \subseteq S$ by $\mathcal{A} = \{A_k \mid k \in \mathbb{N}\}$ and $\mathcal{B} = \{B_k \mid k \in \mathbb{N}\}$.

(a) Show $A_k \subseteq A_{k+1}$ and $B_k \subseteq B_{k+1}$.

(b) Show $\bigcup_k A_k = \mathbb{Q}$ and $\bigcup_k B_k = \{\frac{a}{b} \in \mathbb{Q} \mid b = 2^j,\ j \in \mathbb{Z}_{\geq 0}\} \neq \mathbb{Q}$.

(c) Show that $\mathcal{A}$ and $\mathcal{B}$ are linearly ordered subsets of S and that $\mathcal{A}$ has no upper bound in S while $\mathcal{B}$ has an upper bound in S.

EXERCISE 4.57. Let $S = \{\text{subgroups } H \subseteq (\mathbb{Q}, +) \mid H \neq \mathbb{Q}\}$ be partially ordered by inclusion (cf. Exercise 4.56). This exercise shows S has no maximal element.

(a) By way of proof by contradiction, assume $M \subseteq S$ is maximal and choose $q \in \mathbb{Q} \backslash M$. Show $M + \langle q \rangle = \mathbb{Q}$.

(b) Conclude $\mathbb{Q}/M \cong \langle q \rangle \,/\, (\langle q \rangle \cap M)$ (cf. Exercise 2.167).

(c) Show $\langle q \rangle \cap M \neq \{0\}$. *Hint:* Write $q = \frac{a}{b}$, pick nonzero $\frac{m}{n} \in M$, and look at am.

(d) Since $\langle q \rangle \cong \mathbb{Z}$, conclude $|\mathbb{Q}/M| = k < \infty$.

(e) Show that $k\mathbb{Q} \subseteq M$ and then that $q \in M$ to get a contradiction. *Hint:* $\frac{q}{k}$.

EXERCISE 4.58. Let $R = (x) \subseteq \mathbb{Q}[x]$. For H a subgroup of $(\mathbb{Q}, +)$, let $I_H = Hx + x^2\mathbb{Q}[x]$. If H is proper, this exercise shows I_H is an ideal of R contained in no maximal ideal.

(a) Show R is a commutative ring with no identity and that $I_H \subseteq R$ is an ideal.

(b) If M is an ideal of R containing I_H, show $M = I_J$ for some subgroup J of $\mathbb{Q}$ with $M = R$ if and only if $J = \mathbb{Q}$.

(c) If H is proper, conclude I_H is contained in no maximal ideal of R (cf. Exercise 4.57).

(d) Does this violate Theorem 4.24?

EXERCISE 4.59. Let F be a field and let $\overline{F}$ and $\overline{F}'$ be algebraic closures of F. Show there is an F-isomorphism from $\overline{F}$ to $\overline{F}'$. *Hint:* Let $\mathcal{P} = \{(E, E', \varphi) \mid F \subseteq E \subseteq \overline{F}$ and $F \subseteq E' \subseteq \overline{F}'$ are fields and $\varphi : E \to E'$ is an F-isomorphism$\}$ with partial order $(E_1, E_1', \varphi_1) \leq (E_2, E_2', \varphi_2)$ whenever $E_1 \subseteq E_2$, $E_1' \subseteq E_2'$, and $\varphi_2|_{E_1} = \varphi_1$. Now Zornify.

EXERCISE 4.60 (Cardinality and the Schröder-Bernstein Theorem). If X and Y are nonempty sets, we say $|X| = |Y|$, $|X| \leq |Y|$, or $|X| \geq |Y|$, respectively, if there exists a bijective, injective, or surjective function $f : X \to Y$.

(a) Show $|X| \leq |Y|$ if and only if $|Y| \geq |X|$.

(b) Show $|X| \leq |Y|$ or $|Y| \leq |X|$. This result is called *trichotomy* and turns out to be logically equivalent to Zorn's Lemma. *Hint:* Consider $\{$injective $f : U \to V \mid U \subseteq X$ and $V \subseteq Y\}$. Order by inclusion and compatibility and use Zorn's Lemma to find a maximal element $f : U_0 \to V_0$. Show that either $U_0 = X$ in which case $|X| \leq |Y|$ or that $V_0 = Y$ in which case $|Y| \leq |X|$.

(c) Show $|X| = |Y|$ if and only if $|X| \leq |Y|$ and $|Y| \leq |X|$, called the *Schröder-Bernstein* Theorem. *Hint:* Let $f : X \to Y$ and $g : Y \to X$ be injective. Let $X_\infty = \{x \in X \mid \left(f^{-1} \circ g^{-1}\right)^n (x) \in g(Y)$ and $g^{-1} \circ \left(f^{-1} \circ g^{-1}\right)^n (x) \in f(X)$ for all $n \in \mathbb{Z}_{\geq 0}\}$, $X_X = \{x \in X \mid$for some $m \in \mathbb{Z}_{\geq 0}$, $\left(f^{-1} \circ g^{-1}\right)^n (x) \in g(Y)$, $g^{-1} \circ \left(f^{-1} \circ g^{-1}\right)^{n-1} (x) \in f(X)$, but $\left(f^{-1} \circ g^{-1}\right)^m (x) \notin g(Y)$ for $n \in \mathbb{Z}_{\geq 0}$, $n < m\}$, and $X_Y = \{x \in X \mid$for some $m \in \mathbb{Z}_{\geq 0}$, $\left(f^{-1} \circ g^{-1}\right)^{n+1} (x) \in g(Y)$, $g^{-1} \circ \left(f^{-1} \circ g^{-1}\right)^n (x) \in f(X)$, but $g^{-1} \circ \left(f^{-1} \circ g^{-1}\right)^m (x) \notin f(X)$ for $n \in \mathbb{Z}_{\geq 0}$, $n < m\}$. Similarly, define Y_∞, Y_X, and Y_Y. Now use f to map X_∞ to Y_∞ and X_X to Y_X and use g^{-1} to map X_Y to Y_Y.

(d) Show $|X| < |\mathcal{P}(X)|$, i.e., $|X| \leq |\mathcal{P}(X)|$ but $|X| \neq |\mathcal{P}(X)|$. *Hint:* Show any $f : X \to \mathcal{P}(X)$ cannot be surjective by considering $Y = \{x \in X \mid x \notin f(x)\}$ and checking that $Y \in f(X)$ leads to a contradiction.

EXERCISE 4.61 (Russell's Paradox). Show that the "set of all sets" is actually nonsense. *Hint:* If the set of all sets really made sense as a set, let R be the subset of all sets that do not contain themselves, $R = \{A \mid A \notin A\}$. Now, is $R \in R$ or is $R \notin R$?

EXERCISE 4.62 (The Axiom of Choice and a Few Friends). It is a fact that the following four statements are equivalent (though we do not prove the final equivalence here as it requires much more effort):

(i) *Hausdorff Maximal Principle:* Every partially ordered set has a maximal linearly ordered subset.

(ii) *Zorn's Lemma:* If S is a partially ordered set in which each linearly ordered subset has an upper bound, then S has a maximal element.

(iii) *Well Ordering Principle:* Every nonempty set can be well ordered. (A linearly ordered set for which every nonempty subset has a (unique) minimal element is called *well ordered.* For example, $\mathbb{Z}_{\geq 0}$ is well ordered.)

(iv) *Axiom of Choice:* If $\mathcal{A}$ is a set and X_α is a set for each $\alpha \in \mathcal{A}$, then there exists a function $f : \mathcal{A} \to \bigcup_{\alpha \in \mathcal{A}} X_\alpha$ so that $f(\alpha) \in X_\alpha$ (called a *choice function*). In other words, we can "choose" an element from each X_α.

(a) Show (i) $\Longleftrightarrow$ (ii).

(b) Show (ii) $\Longrightarrow$ (iii). *Hint:* For $E_i \subseteq X$ with $\leq_i$ a well ordering on E_i, say $(E_1, \leq_1) \leq (E_2, \leq_2)$ if $E_1 \subseteq E_2$, $\leq_2 |_{E_1} = \leq_1$, and if elements of E_1 are $\leq_2$ elements of $E_2 \backslash E_1$.

(c) Show (iii) $\Longrightarrow$ (iv). *Hint:* Well order $\bigcup_{\alpha \in \mathcal{A}} X_\alpha$ and let $f(\alpha)$ be the minimal element of X_α.

3. Finite Fields

In this section we classify all finite fields. Remarkably, it turns out that their order is a complete invariant.

To get us started, let R be a ring. Clearly $I = \{k \in \mathbb{Z} \mid kr = 0, \text{ all } r \in R\}$ is an ideal on $\mathbb{Z}$. As $\mathbb{Z}$ is a PID, there is a unique $n \in \mathbb{Z}_{\geq 0}$ that generates I. In other words, the only integers that kill all of R are of the form nk for $k \in \mathbb{Z}$.

DEFINITION 4.30. Let R be a ring. The unique $n \in \mathbb{Z}_{\geq 0}$ satisfying $\langle n \rangle = \{k \in \mathbb{Z} \mid kr = 0, \text{ all } r \in R\}$ is called the *characteristic* of R and is denoted $\operatorname{char} R = n$.

For example, clearly $\operatorname{char} \mathbb{Z} = 0$ and $\operatorname{char} \mathbb{Z}_n = n$ for $n \in \mathbb{N}$.

THEOREM 4.31. **(1)** *Let R be a ring. Then $\langle 1_R \rangle \cong \mathbb{Z}$ if $\operatorname{char} R = 0$ and $\langle 1_R \rangle \cong \mathbb{Z}_n$ if $\operatorname{char} R = n$, $n \in \mathbb{N}$.*

(2) *If F is a field, $\operatorname{char} F$ is either 0 or p for some prime p.*

(3) *If $\operatorname{char} F = 0$, then F contains a subfield isomorphic to $\mathbb{Q}$.*

(4) *If $\operatorname{char} F = p$, then F contains a subfield isomorphic to $\mathbb{Z}_p$.*

(5) *Finite fields have positive characteristic.*

PROOF. First we show that

$$\{k \in \mathbb{Z} \mid kr = 0, r \in R\} = \{k \in \mathbb{Z} \mid k \cdot 1_R = 0, r \in R\}.$$

Clearly it suffices to show that the right-hand side is contained in the left-hand side. For this, suppose $k \cdot 1_R = 0$. Then

$$kr = k \cdot 1_R \cdot r = 0 \cdot r$$

as desired.

For part (1), consider the map $\varphi : \mathbb{Z} \to R$ given by $\varphi(k) = k \cdot 1_R$. It is trivial to check that this map is a homomorphism with image $\langle 1_R \rangle$. Thus $\langle 1_R \rangle \cong \mathbb{Z}/\ker \varphi$. Since $\ker \varphi = \langle \operatorname{char} R \rangle$, we are done. For part (2), use part (1) and the fact that $F \supseteq \langle 1_R \rangle$ has no zero divisors. For part (3), it is straightforward to check that the injective homomorphism $\varphi : \mathbb{Z} \to F$ extends to an injective homomorphism $\varphi : \mathbb{Q} \to F$ by $\varphi(\frac{r}{s}) = \varphi(r)\varphi(s)^{-1}$ (cf. Exercise 3.81). Part (4) is already done by part (1). Part (5) follows from part (1) and the observation that $\mathbb{Z}$ is infinite. $\square$

If F is a finite field, it is common to allow some ambiguity and to simply say $\mathbb{Z}_p \subseteq F$ where $p = \operatorname{char} F$. Of course, the technically correct statement is that $\mathbb{Z}_p \cong \langle 1_R \rangle \subseteq F$, but this is usually too cumbersome to write.

COROLLARY 4.32. *Let F be a finite field with $p = \operatorname{char} F$. Then $|F| = p^n$ where $n = [F : \mathbb{Z}_p]$.*

PROOF. F is a vector space over the subfield $\mathbb{Z}_p$ and is clearly finite dimensional. If $\dim_{\mathbb{Z}_p} F = n$ with $\{\alpha_1, \ldots, \alpha_n\}$ a basis, then each element of F can be written uniquely in the form $c_1\alpha_1 + \cdots + c_n\alpha_n$ with $c_1, \ldots, c_n \in \mathbb{Z}_p$ (cf. Exercise 4.6). Conversely, any choice of $c_1, \ldots, c_n \in \mathbb{Z}_p$ gives rise to an element of F. In particular, there is a bijection from $\mathbb{Z}_p \times \cdots \times \mathbb{Z}_p$ to F. Thus $|F| = |\mathbb{Z}_p \times \cdots \times \mathbb{Z}_p| = p^n$. $\square$

One idea for constructing a field of order p^n is to find an irreducible $f \in \mathbb{Z}_p[x]$ of degree n. Since $\mathbb{Z}_p[x]$ is a Euclidean domain and therefore a PID, $\mathbb{Z}_p[x]/(f)$ is a field of order p^n (Corollary 3.61 and Theorem 3.22). For example, $f = 1+x+x^2 \in \mathbb{Z}_2[x]$ is irreducible so that $\mathbb{Z}_2[x]/(f)$ is a field of order 4. Since the Division Algorithm shows $\mathbb{Z}_2[x]/(f)$ is spanned by $1 + (f)$ and $x + (f)$ over $\mathbb{Z}_2$, it is easy to write a multiplication table for the field. For example, $1 + x + x^2 + (f) = 0 + (f)$ so

$$(x + (f))\,(1 + x + (f)) = x + x^2 + (f) = -1 + (f) = 1 + (f).$$

In general, it is easy to similarly check that

$\cdot$	$0+(f)$	$1+(f)$	$x+(f)$	$1+x+(f)$
$0+(f)$	$0+(f)$	$0+(f)$	$0+(f)$	$0+(f)$
$1+(f)$	$0+(f)$	$1+(f)$	$x+(f)$	$1+x+(f)$
$x+(f)$	$0+(f)$	$x+(f)$	$1+x+(f)$	$1+(f)$
$1+x+(f)$	$0+(f)$	$1+x+(f)$	$1+(f)$	$x+(f)$

If we write 1 for $1 + (f)$, 0 for (f), a for $x + (f)$, and $b = a^{-1}$ for $1 + x + (f)$, the multiplication looks a bit less convoluted:

$\cdot$	0	1	a	b
0	0	0	0	0
1	0	1	a	b
a	0	a	b	1
b	0	b	1	a

The point is that working with a particular finite field is usually remarkably explicit and straightforward. However, to get a uniform existence and uniqueness proof, it turns out to be simpler to work with the idea of splitting fields.

THEOREM 4.33. *Let $p, n \in \mathbb{N}$ with p prime. Up to isomorphism, there exists a unique field $\mathbb{F}_{p^n}$ (alternately called* GF(p^n)*) with $|\mathbb{F}_{p^n}| = p^n$. In fact, $\mathbb{F}_{p^n}$ may be realized as a splitting field of $x^{p^n} - x \in \mathbb{Z}_p[x]$.*

PROOF. Let $f = x^{p^n} - x \in \mathbb{Z}_p[x]$ and choose a splitting field F of f over $\mathbb{Z}_p$. Notice that the formal derivative satisfies $f' = -1$ so that Theorem 3.30 shows all the roots of f are simple. Thus there are distinct $a_k \in F$ so that $f = \prod_{k=1}^{p^n}(x - \alpha_k)$ with $F = \mathbb{Z}_p[\alpha_1, \ldots, \alpha_{p^n}]$.

In fact, we claim that $F = \{\alpha_1, \ldots, \alpha_{p^n}\}$ (the actual set!), which will take care of existence. To see this, let $R = \{\alpha_1, \ldots, \alpha_{p^n}\}$. If we show that R contains $\mathbb{Z}_p$ and that it is a subfield of F, then we are done since then R is clearly the smallest subfield of F containing $\mathbb{Z}_p$ and $\{\alpha_1, \ldots, \alpha_{p^n}\}$, which is the definition of $\mathbb{Z}_p(\alpha_1, \ldots, \alpha_{p^n}) = \mathbb{Z}_p[\alpha_1, \ldots, \alpha_{p^n}]$.

To check the first condition, n iterated applications of Fermat's Little Theorem show that each $\alpha \in \mathbb{Z}_p$ is a root of f so that $\mathbb{Z}_p \subseteq R$ as desired. To check the second condition, we use the *Frobenius endomorphism* $\varphi_{\mathrm{Fr}} : F \to F$ given by

$$\varphi_{\mathrm{Fr}}(c) = c^p$$

for $c \in F$. Since char $F = p$, the Binomial Theorem shows φ_{Fr} is a homomorphism (cf. Exercise 3.43). Since F is a field, $\ker \varphi_{\mathrm{Fr}} = \{0\}$ so that φ_{Fr} is injective. Since F is finite, φ_{Fr} must also be surjective and therefore an automorphism. In particular, n iterated applications of φ_{Fr} show that $\alpha_i - \alpha_j$, $\alpha_i\alpha_j$, and α_i^{-1} (for $\alpha_i \neq 0$) are roots of f and are thus in R. Hence R is a subfield as desired.

For uniqueness, suppose F' is another field of order p^n. As already noted, we may view $\mathbb{Z}_p \subseteq F'$. Since $\left|(F')^\times\right| = p^n - 1$, Corollary 2.29 shows $c^{p^n-1} = 1$ for $c \in F' \backslash \{0\}$. In particular, every element of F' satisfies f. Clearly F' is therefore also a splitting field for f over $\mathbb{Z}_p$. Hence Theorem 4.18 shows F and F' are isomorphic and we are done. □

The above proof realizes $\mathbb{F}_{p^n}$ as the splitting field of $x^{p^n} - x \in \mathbb{Z}_p[x]$. Alternately, since we know that $\mathbb{F}_{p^n}^\times$ is cyclic (Corollary 3.31), $\mathbb{F}_{p^n}^\times$ is actually a simple algebraic extension of $\mathbb{Z}_p$ (just attach the generator). Thus

$$\mathbb{F}_{p^n} \cong \mathbb{Z}_p[x]/(q)$$

for some (any) irreducible $q \in \mathbb{Z}_p[x]$ with $\deg q = n$ (in particular, such a q always exists). Therefore our earlier strategy for explicitly working with a finite field by means of quotients of polynomial rings always works.

3.1. Exercises 4.63–4.81.

EXERCISE 4.63. What is the characteristic of $\mathbb{Z}_{15} \times \mathbb{Z}_6$?

EXERCISE 4.64. Write multiplication tables and find a multiplicative generator for:

(a) $\mathbb{F}_4$,
(b) $\mathbb{F}_8$,
(c) $\mathbb{F}_9$.

EXERCISE 4.65. Show $\mathbb{F}_4 \not\cong \mathbb{Z}_4$.

EXERCISE 4.66. Factor $x^8 - x$ over $\mathbb{Z}$, $\mathbb{Z}_2$, and $\mathbb{F}_8$ (cf. Exercises 3.146 and 3.136).

EXERCISE 4.67. Show $\mathbb{Z}_2[x]/(x^3 + x + 1) \cong \mathbb{Z}_2[x]/(x^3 + x^2 + 1)$.

EXERCISE 4.68. Realize the following as quotients of polynomial rings:
(a) $\mathbb{F}_{16}$,
(b) $\mathbb{F}_{25}$,
(c) $\mathbb{F}_{27}$,
(d) $\mathbb{F}_{125}$.

EXERCISE 4.69. For p prime, show each element of $\mathbb{Z}_p$ has a unique p^{th} root.

EXERCISE 4.70. Show $\mathbb{F}_{p^n} \backslash \{0\}$ is the set of $(p^n - 1)^{\text{th}}$ roots of 1 in $\mathbb{F}_{p^n}$.

EXERCISE 4.71. Let K be a subfield of $\mathbb{F}_{p^n}$. Show $\mathbb{F}_{p^n}$ is a simple algebraic extension of K. *Hint:* Corollary 3.31.

EXERCISE 4.72. **(a)** Let $p \in \mathbb{Z}$ be a prime with $p \equiv 3 \bmod(4)$. Show $\mathbb{Z}[i]/(p) \cong \mathbb{F}_{p^2}$. *Hint:* Theorem 3.67 and Exercise 3.68.

(b) Let $p \in \mathbb{Z}$ be a prime with $p = 2$ or $p \equiv 1 \bmod(4)$ and factor $p = z\bar{z}$ for $z \in \mathbb{Z}[i]$. Show $\mathbb{Z}[i]/(p) \cong \mathbb{Z}[i]/(z) \times \mathbb{Z}[i]/(\bar{z}) \cong \mathbb{F}_p \times \mathbb{F}_p$. (Of course this is not isomorphic to $\mathbb{F}_{p^2}$ since it is not a field.)

EXERCISE 4.73. This exercise shows $x^4 + 1$ is irreducible over $\mathbb{Q}$ though its reduction mod p is reducible for every prime.
(a) Show $x^4 + 1$ is irreducible over $\mathbb{Q}$. *Hint:* $x \to x + 1$.
(b) Show $x^4 + 1$ is reducible over $\mathbb{Z}_2$.

(c) If $p \neq 2$ is prime, show $8 \mid (p^2 - 1)$, deduce $(x^8 - 1) \mid \left(x^{p^2-1} - 1\right)$, and conclude that $x^4 + 1$ splits over $\mathbb{F}_{p^2}$.

(d) If α is a root of $x^4 + 1$, show $[\mathbb{Z}_p[\alpha] : \mathbb{Z}_p] \leq 2$ and conclude $x^4 + 1$ is reducible over $\mathbb{Z}_p$.

EXERCISE 4.74. **(a)** Let $n, m \in \mathbb{N}$ and use the Division Algorithm to write $n = mq + r$ with $0 \leq r < m$. Over any field, show

$$x^n - 1 = (x^m - 1)\left(x^{n-m} + x^{n-2m} + \cdots + x^r\right) + (x^r - 1).$$

(b) Let $a \in \mathbb{Z}$ with $a \geq 2$. Show $(a^m - 1) \mid (a^n - 1)$ if and only if $m \mid n$.

EXERCISE 4.75 (Subfields of $\mathbb{F}_{p^n}$). **(a)** Suppose F, E are subfields of $\mathbb{F}_{p^n}$ with $|F| = |E|$. Show $F = E$. *Hint:* Consider the roots of $x^{|F|} - x$ in $\mathbb{F}_{p^n}$.

(b) If K is a subfield of $\mathbb{F}_{p^n}$, show $K \cong \mathbb{F}_{p^m}$ for some m satisfying $m \mid n$. *Hint:* Use the fact that $\mathbb{F}_{p^n}$ is a vector space over K to see $|K| = p^m$.

(c) If $m \mid n$, show $\mathbb{F}_{p^n}$ has a unique subfield isomorphic to $\mathbb{F}_{p^m}$. *Hint:* Show $x^{p^m} - x$ divides $x^{p^n} - x$ (Exercise 4.74) and look at the splitting field of $x^{p^m} - x$ in $\mathbb{F}_{p^n}$ over $\mathbb{Z}_p$.

EXERCISE 4.76. Show that each element of $\mathbb{F}_{p^n}$ is the sum of two squares. *Hint:* For $\alpha \in \mathbb{F}_{p^n}$, show that the order of both $\{\beta^2 \mid \beta \in \mathbb{F}_{p^n}\}$ and $\{\alpha - \gamma^2 \mid \gamma \in \mathbb{F}_{p^n}\}$ is greater than $\frac{1}{2}|\mathbb{F}_{p^n}|$.

EXERCISE 4.77 (Nonsimple Roots of Irreducible Polynomials). Let F be a field, let $f \in F[x]$ be irreducible, let K be a splitting field for f over F, and let $\alpha \in K$ be a root of f.

(a) If $f' \neq 0$, show f and f' are relatively prime in $F[x]$ and conclude that α is a simple root of f. *Hint:* Theorem 3.30.

(b) Show α is a nonsimple root if and only if $p = \operatorname{char} F > 0$ and if $f(x) = g(x^p)$ for some $g \in F[x]$.

EXERCISE 4.78 (Fixed Points of Frobenius). Let φ_{Fr} be the Frobenius automorphism on $\mathbb{F}_{p^n}$.

(a) Show that the set of fixed points of φ_{Fr} is $\mathbb{F}_p$. *Hint:* Use Fermat's Little Theorem and consider $x^p - x$.

(b) If $1 \leq m \leq n$, show that the number of fixed points of φ_{Fr}^m is at most p^m.

(c) If $m \mid n$, show that the fixed point set of φ_{Fr}^m is $\mathbb{F}_{p^m}$ (cf. Exercise 4.75).

(d) If $1 \leq m \leq n$, show that the fixed point set of φ_{Fr}^m is $\mathbb{F}_{p^{(m,n)}}$.

EXERCISE 4.79 (Automorphisms of $\mathbb{F}_{p^n}$). Choose $\alpha \in \mathbb{F}_{p^n}$ so that $\mathbb{F}_{p^n} = \mathbb{Z}_p[\alpha]$ and let f be the minimal polynomial of α over $\mathbb{Z}_p$ (of degree n).

(a) Show that any automorphism of $\mathbb{F}_{p^n}$ fixes $\mathbb{Z}_p$. *Hint:* $x^p - x$.

(b) Show that an automorphism of $\mathbb{F}_{p^n}$ is completely determined by its value on α.

(c) Write φ_{Fr} for the Frobenius automorphism of $\mathbb{F}_{p^n}$. Show $|\varphi_{\text{Fr}}| = n$ (cf. Exercise 4.78).

(d) Show the roots of f are $\{\varphi_{\text{Fr}}^r(\alpha) \mid 1 \leq r \leq n\}$.

(e) Conclude that the set of automorphisms of $\mathbb{F}_{p^r}$ is $\{\varphi_{\text{Fr}}^r \mid 1 \leq r \leq n\} \cong \mathbb{Z}_n$.

EXERCISE 4.80 (Factoring $x^{p^n} - x$). **(a)** Show that the degree of an irreducible factor of $x^{p^n} - x \in \mathbb{Z}_p[x]$ divides n. *Hint:* Minimal polynomial and $[\mathbb{F}_{p^n} : \mathbb{Z}_p] = n$.

(b) Let $q \in \mathbb{Z}_p[x]$ be irreducible with $\deg q = d$. Show $q \mid \left(x^{p^d} - x\right)$ but $q^2 \nmid \left(x^{p^d} - x\right)$. *Hint:* Note that $\mathbb{Z}_p[x]/(q) \cong \mathbb{F}_{p^d}$ is a normal extension of $\mathbb{Z}_p$ to see that q splits in $\mathbb{F}_{p^d}$.

(c) If $d \mid n$, show $q \mid \left(x^{p^n} - x\right)$ (cf. Exercise 4.75).

(d) Conclude that $x^{p^n} - x$ is the product over all $d \mid n$ of each distinct irreducible monic polynomial of degree d in $\mathbb{Z}_p[x]$.

(e) Use part (d) to inductively list the product of all distinct irreducible monic polynomials of degree d in $\mathbb{Z}_2[x]$ for $1 \leq d \leq 4$ (there exist fast computer algorithms for factoring polynomials over $\mathbb{Z}_p$ which then give an efficient method for computing irreducible polynomials over $\mathbb{Z}_p$).

EXERCISE 4.81 ($\overline{\mathbb{Z}}_p$). If $m \mid n$, let $\iota_{m,n} : \mathbb{F}_{p^m} \hookrightarrow \mathbb{F}_{p^n}$ be the inclusion from Exercise 4.75 set up so that $\iota_{n,r} \circ \iota_{m,n} = \iota_{m,r}$ when $n \mid r$. Define

$$\bigcup_{\rightarrow} \mathbb{F}_{p^n}$$

to be the set of equivalence classes of the disjoint union of $\mathbb{F}_{p^n}$ for $n \in \mathbb{N}$ equipped with the equivalence relation generated by $\alpha \sim \beta$ (and $\beta \sim \alpha$) for $\alpha \in \mathbb{F}_{p^m}$ and $\beta \in \mathbb{F}_{p^n}$ when $m \mid n$ and $\iota_{m,n}(\alpha) = \beta$. For $\gamma \in \mathbb{F}_{p^j}$ and $\delta \in \mathbb{F}_{p^k}$, define binary operations on $\bigcup_{\rightarrow} \mathbb{F}_{p^n}$ by setting $[\gamma] + [\delta] = [\iota_{j,jk}(\gamma) + \iota_{k,jk}(\delta)]$ and $[\gamma] \cdot [\delta] = [\iota_{j,jk}(\gamma) \cdot \iota_{k,jk}(\delta)]$.

(a) Show that the binary operations on $\bigcup_{\rightarrow} \mathbb{F}_{p^n}$ are well defined and yield an infinite field.

(b) Show $\bigcup_{\rightarrow} \mathbb{F}_{p^n}$ is algebraic over $\mathbb{Z}_p$ and that it contains, up to isomorphism, all finite extensions of $\mathbb{Z}_p$.

(c) Conclude that $\bigcup_{\rightarrow} \mathbb{F}_{p^n}$ is the algebraic closure of $\mathbb{Z}_p$.

4. Galois Theory

Galois theory describes a remarkable relationship between automorphisms of a field and its subfields.

4.1. Galois Groups. Recall the notion of an F-automorphism from Definition 4.17.

DEFINITION 4.34. Let $F \subseteq K$ be fields. The *Galois group* of K over F, denoted $\operatorname{Gal}_F K$, is the set of all F-automorphisms of K equipped with the binary operation of composition of functions.

Before plunging into examples, theorems, and other nitty-gritty, let us take a moment to get a panoramic view of where we are heading with the Galois group. For sufficiently well behaved fields $F \subseteq K$ (*Galois* extensions, Definition 4.43), we will eventually show that there is a bijection between the sets

$$\{\text{intermediate fields between } F \text{ and } K\} \longleftrightarrow \{\text{subgroups of } \operatorname{Gal}_F K\}$$

(Corollary 4.45). Of course, making precise and proving this surprising statement takes a fair amount of work. Since there is no time like the present, let us roll up our sleeves and get busy.

Clearly $\operatorname{Gal}_F K$ is a group containing at least the identity map. As a first example, we compute $\operatorname{Gal}_{\mathbb{R}} \mathbb{C}$. This means we need to find all automorphisms σ of $\mathbb{C}$ that fix $\mathbb{R}$. Since every $z \in \mathbb{C}$ can be written as $z = a + ib$ for $a, b \in \mathbb{R}$, it follows that $\sigma(z) = a + \sigma(i)b$. In other words, σ is completely determined by $\sigma(i)$. Since $\iota^2 = -1$, it follows that $\sigma(i)^2 = -1$ so that $\sigma(i) = \pm i$. In fact, either choice gives rise to an $\mathbb{R}$-automorphism of $\mathbb{C}$. The choice of $+i$ yields the identity map and the choice of $-i$ yields complex conjugation, well known to be an $\mathbb{R}$-automorphism of $\mathbb{C}$. Thus $\mathrm{Gal}_{\mathbb{R}}\,\mathbb{C} = \{\,\mathrm{Id}, \sigma\}$ where $\sigma(z) = \overline{z}$. Thus $\mathrm{Gal}_{\mathbb{R}}\,\mathbb{C} \cong \mathbb{Z}_2$.

In the above example of $\mathbb{R} \subseteq \mathbb{C}$, notice that $\mathbb{C} = \mathbb{R}[i]$ and that $\pm i$ (the roots of its minimal polynomial $x^2 + 1$) played a crucial role in determining $\mathrm{Gal}_{\mathbb{R}}\,\mathbb{C}$. Such phenomenon always happens for finite extensions. In particular, Theorem 4.35 below uses these ideas to calculate the Galois group of any simple algebraic extension.

THEOREM 4.35. *Let $F \subseteq K$ be fields.*

(1) *If $\alpha \in K$ is a root of $f \in F[x]$ and if $\sigma \in \mathrm{Gal}_F\, K$, then $\sigma(\alpha)$ is also a root of f.*

(2) *If $K = F(\alpha_1, \ldots, \alpha_n)$ and if $\sigma \in \mathrm{Gal}_F\, K$, then σ is completely determined by the values of $\sigma(\alpha_i)$, $1 \leq i \leq n$. In other words, if $\sigma' \in \mathrm{Gal}_F\, K$ also satisfies $\sigma'(\alpha_i) = \sigma(\alpha_i)$, then $\sigma' = \sigma$.*

(3) *If $K = F[\beta]$ for $\beta \in K$ algebraic over F with minimal polynomial $f_\beta \in F[x]$, let $\{\beta = \beta_1, \beta_2, \ldots, \beta_k\}$ be the distinct roots of f_β appearing in K. Then $\mathrm{Gal}_F\, K = \{\sigma_1, \ldots, \sigma_k\}$ where each F-automorphism σ_i, $1 \leq i \leq k$, is uniquely determined by the condition $\sigma_i(\beta) = \beta_i$.*

PROOF. For part (1), write $f = \sum_k c_k x^k$, $c_k \in F$. Applying σ to the equation $0 = \sum_k c_k \alpha^k$ shows $0 = \sum_k c_k \sigma(\alpha)^k$ so that $\sigma(\alpha)$ is a root of f as desired. For part (2), observe that every element of $F(\alpha_1, \ldots, \alpha_n)$ can be written as a quotient of terms of the form $\sum_{k_1,\ldots,k_n} c_{k_1,\ldots,k_n} \alpha_1^{k_1} \cdots \alpha_n^{k_n}$ for $c_{k_1,\ldots,k_n} \in F$ and $k_i \in \mathbb{Z}_{\geq 0}$. Since σ and σ' are both F-automorphisms and since $\sigma(\alpha_i)^{k_i} = \sigma'(\alpha_i)^{k_i}$, σ and σ' agree on all of $F(\alpha_1, \ldots, \alpha_n)$ so that $\sigma = \sigma'$. For part (3), first observe that each element of $\mathrm{Gal}_F\, K$ must be of the form σ_i for some i by parts (1) and (2). To see that such a σ_i actually exists, use Theorem 4.18, part (1) (with $F_1 = F \subseteq F[\beta]$, $F_2 = F \subseteq F[\beta_i]$, $\varphi = \mathrm{Id}$, and $h = f_\beta$). ☐

Note that $\mathrm{Gal}_F\, K$ is always a finite group for finite extensions $F \subseteq K$ by Theorem 4.35, parts (1) and (2) (Exercise 4.84). Before computing further examples, we add two additional results that are very useful in the setting of splitting fields. Though a bit unconventional, let us view Theorem 4.36 as a continuation of Theorem 4.35 and so continue the numbering from there.

THEOREM 4.36. *Let $F \subseteq K$ be fields.*

(4) *If K is the splitting field of a polynomial over F and if $u, v \in K$, then there exists $\sigma \in \mathrm{Gal}_F\, K$ so that $\sigma(u) = v$ if and only if u and v have the same minimal polynomial over F.*

(5) *If K is the splitting field of $f \in F[x]$ with m distinct roots, then $\sigma \in \mathrm{Gal}_F\, K$ permutes the distinct roots of f and we may identify $\mathrm{Gal}_F\, K$ with a subgroup of S_m.*

PROOF. For part (4), let $f_u \in F[x]$ be the minimal polynomial for u. If $\sigma \in \mathrm{Gal}_F\, K$ exists so that $\sigma(u) = v$, then part (1) shows v is a root of f_u. By irreducibility of the minimal polynomial, it follows that f_u is also the minimal polynomial of v. Conversely suppose u and v have the same minimal polynomial and consider the two extensions $F \subseteq F[u]$ and $F \subseteq F[v]$. Starting with the identity map on F, Theorem 4.18, part (1), shows there exists an isomorphism $\sigma : F[u] \to F[v]$ extending the identity map on F and so that $\sigma(u) = v$. In particular, σ is an F-isomorphism from $F[u]$ to $F[v]$. Since K is a splitting field for some polynomial over F, K remains a splitting field for the same polynomial over $F[u]$ and over $F[v]$. Thus Theorem 4.18, part (2), shows σ can be extended to an isomorphism $\sigma : K \to K$. In particular, σ is an F-automorphism of K (so $\sigma \in \mathrm{Gal}_F K$) satisfying $\sigma(u) = v$ as desired.

For part (5), write $K = F[\alpha_1, \ldots, \alpha_m]$ where the distinct roots of f are $\{\alpha_1, \ldots, \alpha_m\}$ and $f = c\prod_k (x - \alpha_k)^{m_k}$. If $\sigma \in \mathrm{Gal}_F K$, recall that σ extends to an isomorphism $\sigma : K[x] \to K[x]$ fixing $F[x]$ (Exercise 3.92). Thus $f = c\prod_k (x - \sigma(\alpha_k))^{m_k}$. In particular, the roots of f can also be written as $\{\sigma(\alpha_1), \ldots, \sigma(\alpha_m)\}$. Since each α_i is distinct, it follows that σ acts as a permutation on the set of roots of f. Part (2) finishes the proof. □

Theorems 4.35 and 4.36 allow us to compute more examples of Galois groups. For example, consider $\mathbb{Q} \subseteq \mathbb{Q}[2^{\frac{1}{3}}]$. Here $2^{\frac{1}{3}}$ is a root of the irreducible polynomial $f = x^3 - 2$. Since the roots of f in $\mathbb{C}$ are $\{2^{\frac{1}{3}}, 2^{\frac{1}{3}}\omega, 2^{\frac{1}{3}}\omega^2\}$ where $\omega = \frac{-1+i\sqrt{3}}{2}$ and $\omega^2 = \frac{-1-i\sqrt{3}}{2}$, the only root of f in $\mathbb{Q}[2^{\frac{1}{3}}] \subseteq \mathbb{R}$ is $2^{\frac{1}{3}}$. Thus $\mathrm{Gal}_{\mathbb{Q}}\, \mathbb{Q}[2^{\frac{1}{3}}] = \{\mathrm{Id}\}$ by part (3) of Theorem 4.35. Notice that $\mathbb{Q}[2^{\frac{1}{3}}]$ is neither normal nor a splitting field of any polynomial over $\mathbb{Q}$ which (we will soon see) accounts for the small Galois group.

Consider now the case of

$$\mathbb{Q} \subseteq \mathbb{Q}[\sqrt{2}, \sqrt{3}].$$

If $\sigma \in \mathrm{Gal}_{\mathbb{Q}}\, \mathbb{Q}[\sqrt{2}, \sqrt{3}]$, examination of the minimal polynomials $x^2 - 2$ and $x^2 - 3$ shows $\sigma(\sqrt{2}) = \pm\sqrt{2}$ and $\sigma(\sqrt{3}) = \pm\sqrt{3}$. In particular, there are at most 4 elements of the Galois group, $\{\mathrm{Id}, \sigma_2, \sigma_3, \sigma_4\}$, determined by mapping

$$\begin{array}{llll} \sqrt{2} \overset{\mathrm{Id}}{\to} \sqrt{2}, & \sqrt{2} \overset{\sigma_2}{\to} -\sqrt{2}, & \sqrt{2} \overset{\sigma_3}{\to} \sqrt{2}, & \sqrt{2} \overset{\sigma_4}{\to} -\sqrt{2}, \\ \sqrt{3} \overset{\mathrm{Id}}{\to} \sqrt{3}, & \sqrt{3} \overset{\sigma_2}{\to} \sqrt{3}, & \sqrt{3} \overset{\sigma_3}{\to} -\sqrt{3}, & \sqrt{3} \overset{\sigma_4}{\to} -\sqrt{3}. \end{array}$$

A priori, we only know that $\mathrm{Id} \in \mathrm{Gal}_{\mathbb{Q}}\, \mathbb{Q}[\sqrt{2}, \sqrt{3}]$. However, since $\mathbb{Q}[\sqrt{2}, \sqrt{3}]$ is the splitting field of $(x^2-2)(x^2-3)$, Theorem 4.36, part (4), guarantees that there exists an element of the Galois group sending $\sqrt{2} \to -\sqrt{2}$ and one (possibly the same one) sending $\sqrt{3} \to -\sqrt{3}$. In other words, besides Id, we know that either σ_2 and σ_3 (and therefore $\sigma_4 = \sigma_2 \circ \sigma_3$) are in $\mathrm{Gal}_{\mathbb{Q}}\, \mathbb{Q}[\sqrt{2}, \sqrt{3}]$ or that just $\sigma_4 \in \mathrm{Gal}_{\mathbb{Q}}\, \mathbb{Q}[\sqrt{2}, \sqrt{3}]$. To pin down exactly which case is valid, consider the extension $\mathbb{Q}[\sqrt{2}] \subseteq \mathbb{Q}[\sqrt{2}, \sqrt{3}]$ and use Theorem 4.18, part (1) (with $F_1 = F_2 = \mathbb{Q}[\sqrt{2}]$ and $h = x^2 - 3$) to show there exists a $\mathbb{Q}[\sqrt{2}]$-isomorphism of $\mathbb{Q}[\sqrt{2}, \sqrt{3}]$ sending $\sqrt{3} \to -\sqrt{3}$. Since a $\mathbb{Q}[\sqrt{2}]$-isomorphism is simply a $\mathbb{Q}$-isomorphism fixing $\sqrt{2}$, it follows that this isomorphism must be σ_3. Thus $\mathrm{Gal}_{\mathbb{Q}}\, \mathbb{Q}[\sqrt{2}, \sqrt{3}] = \{\mathrm{Id}, \sigma_2, \sigma_3, \sigma_4\}$ and we are done. Note that since $\{\mathrm{Id}, \sigma_2, \sigma_3, \sigma_4\} \cong \{\mathrm{Id}, \sigma_2\} \times \{\mathrm{Id}, \sigma_3\}$, it follows that $\mathrm{Gal}_{\mathbb{Q}}\, \mathbb{Q}[\sqrt{2}, \sqrt{3}] \cong \mathbb{Z}_2 \times \mathbb{Z}_2$.

As a warning, when we view $\mathrm{Gal}_{\mathbb{Q}}\, \mathbb{Q}[\sqrt{2}, \sqrt{3}]$ as subgroup of S_4 by looking at the action of $\mathrm{Gal}_{\mathbb{Q}}\, \mathbb{Q}[\sqrt{2}, \sqrt{3}]$ on $\{\sqrt{2}, -\sqrt{2}, \sqrt{3}, -\sqrt{3}\}$, we get a proper subgroup of S_4 since $|\mathbb{Z}_2 \times \mathbb{Z}_2| = 4$ but $|S_4| = 24$. In other words, not every permutation of roots automatically gives rise to an element of the Galois group.

For a final example, consider the splitting field of $x^3 - 2$ over $\mathbb{Q}$, namely $\mathbb{Q}[2^{\frac{1}{3}}, \omega]$ where $\omega = \frac{-1+i\sqrt{3}}{2}$. Write the three roots of $x^3 - 2$ as $u_k = 2^{\frac{1}{3}}\omega^k$ for $1 \leq k \leq 3$.

By looking at the action of $\mathrm{Gal}_{\mathbb{Q}}\,\mathbb{Q}[2^{\frac{1}{3}},\omega]$ on $\{u_1,u_2,u_3\}$, we view $\mathrm{Gal}_{\mathbb{Q}}\,\mathbb{Q}[2^{\frac{1}{3}},\omega]$ as a subgroup of S_3. Since there is an element of the Galois group mapping u_1 to u_2 by Theorem 4.36, part (4), that element must correspond to either (12) or (123). Similarly, (23) or (123) is in $\mathrm{Gal}_{\mathbb{Q}}\,\mathbb{Q}[2^{\frac{1}{3}},\omega]$. Since it is easy to see (12) and (23) or (12) and (123) generate S_3, it follows that $\mathrm{Gal}_{\mathbb{Q}}\,\mathbb{Q}[2^{\frac{1}{3}},\omega]$ is either all of S_3 or just $\langle(123)\rangle \cong \mathbb{Z}_3$. To decide which case is correct, use Theorem 4.18, part (1) (with $F_1 = F_2 = \mathbb{Q}[2^{\frac{1}{3}}]$ and $h = \left(x^3-2\right)/(x-u_3))$ to show there exists a $\mathbb{Q}[2^{\frac{1}{3}}]$-isomorphism of $\mathbb{Q}[2^{\frac{1}{3}},\omega]$ sending $u_1 \to u_2$ and therefore fixing u_3. In particular, (12) is an element of the Galois group so that $\mathrm{Gal}_{\mathbb{Q}}\,\mathbb{Q}[2^{\frac{1}{3}},\omega] \cong S_3$. In this case we get all possible permutations of the roots.

4.2. Separability. It turns out that the splitting field of a polynomial with simple roots has an especially nicely behaved Galois group. This type of behavior will become even more pronounced in §4.3. As a result, we introduce the following notion.

DEFINITION 4.37. Let $F \subseteq K$ be fields, let nonconstant $f \in F[x]$, and let E be a splitting field for f over F.

(1) We say f is *separable* if f has $\deg f$ distinct roots in E, i.e., if all the roots of f in E are simple.

(2) We say $\alpha \in K$ is *separable* over F if α is algebraic and its minimal polynomial is separable.

(3) We say K is *separable* over F if each element of K is separable over F.

For example, $x^2-2 \in \mathbb{Q}[x]$ is separable since it has two distinct roots in $\mathbb{Q}[\sqrt{2}]$, namely $\pm\sqrt{2}$. As a result, $\sqrt{2}$ is separable over $\mathbb{Q}$ since its minimal polynomial is x^2-2. In fact, it is easy to see $\mathbb{Q}[\sqrt{2}]$ is a separable extension of $\mathbb{Q}$. To verify this, observe that the minimal polynomial for $a \in \mathbb{Q}$ is $x-a$ which is clearly separable and that the minimal polynomial for $a+b\sqrt{2} \in \mathbb{Q}[\sqrt{2}]$, $b \neq 0$, is $\left(x-a-b\sqrt{2}\right)\left(x-a+b\sqrt{2}\right) = x^2-2ax+a^2-2b^2$ which is also separable. For an example of a nonseparable polynomial, consider $(x^2-2)^2 \in \mathbb{Q}[x]$ which has repeated roots by construction. Finding an example of an irreducible nonseparable polynomial is more difficult (Exercise 4.100).

THEOREM 4.38. *Let F be a field and let $f \in F[x]$ be irreducible.*

(1) *If* $\operatorname{char} F = 0$, *then f is separable.*

(2) *If* $\operatorname{char} F = p > 0$, *then f is separable if and only if $f(x)$ is not of the form $g(x^p)$ for some $g \in F[x]$.*

PROOF. Let α be a root of f in a splitting field. By Theorem 3.30, a is simple if and only if a is not a root of f'. Since f is irreducible, we in fact claim that α is simple if and only if $f' \neq 0$. To prove this, first consider the case of $f' \neq 0$. In this situation, f and f' are relatively prime in $F[x]$ since f is irreducible and f' is of lower degree (but not zero). Since $F[x]$ is a PID, we can then write $1 = fg + f'h$ for some $g, h \in F[x]$. In particular, since $1 \neq 0$, evaluating at α shows α cannot be a root of f' so that α is simple. On the other hand, if $f' = 0$, clearly α is a root of f' so that α is not simple.

Of course, when $\operatorname{char} F = 0$, $f' \neq 0$ by looking at the leading coefficient. This finishes part (1). If $\operatorname{char} F = p$, write $f = \sum_k c_k x^k$. Then by definition of the derivative, $f' = 0$ if and only if $c_k = 0$ when $p \nmid k$ if and only if $f = \sum_j c_{pj} x^{pj}$. Using $g = \sum_j c_{pj} x^j$ finishes part (2). $\square$

As an immediate corollary, minimal polynomials are separable when $\operatorname{char} F = 0$. Therefore we get the following.

COROLLARY 4.39. *If* $\operatorname{char} F = 0$, *algebraic extensions of* F *are separable.*

It turns out that the above corollary is also true for finite fields (Exercise 4.98), though it can be false for infinite fields of characteristic p (Exercise 4.100).

As a final result on separable extensions, we show the remarkable fact that every finite separable extension is actually simple. For instance, this means there exists a single $\alpha \in \mathbb{Q}[\sqrt{2}, e^{\frac{2\pi i}{17}}, 4^{\frac{1}{3}} + \sqrt{5}]$ so that $\mathbb{Q}[\sqrt{2}, e^{\frac{2\pi i}{17}}, 4^{\frac{1}{3}} + \sqrt{5}] = \mathbb{Q}[\alpha]$.

THEOREM 4.40 (Primitive Element Theorem). *Let* K *be a finite separable extension of* F. *Then* K *is a simple extension.*

PROOF. It turns out that the method of proof depends on whether F is finite or infinite. In the case where F is finite, then so is K. Hence Corollary 3.31 shows $K^\times$ is cyclic. In particular, if α is a generator of $K^\times$, then clearly $K = F[\alpha]$ and the theorem is finished.

Assume now that F is infinite. By Theorem 4.13, $K = F[\alpha_1, \ldots, \alpha_n]$ with α_i algebraic over F. By induction on n and writing

$$F[\alpha_1, \ldots, \alpha_n] = (F[\alpha_1, \ldots, \alpha_{n-1}])\,[\alpha_n],$$

it suffices to prove the theorem in the case of $n = 2$. For $c \in F$, consider $\beta = \alpha_1 + c\alpha_2$. We hope to choose c so that $K = F[\beta]$. Clearly $F[\beta] \subseteq F[\alpha_1, \alpha_2]$. Since $\alpha_1 = \beta - c\alpha_2$, it follows that $F[\alpha_1, \alpha_2] \subseteq F[\beta]$ if and only if $\alpha_2 \in F[\beta]$. Let $f_i \in F[x]$ be the minimal polynomial of α_i over F and let $g \in (F[\beta])[x]$ be the minimal polynomial of α_2 over $F[\beta]$. It therefore suffices to find a c so that $\deg g = 1$.

Clearly $g \mid f_2$. The trick is to look at $h_1(x) = f_1(\beta - cx)$. Since $h_1(\alpha_2) = f_1(\alpha_1) = 0$, we also have $g \mid h_1$. Choose a splitting field for $f_1 f_2 g$ and write the roots of f_i as $\alpha_{i,j}$ and the roots of g as γ_k. Each γ_k is simultaneously a root of f_2 and a root of h_1. Thus $\gamma_k = \alpha_{2,i}$ and $\beta - c\gamma_k = \alpha_{1,j}$ for some i, j depending on k. Substituting $\beta = \alpha_1 + c\alpha_2$ and $\gamma_k = \alpha_{2,i}$ into the last equation shows $(\alpha_1 + c\alpha_2) - c\alpha_{2,i} = \alpha_{1,j}$. In turn, this requires either $\alpha_2 = \alpha_{2,i}$ (and $\alpha_1 = \alpha_{1,j}$, not that we care) or

$$c = \frac{\alpha_{1,j} - \alpha_1}{\alpha_2 - \alpha_{2,i}}$$

when $\alpha_2 \neq \alpha_{2,i}$. Since F is infinite, we can choose c so that $c \neq \frac{\alpha_{1,j} - \alpha_1}{\alpha_2 - \alpha_{2,i}}$ for any i, j when $\alpha_2 \neq \alpha_{2,i}$. When this is done, it follows that α_2 is the only root of g. Since K is separable, α_2 is a simple root of f_2 and therefore a simple root of g. Thus $\deg g = 1$ as desired. □

If K is a finite separable extension of F and if we write $K = F[\beta]$ for $\beta \in K$, the element β is sometimes called a *primitive element* for K over F.

4.3. Galois Correspondence.

DEFINITION 4.41. Let $F \subseteq K$ be fields. Write $G = \operatorname{Gal}_F K$.

(1) If H is a subgroup of G, the *fixed field* of H is

$$K_H = \{k \in K \mid \sigma(k) = k \text{ for all } \sigma \in H\}.$$

(2) If E is an intermediate field between F and K, $F \subseteq E \subseteq K$, the subgroup of G fixing E is

$$G^E = \{\sigma \in G \mid \sigma(\alpha) = \alpha \text{ for all } \alpha \in E\}.$$

By construction, observe that we may alternately write

$$G^E = \operatorname{Gal}_E K$$

(Exercise 4.107).

(3) The *Galois correspondence* is given by the maps

$$\{\text{intermediate fields } E \text{ between } F \text{ and } K\} \underset{\Psi_{\text{group}}}{\overset{\Psi_{\text{field}}}{\rightleftarrows}} \{\text{subgroups } H \text{ of } \operatorname{Gal}_F K\}$$

where

$$\Psi_{\text{field}}(E) = G^E,$$
$$\Psi_{\text{group}}(H) = K_H.$$

It is easy to see that the fixed field K_H really is a subfield of K containing F and that the G^E really is a subgroup of G (Exercise 4.107). In nice cases (*Galois* extensions, Definition 4.43), it will turn out that Ψ_{field} and Ψ_{group} are inverses to each other so that there is a bijection from the set of intermediate fields between F and K and the set of subgroups of $\operatorname{Gal}_F K$. However, for arbitrary finite extensions, we can only guarantee that Ψ_{field} is surjective (Theorem 4.44). When we toss in the adjectives normal and separable, we will then get injectivity (Corollary 4.45).

As an example, consider again the extension $\mathbb{Q} \subseteq \mathbb{Q}[\sqrt{2}, \sqrt{3}]$. There are five easy-to-spot intermediate fields between $\mathbb{Q}$ and $\mathbb{Q}[\sqrt{2}, \sqrt{3}]$, namely $\mathbb{Q}$, $\mathbb{Q}[\sqrt{2}]$, $\mathbb{Q}[\sqrt{3}]$, $\mathbb{Q}[\sqrt{6}]$, and $\mathbb{Q}[\sqrt{2}, \sqrt{3}]$. Though not at all obvious at this point in time, in fact these are all the intermediate fields. Drawing an arrow to represent inclusion, the set of intermediate fields is pictorially represented as

$$\begin{array}{ccccc} & & \mathbb{Q}[\sqrt{2}, \sqrt{3}] & & \\ & \nearrow & \uparrow & \nwarrow & \\ \mathbb{Q}[\sqrt{2}] & & \mathbb{Q}[\sqrt{6}] & & \mathbb{Q}[\sqrt{3}] \\ & \nwarrow & \uparrow & \nearrow & \\ & & \mathbb{Q} & & \end{array}.$$

Turning to the group side, recall our computation in §4.1 that $\operatorname{Gal}_{\mathbb{Q}} \mathbb{Q}[\sqrt{2}, \sqrt{3}] = \{\mathrm{Id}, \sigma_2, \sigma_3, \sigma_4\}$. It is an easy exercise to see that $\operatorname{Gal}_{\mathbb{Q}} \mathbb{Q}[\sqrt{2}, \sqrt{3}] \cong \mathbb{Z}_2 \times \mathbb{Z}_2$ has exactly five subgroups, namely $\{\mathrm{Id}\}$, $\{\mathrm{Id}, \sigma_2\}$, $\{\mathrm{Id}, \sigma_3\}$, $\{\mathrm{Id}, \sigma_4\}$, and $\{\mathrm{Id}, \sigma_2, \sigma_3, \sigma_4\}$. Drawing an arrow to represent inclusion (going down in this picture for reasons soon to be made apparent), the set of subgroups is pictorially represented as

$$\begin{array}{ccccc} & & \{\mathrm{Id}\} & & \\ & \swarrow & \downarrow & \searrow & \\ \{\mathrm{Id}, \sigma_3\} & & \{\mathrm{Id}, \sigma_4\} & & \{\mathrm{Id}, \sigma_2\} \\ & \searrow & \downarrow & \swarrow & \\ & & \{\mathrm{Id}, \sigma_2, \sigma_3, \sigma_4\} & & \end{array}.$$

Next we calculate Ψ_{field}. To simplify notation, write $G = \operatorname{Gal}_{\mathbb{Q}} \mathbb{Q}[\sqrt{2}, \sqrt{3}]$. Recall that Ψ_{field} maps each intermediate field E to G^E, the subgroup of elements fixing everything in E. For instance, if $E = \mathbb{Q}[\sqrt{2}]$, it is clear that $\sigma \in G$ fixes each element of $\mathbb{Q}[\sqrt{2}]$ if and only if $\sigma(\sqrt{2}) = \sqrt{2}$. By inspection, this is $\{\mathrm{Id}, \sigma_3\}$. Thus $\Psi_{\text{field}}\left(\mathbb{Q}[\sqrt{2}]\right) = \{\mathrm{Id}, \sigma_3\}$. Similarly, the set of $\sigma \in G$ fixing $\sqrt{6}$ is $\{\mathrm{Id}, \sigma_4\}$ and the set of $\sigma \in G$ fixing $\sqrt{3}$ is $\{\mathrm{Id}, \sigma_2\}$. Thus $\Psi_{\text{field}}\left(\mathbb{Q}[\sqrt{6}]\right) = \{\mathrm{Id}, \sigma_4\}$ and $\Psi_{\text{field}}\left(\mathbb{Q}[\sqrt{3}]\right) = \{\mathrm{Id}, \sigma_2\}$. If $E = \mathbb{Q}[\sqrt{2}, \sqrt{3}]$, then $\sigma \in G$ is in G^E if and only if $\sigma(\sqrt{2}) = \sqrt{2}$ and $\sigma(\sqrt{3}) = \sqrt{3}$. By inspection, only the identity element does this. Thus $\Psi_{\text{field}}\left(\mathbb{Q}[\sqrt{2}, \sqrt{3}]\right) = \{\mathrm{Id}\}$. Finally, all $\sigma \in G$ fix $\mathbb{Q}$ so that $\Psi_{\text{field}}(\mathbb{Q}) = \{\mathrm{Id}, \sigma_2, \sigma_3, \sigma_4\}$. In summary,

$$\begin{array}{ccccc} & & \Psi_{\text{field}}\left(\mathbb{Q}[\sqrt{2}, \sqrt{3}]\right) = \{\mathrm{Id}\} & & \\ & \swarrow & \downarrow & \searrow & \\ \Psi_{\text{field}}\left(\mathbb{Q}[\sqrt{2}]\right) = \{\mathrm{Id}, \sigma_3\} & & \Psi_{\text{field}}\left(\mathbb{Q}[\sqrt{6}]\right) = \{\mathrm{Id}, \sigma_4\} & & \Psi_{\text{field}}\left(\mathbb{Q}[\sqrt{3}]\right) = \{\mathrm{Id}, \sigma_2\} \\ & \searrow & \downarrow & \swarrow & \\ & & \Psi_{\text{field}}(\mathbb{Q}) = \{\mathrm{Id}, \sigma_2, \sigma_3, \sigma_4\} & & \end{array}$$

so that Ψ_{field} is a(n) (inclusion order reversing) bijection from the set of intermediate fields to the set of subgroups of G.

Next, we calculate Ψ_{group}. To simplify notation, write $K = \mathbb{Q}[\sqrt{2}, \sqrt{3}]$ and recall that Ψ_{group} maps a subgroup H of G to its fixed field K_H. It is easy to see that a basis for K over $\mathbb{Q}$ is given by $\{1, \sqrt{2}, \sqrt{6}, \sqrt{3}\}$. Thus every element of K can be uniquely written as $a + b\sqrt{2} + c\sqrt{2}\sqrt{3} + d\sqrt{3}$ for $a, b, c, d \in \mathbb{Q}$ and it will be easy to explicitly find all $\alpha \in K$ so that $\sigma(\alpha) = \alpha$ for every $\sigma \in H$. For instance, if $H = \{\mathrm{Id}, \sigma_3\}$, it follows that $\alpha \in K_H$ if and only if $\sigma_3(\alpha) = \alpha$. By inspection, the fixed point set of σ_3 is $\mathbb{Q}[\sqrt{2}]$. Thus $\Psi_{\text{group}}(\{\mathrm{Id}, \sigma_3\}) = \mathbb{Q}[\sqrt{2}]$. Similarly, the fixed point set of σ_4 is $\mathbb{Q}[\sqrt{6}]$ and the fixed point set of σ_2 is $\mathbb{Q}[\sqrt{3}]$ so we get $\Psi_{\text{group}}(\{\mathrm{Id}, \sigma_4\}) = \mathbb{Q}[\sqrt{6}]$ and $\Psi_{\text{group}}(\{\mathrm{Id}, \sigma_2\}) = \mathbb{Q}[\sqrt{3}]$. Also, by inspection, the only α fixed by all of $\{\mathrm{Id}, \sigma_2, \sigma_3, \sigma_4\}$ is $\mathbb{Q}$. Thus $\Psi_{\text{group}}(\{\mathrm{Id}, \sigma_2, \sigma_3, \sigma_4\}) = \mathbb{Q}$. Finally, it is obvious that $\Psi_{\text{group}}(\{\mathrm{Id}\}) = K$. In summary,

$$\begin{array}{ccccc} & & \Psi_{\text{group}}(\{\mathrm{Id}\}) = \mathbb{Q}[\sqrt{2}, \sqrt{3}] & & \\ & \nearrow & \uparrow & \nwarrow & \\ \Psi_{\text{group}}(\{\mathrm{Id}, \sigma_3\}) = \mathbb{Q}[\sqrt{2}] & & \Psi_{\text{group}}(\{\mathrm{Id}, \sigma_4\}) = \mathbb{Q}[\sqrt{6}] & & \Psi_{\text{group}}(\{\mathrm{Id}, \sigma_2\}) = \mathbb{Q}[\sqrt{3}] \\ & \nwarrow & \uparrow & \nearrow & \\ & & \Psi_{\text{group}}(\{\mathrm{Id}, \sigma_2, \sigma_3, \sigma_4\}) = \mathbb{Q} & & \end{array}$$

so that Ψ_{group} is an inclusion order reversing bijection from the set of subgroups of G to the set of intermediate fields. Moreover, we see that $\Psi_{\text{group}} = \Psi_{\text{field}}^{-1}$.

In order to investigate how much of this generalizes, we need the following lemma.

LEMMA 4.42. *Let K be a finite extension of F and let H be a subgroup of $\mathrm{Gal}_F K$. Then K is a separable, simple, normal extension of K_H with $[K : K_H] \leq |H|$.*

PROOF. For $\alpha \in K$, let $f \in K_H[x]$ be its minimal polynomial over K_H. Let $\mathcal{O}_\alpha$ be the orbit of α under the action of H; i.e., $\mathcal{O}_\alpha = \{\sigma(\alpha) \mid \sigma \in H\}$. Noting that $\mathrm{Gal}_F K$ is finite by Theorem 4.35, parts (1) and (2) (Exercise 4.84), write $\mathcal{O}_\alpha = \{\alpha = \alpha_1, \alpha_2, \ldots, \alpha_n\}$ for distinct α_i. It follows that $n \leq |H|$ and that $\sigma \in H$ maps $\mathcal{O}_\alpha$ into itself. As σ is an automorphism, σ acts as a permutation on $\mathcal{O}_\alpha$.

Let $g = \prod_i (x - \alpha_i) \in K[x]$. Extending σ to an isomorphism $\sigma : K[x] \to K[x]$ fixing $K_H[x]$ (Exercise 3.92), it follows that $\sigma g = \sigma \prod_i (x - \alpha_i) = \prod_i (x - \sigma(\alpha_i)) = g$. Thinking of g now as a sum, this means every coefficient of x^j in g is fixed by each

$\sigma \in H$. Thus $g \in K_H[x]$. Since $g(\alpha) = 0$, $f \mid g$ and we see that f is separable with $\deg f \leq n$. In particular, K is a separable extension of K_H.

As K is a finite extension of F, it is also a finite extension of K_H. Thus Theorem 4.40 shows K is a simple extension of K_H. Write $K = K_H[\alpha]$ for some $\alpha \in K$. From the above paragraph, the minimal polynomial f for α over K_H splits so that $K = K_H[\alpha]$ is a splitting field for f. Theorem 4.21 establishes normality and observing that $[K : K_H] = \deg f$ finishes the proof. □

The next theorem (Theorem 4.44) shows that Ψ_{field} is always surjective for finite extensions but that Ψ_{group} can fail to be surjective in the absence of normality and separability. As a result, we introduce the following notation.

DEFINITION 4.43. Let $F \subseteq K$ be fields. We say K is *Galois* over F if K is a finite, normal, separable extension of F.

In the case of $\operatorname{char} F = 0$, note that K is Galois if and only if K is the splitting field of some $f \in F[x]$ by Theorem 4.21 and Corollary 4.39. In the case of an arbitrary characteristic, it is a fact that K is Galois over F if and only if K is the splitting field of some separable $f \in F[x]$ (Exercise 4.129).

THEOREM 4.44. *Let K be a finite extension of F.*

(1)

$$\Psi_{\text{field}} \circ \Psi_{\text{group}} = \operatorname{Id}_{\{\textit{subgroups of } \operatorname{Gal}_F K\}}$$

In other words, if H is a subgroup of $\operatorname{Gal}_F K$, then

$$\operatorname{Gal}_{K_H} K = H.$$

In particular, Ψ_{field} is always surjective.

(2) *Moreover,*

$$[K : K_H] = |H|.$$

(3) *The range of Ψ_{group} is*

$$\mathcal{IF}_{\text{Gal}} = \{\textit{intermediate fields } E \textit{ between } F \textit{ and } K \mid K \textit{ is Galois over } E\}.$$

Restricting Ψ_{field} to $\mathcal{IF}_{\text{Gal}}$,

$$\Psi_{\text{group}} \circ (\Psi_{\text{field}})|_{\mathcal{IF}_{\text{Gal}}} = \operatorname{Id}_{\mathcal{IF}_{\text{Gal}}}.$$

In other words, if $E \in \mathcal{IF}_{\text{Gal}}$, then

$$K_{\operatorname{Gal}_E K} = E.$$

(4) *$(\Psi_{\text{field}})|_{\mathcal{IF}_{\text{Gal}}}$ and Ψ_{group} each give a bijection*

$$\mathcal{IF}_{\text{Gal}} \longleftrightarrow \{\textit{subgroups of } \operatorname{Gal}_F K\}$$

on which

$$(\Psi_{\text{group}})^{-1} = (\Psi_{\text{field}})|_{\mathcal{IF}_{\text{Gal}}}.$$

PROOF. First we show that $\operatorname{Gal}_{K_H} K = H$. Recall that Lemma 4.42 shows that K is a simple, normal, separable extension of K_H with $[K : K_H] \leq |H|$. Write $K = K_H[\alpha]$ for some $\alpha \in K$ and let $f \in K_H[x]$ be the minimal polynomial of α over K_H so that $n = \deg f = [K : K_H] \leq |H|$. Since K is a normal, separable extension of K_H, K has n distinct roots in K. Thus Theorem 4.35, part (3), shows

$|\mathrm{Gal}_{K_H} K| = n$. Thus $|\mathrm{Gal}_{K_H} K| \leq |H|$. On the other hand, clearly $H \subseteq \mathrm{Gal}_{K_H} K$ by definition. Hence $\mathrm{Gal}_{K_H} K = H$ and parts (1) and (2) are complete.

For part (3), Lemma 4.42 shows that the range of Ψ_{group} is contained in $\mathcal{IF}_{\text{Gal}}$. To finish the remaining statements, we must show $K_{\mathrm{Gal}_E K} = E$ when $E \in \mathcal{IF}_{\text{Gal}}$. By definition, clearly $E \subseteq K_{\mathrm{Gal}_E K}$. It remains to see that $K_{\mathrm{Gal}_E K} \subseteq E$. We prove the contrapositive. Namely, if $\alpha \in K$ with $\alpha \notin E$, we show $\alpha \notin K_{\mathrm{Gal}_E K}$. In other words, we need to find a $\sigma \in \mathrm{Gal}_E K$ so that $\sigma(\alpha) \neq \alpha$. Let $f \in E[x]$ be the minimal polynomial for α over E. By construction, $\deg f \geq 2$. By normality and separability, f splits as a product of distinct roots over K. In particular, there exists a root $\beta \in K$ of f with $\alpha \neq \beta$. Theorems 4.21 and 4.36, part (4), show there exists $\sigma \in \mathrm{Gal}_E K$ so that $\sigma(\alpha) = \beta$. Thus $\alpha \notin K_{\mathrm{Gal}_E K}$ as desired and part (3) is complete. Of course part (4) follows from parts (1), (2), and (3). □

COROLLARY 4.45 (Fundamental Theorem of Galois Theory). *Let K be a Galois extension of F.*

(1) *Ψ_{field} and Ψ_{group} each give a bijection*

$$\{\textit{intermediate fields between } F \textit{ and } K\} \longleftrightarrow \{\textit{subgroups of } \mathrm{Gal}_F K\}$$

on which

$$(\Psi_{\text{group}})^{-1} = (\Psi_{\text{field}}).$$

(2) *Each bijection reverses the inclusion ordering; i.e., if $E \subseteq E'$ are intermediate fields, then $\Psi_{\text{field}}(E) \supseteq \Psi_{\text{field}}(E')$, and if $H \subseteq H'$ are subgroups of $\mathrm{Gal}_F K$, then $\Psi_{\text{group}}(H) \supseteq \Psi_{\text{group}}(H')$.*

(3)

$$[K : E] = |\mathrm{Gal}_E K|,$$
$$[E : F] = |\mathrm{Gal}_F K| \,/\, |\mathrm{Gal}_E K|.$$

(4) *E is a normal extension of F if and only if $\mathrm{Gal}_E K$ is a normal subgroup of $\mathrm{Gal}_F K$. In this case,*

$$\mathrm{Gal}_F E \cong (\mathrm{Gal}_F K) \,/\, (\mathrm{Gal}_E K)$$

and E is Galois over F.

PROOF. Observe that if K is Galois over F and if E is an intermediate field, then K is still Galois over E (Exercise 4.108). Therefore Theorem 4.44 establishes part (1). For part (2), it follows trivially from the definitions that $E \subseteq E'$ implies $\mathrm{Gal}_E K \supseteq \mathrm{Gal}_{E'} K$ and that $H \subseteq H'$ implies $K_H \supseteq K_{H'}$. For part (3), write $H = \Psi_{\text{group}}(E)$ so that $K_H = E$ and $\mathrm{Gal}_E K = H$. Thus Theorem 4.44, part (1), shows $[K : E] = |\mathrm{Gal}_E K|$. Theorem 4.9 handles the second equation.

For part (4), first assume E is a normal extension of F. Using Theorem 4.21, choose $f \in F[x]$ so that E is a splitting field of f. By Theorem 4.35, part (1), any $\sigma \in \mathrm{Gal}_F K$ maps a root of f to a root of f and therefore preserves E. Hence the map $\rho : \mathrm{Gal}_F K \to \mathrm{Gal}_F E$ given by $\rho(\sigma) = \sigma|_E$ is well defined. Obviously ρ is a homomorphism with $\ker \rho = \mathrm{Gal}_E K$. In particular, $\mathrm{Gal}_E K$ is normal. To see $\mathrm{Gal}_F E \cong (\mathrm{Gal}_F K) \,/\, (\mathrm{Gal}_E K)$, it remains only to show ρ is surjective. For this, note that K is a splitting field over E (with respect to the same polynomial for which it is a splitting field over F, Exercise 4.48). Therefore Theorem 4.18, part (2), shows any element of $\mathrm{Gal}_F E$ has an extension to $\mathrm{Gal}_F K$ so that ρ is surjective.

Conversely, assume $\mathrm{Gal}_E K$ is a normal subgroup of $\mathrm{Gal}_F K$. If $g \in F[x]$ is irreducible and if it has a root $\alpha \in E$, it suffices to show that g splits in E. Of course, g splits in K so it is enough to show that any root $\beta \in K$ of g actually lies in E. Use Theorem 4.36, part (4), to find $\sigma \in \mathrm{Gal}_F K$ with $\sigma(\alpha) = \beta$. If $\tau \in \mathrm{Gal}_E K$, then $\tau' = \sigma^{-1}\tau\sigma \in \mathrm{Gal}_E K$ so that

$$\tau(\beta) = \tau\sigma(\alpha) = \sigma\tau'(\alpha) = \sigma(\tau'(\alpha)) = \sigma(\alpha) = \beta.$$

Hence $\beta \in K_{\mathrm{Gal}_E K} = E$ as desired. □

For example, $\mathbb{Q} \subseteq \mathbb{Q}[2^{\frac{1}{3}}, \omega]$, where $\omega = \frac{-1+i\sqrt{3}}{2}$, is the splitting field for $x^3 - 2$ over $\mathbb{Q}$ and is therefore Galois over $\mathbb{Q}$. Writing the three roots of x^3-2 as $u_k = 2^{\frac{1}{3}}\omega^k$ for $1 \leq k \leq 3$, we have already seen that the action of $\mathrm{Gal}_{\mathbb{Q}}\,\mathbb{Q}[2^{\frac{1}{3}}, \omega]$ on $\{u_1, u_2, u_3\}$ allows us to identify $\mathrm{Gal}_{\mathbb{Q}}\,\mathbb{Q}[2^{\frac{1}{3}}, \omega]$ with S_3. In particular, we can use Corollary 4.45 to calculate all the intermediate fields of $\mathbb{Q} \subseteq \mathbb{Q}[2^{\frac{1}{3}}, \omega]$. Since it is easy to see that the only proper subgroups of S_3 are $\langle(12)\rangle$, $\langle(23)\rangle$, and $\langle(13)\rangle$, each of order 2, and $\langle(123)\rangle$ of order 3, the inclusion diagram for subgroups of S_3 is

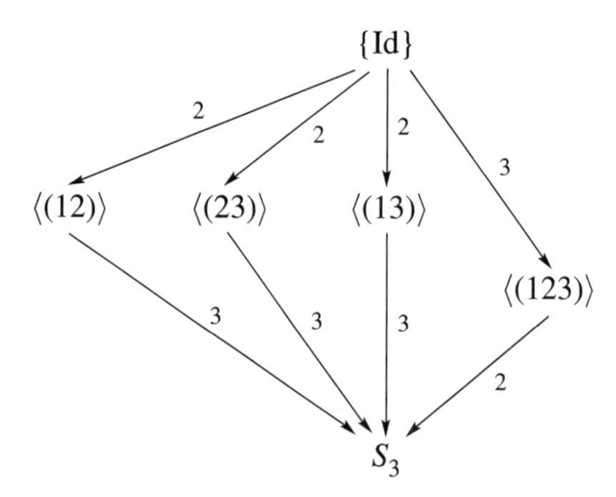

where the integer above the inclusion arrows indicates the index of the two groups. Note that the only proper normal subgroup is $\langle(123)\rangle$.

If H is a subgroup of S_3, the corresponding intermediate field is $\mathbb{Q}[2^{\frac{1}{3}}, \omega]_H$. For example, it is obvious that $\mathbb{Q}[u_3 = 2^{\frac{1}{3}}] \subseteq \mathbb{Q}[2^{\frac{1}{3}}, \omega]_{\langle(12)\rangle}$. To see equality, first use Theorem 4.44, part (2), to verify that $[\mathbb{Q}[2^{\frac{1}{3}}, \omega] : \mathbb{Q}[2^{\frac{1}{3}}, \omega]_{\langle(12)\rangle}] = 2$. Since $[\mathbb{Q}[2^{\frac{1}{3}}, \omega] : \mathbb{Q}[2^{\frac{1}{3}}]] = 2$, it follows that $\mathbb{Q}[2^{\frac{1}{3}}] = \mathbb{Q}[2^{\frac{1}{3}}, \omega]_{\langle(12)\rangle}$. Note that since $\langle(12)\rangle$ is not a normal subgroup, the corresponding intermediate field $\mathbb{Q}[2^{\frac{1}{3}}]$ is not a normal extension of $\mathbb{Q}$ and therefore is not Galois over $\mathbb{Q}$. Similarly, it is straightforward to calculate that $\mathbb{Q}[2^{\frac{1}{3}}, \omega]_{\langle(23)\rangle} = \mathbb{Q}[2^{\frac{1}{3}}\omega]$, $\mathbb{Q}[2^{\frac{1}{3}}, \omega]_{\langle(13)\rangle} = \mathbb{Q}[2^{\frac{1}{3}}\omega^2]$, and $\mathbb{Q}[2^{\frac{1}{3}}, \omega]_{\langle(123)\rangle} = \mathbb{Q}[\omega]$. The last field is a normal extension of $\mathbb{Q}$ since $\langle(123)\rangle$ is a normal subgroup (in fact, it is the splitting field of $x^2 + x + 1$). As a result, the inclusion diagram for the intermediate fields is

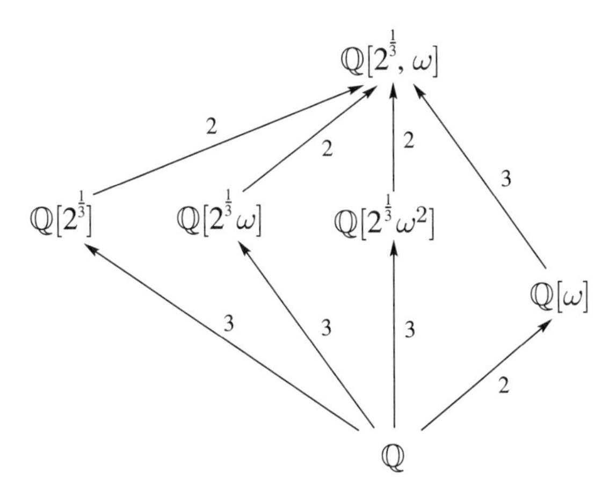

where the integers above the inclusion arrows indicate the order of the extension.

4.4. Exercises 4.82—4.135.

4.4.1. *Galois Groups.*

EXERCISE 4.82. Calculate the Galois group for the following field extensions:
(a) $\mathbb{Q} \subseteq \mathbb{Q}[\sqrt{3}]$.
(b) $\mathbb{Q} \subseteq \mathbb{Q}[\sqrt{3}, \sqrt{5}]$.
(c) $\mathbb{Q} \subseteq \mathbb{Q}[\sqrt{3}, \sqrt{5}, \sqrt{7}]$.
(d) $\mathbb{Q} \subseteq \mathbb{Q}[11^{\frac{1}{3}}]$.
(e) $\mathbb{Q} \subseteq \mathbb{Q}[11^{\frac{1}{4}}]$.
(f) $\mathbb{Q} \subseteq \mathbb{Q}[\sqrt{2}, i]$.
(g) $\mathbb{Q} \subseteq \mathbb{Q}[e^{\frac{2\pi i}{3}}]$.
(h) $\mathbb{Q} \subseteq \mathbb{Q}[7^{\frac{1}{3}}, e^{\frac{2\pi i}{3}}]$.
(i) $\mathbb{Q} \subseteq \mathbb{Q}[7^{\frac{1}{4}}, i]$. (N.B.: Show that the Galois group is isomorphic to D_4.)
(j) $\mathbb{Q} \subseteq \mathbb{R}$ (cf. Exercise 4.46).
(k) $\mathbb{Q} \subseteq \mathbb{Q}[\sqrt[3]{1+\sqrt{3}}]$.
(l) $\mathbb{Q} \subseteq \mathbb{Q}[\sqrt{1+\sqrt{1+\sqrt{2}}}]$.
(m) $\mathbb{Q}[e^{\frac{2\pi i}{3}}] \subseteq \mathbb{Q}[2^{\frac{1}{6}}, e^{\frac{2\pi i}{6}}]$.

EXERCISE 4.83. Let K be an extension field of $\mathbb{Q}$. Show that an automorphism of K is the same as a $\mathbb{Q}$-automorphism of K. *Hint:* If σ is an automorphism of K, start with $\sigma(1) = 1$ to show $\sigma(n) = n$ for $n \in \mathbb{Z}$.

EXERCISE 4.84. Let K be a finite extension of the field F. Show $\mathrm{Gal}_F K$ is a finite group.

EXERCISE 4.85. If $F \subseteq K$ are fields with $[K : F] = 2$, show $\mathrm{Gal}_F K \cong \mathbb{Z}_2$. *Hint:* Exercise 4.40.

EXERCISE 4.86. If $p_1, \ldots, p_n$ are distinct positive primes, show that $\mathrm{Gal}_{\mathbb{Q}} \mathbb{Q}[\sqrt{p_1}, \ldots, \sqrt{p_n}]$ is abelian.

EXERCISE 4.87. **(a)** For $n \in \mathbb{N}$, let $\omega = e^{\frac{2\pi i}{n}}$. Show $\mathrm{Gal}_{\mathbb{Q}} \mathbb{Q}[\omega]$ is abelian.
(b) For $p > 2$ prime, show $\mathrm{Gal}_{\mathbb{Q}} \mathbb{Q}[\omega] \cong U_p$.

EXERCISE 4.88. Let K be a splitting field for $x^p - 2$ over $\mathbb{Q}$ where p is prime. Show $\operatorname{Gal}_{\mathbb{Q}} K \cong \left\{ \begin{pmatrix} a & b \\ 0 & 1 \end{pmatrix} \mid a, b \in \mathbb{Z}_p \text{ with } a \neq 0 \right\}$.

EXERCISE 4.89. Show $\operatorname{Gal}_{\mathbb{F}_p} \mathbb{F}_{p^n} \cong \mathbb{Z}_n$. *Hint:* Theorem 4.35 or Exercise 4.79.

EXERCISE 4.90. **(a)** Let F be a field and let $F(t)$ be the field of rational functions. Show that $\operatorname{Gal}_F F(t)$ consists of all *linear fractional transformations*, i.e., all automorphisms determined by mapping $t \to \frac{at+b}{ct+d}$ for $a, b, c, d \in F$ with $ad - bc \neq 0$.

(b) Show $\operatorname{Gal}_F F(t) \cong PGL(2, F)$ (cf. Exercise 2.169).

EXERCISE 4.91 (Independence of Automorphisms). Let $F \subseteq K$ be fields. Show that the elements of $\operatorname{Gal}_F K$ are linearly independent as functions on K. *Hint:* Exercise 4.12.

4.4.2. *Separability.*

EXERCISE 4.92. Let F be a field and let $f \in F[x]$ be separable of degree n. If K is a splitting field for f, show $|\operatorname{Gal}_F K| \mid n!$.

EXERCISE 4.93. Let $F \subseteq E \subseteq K$ be fields and let $f \in F[x]$.

(a) If f is separable over F, show that f is separable over K.

(b) If K is separable over F, show that K is separable over E and that E is separable over F.

EXERCISE 4.94. **(a)** Let K be a finite separable extension of F. Show $|\operatorname{Gal}_F K| \leq [K : F]$. *Hint:* Write $K = F[\alpha]$.

(b) If K is also a splitting field for some $f \in F[x]$, show $|\operatorname{Gal}_F K| = [K : F]$.

(c) If $\operatorname{char} F = 0$ and if K is a splitting field for some $f \in F[x]$, show $|\operatorname{Gal}_F K| = [K : F]$.

EXERCISE 4.95. For distinct primes $p, q, r \in \mathbb{N}$, use the proof of Theorem 4.40 to show:

(a) $\mathbb{Q}[\sqrt{p}, \sqrt{q}] = \mathbb{Q}[\sqrt{p} + \sqrt{q}]$.

(b) $\mathbb{Q}[\sqrt{p}, \sqrt{q}, \sqrt{r}] = \mathbb{Q}[\sqrt{p} + \sqrt{q} + \sqrt{r}]$.

EXERCISE 4.96. Let $F \subseteq K$ be fields. If there are $\alpha_1, \ldots, \alpha_n \in K$ algebraic over F so that $K = F[\alpha_1, \ldots, \alpha_n]$ and so that $\alpha_2, \ldots, \alpha_n$ are separable over F, show that $K = F[\beta]$ for some $\beta \in K$. *Hint:* Proof of Theorem 4.40.

EXERCISE 4.97. Let F be a field and let $f \in F[x]$. Show f is separable if and only if f and f' are relatively prime (cf. Exercise 4.45).

EXERCISE 4.98 (Perfect Fields). A field F is called *perfect* if every irreducible polynomial in $F[x]$ is separable.

(a) Show F is perfect if and only if every algebraic extension of F is separable if and only if every finite extension of F is separable.

(b) If $\operatorname{char} F = 0$, show F is perfect.

(c) If F is finite, show F is perfect. *Hint:* For $\mathbb{F}_{p^n}$, use the Frobenius automorphism to show $f(x^p)$ is not irreducible by showing it is a pth power.

(d) If F is perfect and K is an algebraic extension of F, show K is perfect. *Hint:* If $g \in K[x]$ is irreducible, L is a splitting field for g over K, and $\alpha \in L$ is a root of g, look at the minimal polynomial of α over F (cf. Exercise 4.25) and show $g \mid f$.

Exercise 4.99 (Minimal Polynomials over Finite Fields). Let K be a finite extension of $\mathbb{F}_{p^m}$ and let $\alpha \in K$ with minimal polynomial $f \in \mathbb{F}_{p^m}[x]$ with $\deg f = k$. Show $f = \prod_{j=0}^{k-1}(x - \alpha^{p^{mj}})$. *Hint:* Exercise 4.89 and Corollary 4.45 or Exercises 4.75 and 4.78.

Exercise 4.100. Consider the field of rational functions $F = \mathbb{Z}_2(t)$ and let $f \in F[x]$ be given by $f = x^2 - t$. This exercise shows F is not a perfect field (cf. Exercise 4.98).

(a) Show that $\operatorname{char} F = 2$ and that F is infinite.

(b) Show f is irreducible. *Hint:* If $\alpha \in F$ is a root of f, write $\alpha = \frac{g}{h}$ for $g, h \in \mathbb{Z}_2(t)$ and look for a contradiction while reducing to $\mathbb{Z}_2[t]$.

(c) Show f is not separable. *Hint:* Use f'.

Exercise 4.101 (Frobenius Automorphism). Let F be a field with $\operatorname{char} F = p > 0$.

(a) The *Frobenius endomorphism* $\varphi_{\mathrm{Fr}} : F \to F$ is given by $\varphi_{\mathrm{Fr}}(c) = c^p$ for $c \in F$. Show φ_{Fr} is an injective homomorphism (cf. the proof of Theorem 4.33).

(b) For $a \in F$, let $f = x^p - a$. If β is a root of f in a splitting field, show $f = (x - \beta)^p$.

(c) Show F is perfect (Exercise 4.98) if and only if each $a \in F$ can be written as $a = b^p$ for some $b \in F$. *Hint:* If F is perfect, look at an irreducible factor of $x^p - a$. For the converse, see the hint in Exercise 4.98, part (c).

(d) Conclude F is perfect if and only if φ_{Fr} is surjective and therefore an automorphism.

Exercise 4.102. Let F be a field with $\operatorname{char} F = p > 0$ with $f \in F[x]$ irreducible. Show there is a unique $k \in \mathbb{Z}_{\geq 0}$ and irreducible, separable $f_{\mathrm{sep}} \in F[x]$ so that $f(x) = f_{\mathrm{sep}}(x^{p^k})$. The degree of f_{sep} is called the *separable degree* of f and the k is called the *inseparable degree* of f. *Hint:* If f is not separable, then $f(x) = f_1(x^p)$. If f_1 is separable, then $k = 1$. Otherwise, repeat.

Exercise 4.103. Let F be a field with $\operatorname{char} F = p > 0$ and let K be an extension field of F. Let $\mathrm{Sep}_F(K) = \{\alpha \in K \mid \alpha \text{ is separable over } F\}$, called the *separable closure* of F in K. Show S is the maximal separable subfield of K containing F.

Exercise 4.104. Let $F \subseteq E \subseteq K$ be fields with K a separable extension of F. Show that K is separable over E and that E is separable over F.

Exercise 4.105 (Separable Extensions of Separable Extensions). Let $F \subseteq E \subseteq K$ be fields with E a separable extension of F and let K be a separable extension of E. This exercise shows that K is a separable extension of F.

(a) Reduce to the case where $\operatorname{char} F = p$. Let $\alpha \in K$ and show there is a $k \in \mathbb{Z}_{\geq 0}$ so that $\alpha^{p^k} \in \mathrm{Sep}_F(K)$ (cf. Exercises 4.102 and 4.103).

(b) Let $f \in (\mathrm{Sep}_F(K))[x]$ be the minimal polynomial of α over $\mathrm{Sep}_F(K)$. Show f is separable (cf. Exercise 4.93).

(c) Show that $f \mid \left(x^{p^k} - \alpha^{p^k}\right)$ and that α is the only root of f in a splitting field for f. Conclude that α is separable over F. *Hint:* Frobenius endomorphism (Exercise 4.101).

Exercise 4.106. If $F \subseteq K$ are fields with $\alpha_1, \ldots, \alpha_n \in K$ separable over F, show $F[\alpha_1, \ldots, \alpha_n]$ is a separable extension of F. *Hint:* Look at the splitting field for the minimal polynomials of $\alpha_1, \ldots, \alpha_n$ and use the ideas of Exercise 4.129 or use Exercise 4.105.

4.4.3. *Galois Correspondence.*

EXERCISE 4.107. Let $F \subseteq K$ be fields and let $G = \mathrm{Gal}_F K$.

(a) Show $G^E = \{\sigma \in G \mid \sigma(\alpha) = \alpha \text{ for all } \alpha \in E\}$ coincides with $\mathrm{Gal}_E K$.

(b) If E is an intermediate field between F and K, show $\Psi_{\text{group}}(E) = G^E = \mathrm{Gal}_E K$ is a subgroup of G.

(c) If H is a subgroup of G, show $\Psi_{\text{field}}(H) = K_H$ is an intermediate field between F and K.

EXERCISE 4.108. Let $F \subseteq E \subseteq K$ be fields with K Galois over F. Show K is Galois over E. *Hint:* Theorem 4.21 and Exercise 4.93.

EXERCISE 4.109. If K is Galois over F, show there are only a finite number of intermediate fields.

EXERCISE 4.110. Let $F \subseteq K$ be fields and let H be a finite subgroup of $\mathrm{Gal}_F K$. If $\alpha \in K$, show $\mathrm{tr}_H(\alpha) = \sum_{\sigma \in H} \sigma(\alpha)$ and $n_H(\alpha) = \prod_{\sigma \in H} \sigma(\alpha)$ are both in K_H.

EXERCISE 4.111. Show the following extensions are Galois, calculate the Galois correspondence and all intermediate fields, put the intermediate fields in a diagram showing their inclusion ordering, and determine which intermediate fields are normal over $\mathbb{Q}$:

(a) $\mathbb{Q} \subseteq \mathbb{Q}[\sqrt{3}, \sqrt{5}]$.

(b) $\mathbb{Q} \subseteq \mathbb{Q}[\sqrt{3}, \sqrt{5}, \sqrt{7}]$.

(c) $\mathbb{Q} \subseteq \mathbb{Q}[7^{\frac{1}{3}}, e^{\frac{2\pi i}{3}}]$.

(d) $\mathbb{Q} \subseteq \mathbb{Q}[11^{\frac{1}{4}}, i]$.

(e) $\mathbb{Q} \subseteq \mathbb{Q}[\sqrt{2}, i]$.

(f) $\mathbb{Q}[e^{\frac{2\pi i}{3}}] \subseteq \mathbb{Q}[2^{\frac{1}{6}}, e^{\frac{2\pi i}{6}}]$.

EXERCISE 4.112. Consider the extension $\mathbb{Q} \subseteq \mathbb{Q}[2^{\frac{1}{3}}]$. Show $\Psi_{\text{field}}(\mathbb{Q}) = \Psi_{\text{field}}(\mathbb{Q}[2^{\frac{1}{3}}])$ and calculate the range of Ψ_{group}. Does this violate Corollary 4.45?

EXERCISE 4.113. If K is a normal extension of $\mathbb{Q}$ with $[K : \mathbb{Q}] = p$ prime, show $\mathrm{Gal}_{\mathbb{Q}} K \cong \mathbb{Z}_p$.

EXERCISE 4.114. Let K be a Galois extension of F so that $\mathrm{Gal}_F K$ is abelian of order 15. Calculate the number of intermediate fields and give the dimension of each over F.

EXERCISE 4.115. **(a)** If K is a Galois extension of F with $\mathrm{Gal}_F K$ abelian and E an intermediate field, show E is Galois over F and that

$$\mathrm{Gal}_F E \cong (\mathrm{Gal}_F K) / (\mathrm{Gal}_E K).$$

(b) If $\mathrm{Gal}_F K \cong \mathbb{Z}_n$ for $n \in \mathbb{N}$, show that the intermediate fields are in one-to-one correspondence with $\{k \in \mathbb{N} \mid k \mid n\}$ and that if k corresponds to E, then $\mathrm{Gal}_E K \cong \mathbb{Z}_{\frac{n}{k}}$ and $\mathrm{Gal}_F E \cong \mathbb{Z}_k$. *Hint:* Exercises 2.63 and 2.156.

EXERCISE 4.116. Let E and E' be intermediate fields of $F \subseteq K$. We say E and E' are *conjugate* if there exists $\sigma \in \mathrm{Gal}_F K$ so that $\sigma(E) = E'$. Show E and E' are conjugate if and only if $\mathrm{Gal}_E K$ and $\mathrm{Gal}_{E'} K$ are conjugate subgroups of $\mathrm{Gal}_F K$ (cf. Exercise 2.71).

EXERCISE 4.117. Let K be Galois over F. If $p^k \mid [K : F]$ for a prime p, show there is an intermediate field E so that $[K : E] = p^k$. *Hint:* Theorem 2.69.

EXERCISE 4.118. Let E and E' be intermediate fields of $F \subseteq K$ and let H and H' be subgroups of $\operatorname{Gal}_F K$.

(a) Show $\Psi_{\text{field}}(F(E, E')) = \Psi_{\text{field}}(E) \cap \Psi_{\text{field}}(E')$.

(b) Show $\Psi_{\text{group}}(\langle H, H' \rangle) = \Psi_{\text{group}}(H) \cap \Psi_{\text{group}}(H')$.

EXERCISE 4.119. **(a)** Suppose that $f \in \mathbb{Q}[x]$ is irreducible, that it has exactly two non-real roots, and that $\deg f = p > 2$ is prime. If K is a splitting field for f, show $G = \operatorname{Gal}_{\mathbb{Q}} K \cong S_p$. *Hint:* If α is a root of f, look at $\mathbb{Q}[\alpha]$ to show $p \mid [K : \mathbb{Q}]$ and $p \mid |G|$. Then use Cauchy's Theorem and Exercise 2.200.

(b) Show such an f exists. *Hint:* For instance, first find a polynomial with $p - 2$ distinct nonzero real roots, multiply by x^2, and then try adding a small rational constant to make it irreducible but possessing exactly $p - 2$ real roots.

EXERCISE 4.120. Let $F \subseteq E \subseteq K$ be fields with E normal over F. If $\sigma \in \operatorname{Gal}_F K$, then $\sigma|_E \in \operatorname{Gal}_F E$.

EXERCISE 4.121. Let $F \subseteq E \subseteq K$ be fields and let $H = \{\sigma \in \operatorname{Gal}_F K \mid \sigma(E) = E\}$.

(a) Show H is a subgroup of $\operatorname{Gal}_F K$.

(b) If K is a splitting field for some polynomial over F, show $\operatorname{Gal}_F E \cong H / \operatorname{Gal}_E K$. *Hint:* See the proof of Corollary 4.45.

(c) If E is a splitting field for some polynomial over F, show $H = \operatorname{Gal}_F K$.

EXERCISE 4.122. Let K be Galois over F. If $\alpha \in K$, let

$$f = \prod_{\sigma \in \operatorname{Gal}_F K} (x - \sigma(\alpha)) \in E[x].$$

Show that $f \in F[x]$ and that f is a power of the minimal polynomial of α over F.

EXERCISE 4.123. Suppose that K is Galois over F and that $f \in F[x]$ is irreducible. If there are irreducible $g_1, \ldots, g_m \in K[x]$ so that $f = \prod_i g_i$, show each g_i has the same degree. *Hint:* Look at, say, $h = \prod_{\sigma \in \operatorname{Gal}_F K} \sigma(g_1)$ and then look at a root in a splitting field to show $f \mid h$.

EXERCISE 4.124. Let $F \subseteq K$ be fields with $\operatorname{char} F \neq 2$. Show $K = F[\sqrt{\alpha_1}, \ldots, \sqrt{\alpha_n}]$ for some $\sqrt{\alpha_i} \in K$ satisfying $\left(\sqrt{\alpha_i}\right)^2 = \alpha_i \in F$ if and only if K is Galois over F with $\sigma^2 = \operatorname{Id}$ for all $\sigma \in \operatorname{Gal}_F K$. *Hint:* Exercise 2.45.

EXERCISE 4.125. Let $F_i \subseteq E_i$ be fields, let $\varphi : F_1 \to F_2$ be an isomorphism of fields, and let $f \in F_1[x]$ be separable. Suppose that E_1 is a splitting field for f over F_1 and that E_2 is a splitting field for $\varphi(f)$ over F_2. Show there exist exactly $[E_1 : F_1]$ isomorphisms $\Phi : E_1 \to E_2$ extending φ. *Hint:* Use Theorem 4.18 to reduce to the case of $F_1 = F_2$, $E_1 = E_2$, and $\varphi = \operatorname{Id}$.

EXERCISE 4.126. Show a Galois extension of a Galois extension need not be Galois. *Hint:* $\mathbb{Q} \subseteq \mathbb{Q}[\sqrt{2}] \subseteq \mathbb{Q}[\sqrt[4]{2}]$.

EXERCISE 4.127 (Alternate Definition of Galois Extension, I). Let $F \subseteq K$ be fields. Show K is Galois if and only if K is a finite extension and $K_{\operatorname{Gal}_F K} = F$.

EXERCISE 4.128 (Alternate Definition of Galois Extension, II). Let $F \subseteq K$ be fields. Show K is Galois if and only if K is a finite extension and $|\mathrm{Gal}_F K| = [K : F]$. *Hint:* Theorem 4.44, part (1), and Exercise 4.127.

EXERCISE 4.129 (Alternate Definition of Galois Extension, III). Let $F \subseteq K$ be fields. This exercise shows K is a Galois extension of F if and only if K is the splitting field of a separable polynomial over F.

(a) If K is a Galois extension of F, show K is the splitting field of a separable polynomial over F.

(b) Conversely, suppose K is the splitting field of a separable polynomial over F for the remainder. Show $K = F[\alpha]$ for some separable $\alpha \in K$. *Hint:* Exercise 4.96.

(c) Let $f \in F[x]$ be the minimal polynomial of α with $\deg f = n$. Show f has distinct roots $\alpha = \alpha_1, \ldots, \alpha_n \in K$ and that there exists $\sigma_i \in \mathrm{Gal}_F K$ so that $\sigma_i(\alpha) = \alpha_i$.

(d) Suppose $\beta \in K_{\mathrm{Gal}_F K}$ and write $\beta = \sum_{j=0}^{n-1} c_j \alpha^j$ for $c_j \in F$. Let $h = (c_0 - \beta) + \sum_{j=1}^{n-1} c_j x^j \in (K[\alpha])[x]$. Show each α_i is a root of h and conclude $K_{\mathrm{Gal}_F K} = F$. *Hint:* Look at $\sigma_i(\beta)$ and then at $\deg h$.

(e) Conclude that K is Galois over F.

EXERCISE 4.130 (Dedekind-Artin Theorem). Let K be a field and let G be a finite group of automorphisms of K. This exercise shows $[K : K_G] = |G|$.

(a) Show K is a separable, algebraic extension of K_G for which each minimal polynomial has degree at most $n = |G|$. *Hint:* See the proof of Lemma 4.42.

(b) Show K is a finite extension of $F = K_G$ with $[K : F] \leq n$. *Hint:* If the extension is not finite, choose $u_1, \ldots, u_{n+1}$ in K independent over F. Now apply part (a) and Theorem 4.40 to $K[u_1, \ldots, u_{n+1}]$.

(c) Show $n \leq |\mathrm{Gal}_F K| \leq [K : F]$. *Hint:* Show normality as in Lemma 4.42 and then use Theorems 4.40 and 4.35, part (3).

EXERCISE 4.131 (All Finite Groups Are Galois Groups). This exercise shows that if G is a finite group, then there exists a Galois extension K of F so that $\mathrm{Gal}_F K \cong G$.

(a) Let E be any field and let $n = |G|$. Define $s_0, s_1, \ldots, s_n$ to be the *elementary symmetric polynomials* (cf. Exercise 2.243) in $E[x_1, \ldots, x_n]$ by expanding

$$\prod_{j=1}^{n} (t - x_j) = \sum_{k=0}^{n} (-1)^k s_k \, t^{n-k}$$

in $E[t, x_1, \ldots, x_n]$. Show $\deg s_k = k$ and

$$s_k = \sum_{1 \leq j_1 < j_2 < \cdots < j_k \leq n} x_{j_1} x_{j_2} \cdots x_{j_k}.$$

(b) Show that $K = E(x_1, \ldots, x_n)$ is a splitting field for

$$\prod_{j=1}^{n} (t - x_j) \in (E(s_1, \ldots, s_n))\,[t]$$

over $L = E(s_1, \ldots, s_n)$ and is therefore Galois (cf. Exercise 4.129). The polynomial $\prod_{j=1}^{n} (t - x_j)$ is called the *general polynomial* over E.

(c) Define an action of S_n on $E[x_1, \ldots, x_n]$ by

$$\sigma \cdot \left(\sum_{i=1}^{N} c_i \, x_1^{k_{1,i}} \cdots x_n^{k_{n,i}} \right) = \sum_{i=1}^{N} c_i \, x_{\sigma(1)}^{k_{1,i}} \cdots x_{\sigma(n)}^{k_{n,i}}$$

for $\sigma \in S_n$ where $N \in \mathbb{N}$, $c_i \in \mathbb{R}$, and $k_{j,i} \in \mathbb{Z}_{\geq 0}$. Extend the action to K by a similar formula in the denominator. Show $\sigma \in \operatorname{Gal}_L K$.

(d) Conclude $\operatorname{Gal}_L K \cong S_n$. *Hint:* Recall $[K : L] \mid n!$.

(e) Use Cayley's Theorem to show there is an intermediate field F between L and K so that $\operatorname{Gal}_F K \cong G$.

EXERCISE 4.132 (Fundamental Theorem on Symmetric Functions). Continue the notation from Exercise 4.131. A rational function $f \in E(x_1, \ldots, x_n)$ is called *symmetric* if $\sigma \cdot f = f$ for all $\sigma \in S_n$.

(a) Show that the set of symmetric rational functions is exactly $E(s_1, \ldots, s_n)$, i.e., the set of rational functions in the elementary symmetric functions. *Hint:* Use $\operatorname{Gal}_{E(s_1,\ldots,s_n)} E(x_1, \ldots, x_n) \cong S_n$.

(b) Give a constructive proof showing that the set of symmetric polynomials is exactly $E[s_1, \ldots, s_n]$, i.e., the set of polynomials in the elementary symmetric functions. *Hint:* Order monomials *lexicographically*; i.e., write $x_1^{i_1} \cdots x_n^{i_n} > x_1^{j_1} \cdots x_n^{j_n}$ if the first nonzero term in the sequence $i_1 - j_1, \ldots, i_n - j_n$ is positive. If $a x_1^{i_1} \cdots x_n^{i_n}$, $a \in E^\times$, is the largest monomial appearing in a symmetric polynomial f, show $i_1 \geq i_2 \geq \cdots \geq i_n$. Now consider $f - a s_1^{i_1 - i_2} s_2^{i_2 - i_3} \cdots s_{n-1}^{i_{n-1} - i_n} s_n^{i_n}$ and induct.

EXERCISE 4.133. Continue the notation from Exercise 4.131. The *discriminant* of $x_1, \ldots, x_n$ is $\Delta = \prod_{i<j} (x_i - x_j)^2$. Let $\sqrt{\Delta} = \prod_{i<j} (x_i - x_j)$ and identify $\operatorname{Gal}_{E(s_1,\ldots,s_n)} E(x_1, \ldots, x_n)$ with S_n.

(a) Show $\Delta \in E(s_1, \ldots, s_n)$.

(b) For $\sigma \in S_n$, show $\sigma(\sqrt{\Delta}) = \operatorname{sgn}(\sigma)\sqrt{\Delta}$. For $\operatorname{char} E \neq 2$, conclude $\sigma \in A_n$ if and only if $\sigma(\sqrt{\Delta}) = \sqrt{\Delta}$.

(c) If $n \geq 5$, show that there is exactly one normal intermediate field between $E(s_1, \ldots, s_n)$ and $E(x_1, \ldots, x_n)$, namely the fixed field of A_n, and that it is an extension of degree 2.

(d) If $\operatorname{char} E \neq 2$ and $n \geq 5$, show $\sqrt{\Delta}$ generates this extension.

EXERCISE 4.134. Continue the notation from Exercise 4.133. Let E_f be a splitting field for $f \in E[x]$ with (possibly repeated) roots $\alpha_1, \ldots, \alpha_n \in E_f$. Define the *Galois group* of f to be $\operatorname{Gal}_E E_f$ and the *discriminant* of f to be $\Delta_f = \prod_{i<j} (\alpha_i - \alpha_j)^2$ when f is monic.

(a) Show $\Delta_f \neq 0$ if and only if f is separable.

(b) Show Δ_f is a polynomial in the coefficients of f. *Hint:* Use Exercise 4.132 and then the original definition of s_k in Exercise 4.131 to see the relation to the coefficients of f.

(c) If $f = x^2 + bx + c$, show $\Delta_f = b^2 - 4c$.

(d) For separable f and $\operatorname{char} E \neq 2$, show $\operatorname{Gal}_E E_f \subseteq A_n$ if and only if Δ_f is the square of an element in E. *Hint:* Consider $\sqrt{\Delta_f} = \prod_{i<j} (\alpha_i - \alpha_j)$ and see Exercise 4.133, part (b).

(e) For irreducible, separable f and $\operatorname{char} E \neq 2$ with $\deg f = 3$, show $\operatorname{Gal} f$ is isomorphic to either $\mathbb{Z}_3$ or S_3.

EXERCISE 4.135 (Proof of the Fundamental Theorem of Algebra). This exercise outlines a proof of the Fundamental Theorem of Algebra. We assumes that it is known that every polynomial of odd degree with real coefficients has a real root (a fact that follows from the Intermediate Value Theorem which is proved in an analysis course).

(a) Show it suffices to see that $\mathbb{C}$ has no proper finite field extensions.

(b) By way of contradiction, suppose $\mathbb{C}$ has a proper finite field extension, $K \supseteq \mathbb{C}$. Show that it is possible to take K to be a Galois extension of $\mathbb{R}$.

(c) Show $\operatorname{Gal}_{\mathbb{R}} K$ has a 2-Sylow subgroup, H. Conclude that K_H has odd degree over $\mathbb{R}$.

(d) Show $K_H = \mathbb{R}$. Conclude $[K : \mathbb{C}]$ is a power of 2.

(e) Show $\operatorname{Gal}_{\mathbb{C}} K$ contains a subgroup H' of index 2 (cf. Exercise 2.254). Conclude that K contains a subextension of $\mathbb{C}$ of degree 2.

(f) Obtain a contradiction by showing that $\mathbb{C}$ has no extension of degree 2.

5. Famous Impossibilities

5.1. Compass and Straightedge Constructions. The ancient Greeks were interested in the geometry that is constructible using only a compass and a straightedge. While the ancient Greeks were very clever, they were unable to answer questions such as: Is it possible to trisect every angle? Is it always possible to construct a cube with twice the volume of some other cube (classically called *duplicating the cube*)? Is it always possible to construct a square with the same area as a given circle (classically called *squaring the circle* or the *quadrature of the circle*)? While these questions appear to be very straightforward geometrical questions, it took until the 19th century for them to be answered. It turns out that the missing ingredient is algebra.

The reader will surely recall from geometry that it is possible to construct a perpendicular line through a point on another line using only a compass and straightedge. Starting with a pair of arbitrary perpendicular lines in the plane, view one as the x-axis and the other as the y-axis. Their intersection is called the origin. Next choose any fixed radius (called the *unit distance*) and use the compass to mark off points that are an integer multiple of this radius away from the origin on the x-axis and y-axis. For the sake of labeling, we denote these points as $(n, 0)$ or $(0, n)$ in the obvious way. Finally, we can uniquely label every point in the plane by a pair of real numbers as usual.

Looking at the intersection of perpendicular lines through $(n, 0)$ and $(0, m)$, we can clearly construct, using only our compass and straightedge, all points of the form $\{(n, m) \mid n, m \in \mathbb{Z}\}$, called the *integer points*. As a result, the integer points are called *constructible*. Furthermore, any point that is the intersection of circles or lines constructed by a compass or straightedge using integer points is also called *constructible*. In general, we inductively say that a point in the plane is *constructible* if it is the intersection of circles or lines constructed by a compass or straightedge using previously obtained constructible points. We also say that $r \in \mathbb{R}$ is *constructible* if it is the x-coordinate of a constructible point (equivalently, use y-coordinates).

LEMMA 4.46. **(1)** *If a, b are constructible numbers, then so are $a + b$, $a - b$, ab, $\frac{a}{b}$ (for $b \neq 0$), and $\sqrt{a}$ (for $a > 0$).*

(2) *The set of constructible numbers forms an intermediate field between* $\mathbb{Q}$ *and* $\mathbb{R}$.

PROOF. For part (1) first observe that $a+b$ and $a-b$ are constructible by using the compass to copy the length of b and placing a copy of that length on the appropriate side of $(a,0)$. Now write $B=(b,0)$, write U for a point a unit distance from the origin, O, but not on either axis, and write A for the point on the resulting line from O to A a distance a from the origin. The constructions of ab and $\frac{a}{b}$ follow from the pictures

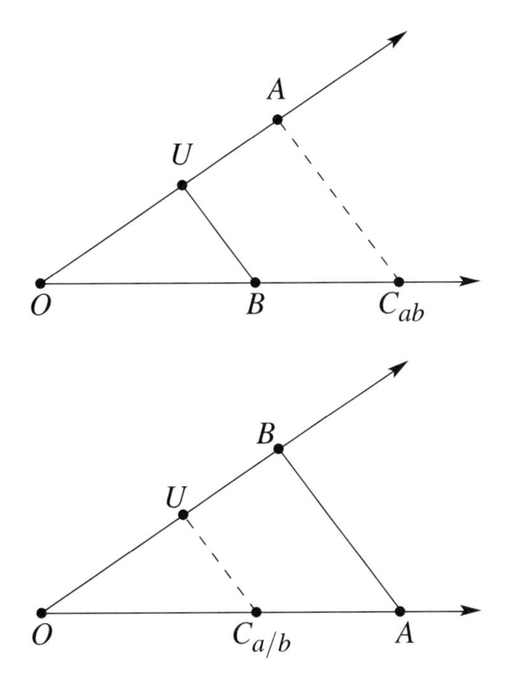

and use of similar triangles (these pictures are for $a>1$ and $b>0$). Using similar notation, the construction of $\sqrt{a}$ follows from the picture

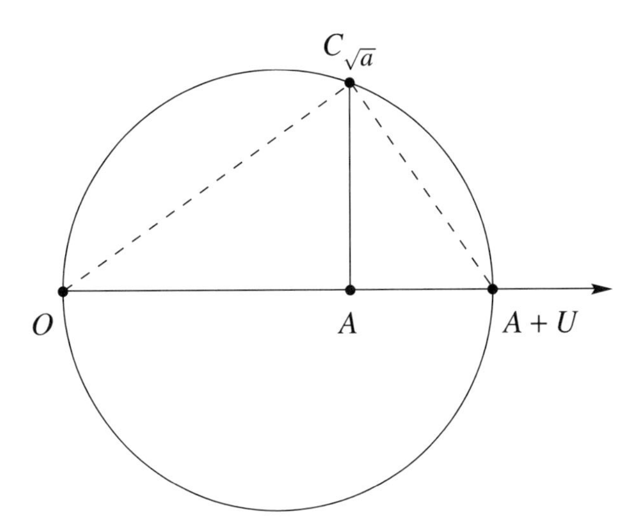

and use of similar triangles. Note that the angle between the dashed lines is a right angle. Thus part (1) is complete and clearly part (2) follows from part (1). □

THEOREM 4.47. *If $\alpha \in \mathbb{R}$ is constructible, then it is algebraic over $\mathbb{Q}$ and $[\mathbb{Q}[\alpha] : \mathbb{Q}] = 2^n$ for some $n \in \mathbb{Z}_{\geq 0}$.*

PROOF. Write $C = \{\alpha \in \mathbb{R} \mid \alpha \text{ is constructible}\}$. By definition, any constructible point P is constructed from the integer points by, say, n iterated compass and straightedge constructions. This means there are constructible points $P_1, P_2, \ldots, P_n$ with $P = P_n$ and P_j constructible by a single compass and straightedge construction from the integer points and $P_1, P_2, \ldots, P_{j-1}$ for $1 \leq j \leq n$. Write $P_j = (x_j, y_j)$ and let $C_j = \mathbb{Q}(x_1, \ldots, x_j, y_1, \ldots, y_j)$. Note that C is the union over all constructible P of the corresponding C_n.

Again by definition, P_j is the intersection of (i) two lines, each passing through two points with coordinates in C_{j-1}, (ii) the intersection of two circles, each having centers with coordinates in C_{j-1} and radii a square root of C_{j-1}, or (iii) the intersection of such a line and circle. It is straightforward to see that this implies that the equation of such lines takes the form $c_1x + c_2y = c_3$ and the equation of such circles takes the form $x^2 + y^2 + c_4x + c_5y = c_6$ for $c_i \in C_{j-1}$. Using simple algebra and the quadratic formula if necessary, it follows that x_j and y_j both either lie in C_{j-1} or in a quadratic extension of C_{j-1}. Therefore $[C_j : C_{j-1}]$ is either 1 or 2. Looking at the chain $\mathbb{Q} \subseteq C_1 \subseteq \cdots \subseteq C_n$, it follows that $[C_n : \mathbb{Q}] = 2^m$ for some $0 \leq m \leq n$. In particular, since $\mathbb{Q} \subseteq \mathbb{Q}[x_n] \subseteq C_n$, the proof is complete. □

The above restriction on constructible numbers is rather surprising and allows us to show the impossibility of certain types of classical constructions.

COROLLARY 4.48. *Using only the compass and straightedge:*

(1) *There exist angles that cannot be trisected.*

(2) *There exist cubes for which a cube with twice the volume cannot be constructed.*

(3) *There exist circles for which a square with the same area cannot be constructed.*

PROOF. For part (1), construct an equilateral triangle with base at $(0, 0)$ and $(1, 0)$. If the angle $\frac{\pi}{3}$ could be trisected using a compass and straightedge, intersecting that line with the unit circle and dropping a perpendicular would show that $\alpha = \cos\frac{\pi}{9}$ would be a constructible number. However, using the trigonometric identity $\cos 3\theta = 4\cos^3\theta - 3\cos\theta$, it follows that $\frac{1}{2} = 4\alpha^3 - 3\alpha$. Since $f = 8x^3 - 6x - 1 \in \mathbb{Q}[x]$ has no root in $\mathbb{Q}$ by the Rational Root Test, f is irreducible. Therefore $[\mathbb{Q}[\alpha] : \mathbb{Q}] = 3$ and therefore α is not constructible by Theorem 4.47.

For part (2), start with a cube of side length 1. Constructing a cube with twice the volume means that $\alpha = \sqrt[3]{2}$ would be constructible. However, looking at $x^3 - 2$ shows $[\mathbb{Q}[\alpha] : \mathbb{Q}] = 3$ so that α is not constructible.

For part (3), start with a circle of radius 1. Since it is known that π is not algebraic, certainly $\sqrt{\pi}$ is not algebraic and therefore is not constructible. As a result, a square with area π cannot be constructed. □

Note that if we allow more versatility in our definition of constructible, then some of these impossibilities become possible. For instance, it is known that if we also allow a markable ruler, then it is possible to trisect angles and duplicate cubes, though it remains impossible to square the circle.

5.2. Solvability of Polynomials. Everybody knows and loves the quadratic formula. Namely, the solution to the general quadratic equation $ax^2 + bx + c = 0$ over $\mathbb{C}$, $a \neq 0$, is given by

$$x = \frac{-b \pm \sqrt{b^2 - 4ac}}{2a}.$$

Though most people have not seen them because of their ugliness, there are similar (though much more complicated) formulas for solving the general 3rd degree equation, $ax^3 + bx^2 + cx + d = 0$, and the general 4th degree equation, $ax^4 + bx^3 + cx^2 + dx + e = 0$ (Exercises 3.125 and 3.126). However, though people searched for hundreds of years, no one was able to find the solution to the general 5th degree equation, $ax^5 + bx^4 + cx^3 + dx^2 + ex + f = 0$. It turns out that is because no general formula is possible, at least not using the four arithmetic operations and extraction of roots. We make these ideas precise in this section. First we deal with what it means to solve a polynomial equation with roots.

DEFINITION 4.49. **(1)** If $F \subseteq E$ are fields, we say E is a *radical extension* if there exists a chain of fields $F = E_0 \subseteq E_1 \subseteq \cdots \subseteq E_n = E$ so that $E_j = E_{j-1}[\alpha_j]$ for $\alpha_j \in E_j$ satisfying $\alpha_j^{k_j} \in E_{j-1}$ for some $k_j \in \mathbb{N}$, $1 \leq j \leq n$.

(2) If $f \in F[x]$, we say f is *solvable* over F if a splitting field for f over F lies inside a radical extension of F.

Radical extensions are precisely those obtained by successively attaching various k^{th} roots to the field. For instance, the quadratic equation says that only the four arithmetic operations and square roots are needed to solve a second degree polynomial. In particular, the roots are contained in a radical extension and so any second degree polynomial is solvable. Examination of the solution to the 3rd degree equation (Exercise 3.125) shows that only cube roots of square roots are needed. Therefore any 3rd degree polynomial is also solvable. Similarly, roots of the 4th degree equation (Exercise 3.126) require square roots of cube roots of square roots so that any 4th degree polynomial is also solvable.

Stunningly, it turns out that this wonderful pattern stops and that some 5th degree equations are not solvable. We will see that there is *no formula* for the roots of a general 5th degree equation using only field operations and extraction of roots (Corollary 4.56). This is a mind-blowing statement! Of course, this result says nothing about trying to solve a general 5th degree equation using fancier techniques than arithmetic and extraction of roots. In fact such ideas are possible, but they require tools from analysis.

We start our examination of solving polynomial equations by proving a lemma that is useful for improving the behavior of a radical extension.

LEMMA 4.50. *Let $F \subseteq E$ be a radical extension. There exists a field K with $E \subseteq K$ so that K is simultaneously a radical extension of F and a splitting field for a polynomial over F.*

PROOF. Write $E = F[\alpha_1, \ldots, \alpha_n]$ with $\alpha_j^{k_j} \in F[\alpha_1, \ldots, \alpha_{j-1}]$. Let f_j be the minimal polynomial of α_j over F and choose K to be a splitting field for $f = f_1 \cdots f_n$ over E. Since $f \in F[x]$ and $E = F[\alpha_1, \ldots, \alpha_n]$, K is also a splitting field for f over F. By Theorem 4.36, part (4), the roots of f_j are $R_j = \{\sigma(\alpha_j) \mid$

$\sigma \in \operatorname{Gal}_F K\}$. Noting that $\sigma(\alpha_j)^{k_j} \in F[\sigma(\alpha_1), \ldots, \sigma(\alpha_{j-1})]$, it is then easy to see K is a radical extension of F by attaching, one at a time, the elements of R_1, then R_2, ..., then R_n. $\square$

Having already developed Galois theory, we expect that the notion of a solvable extension should have an analogue on the group side (cf. Exercise 2.301). It is given in the following definition.

DEFINITION 4.51. A group G is said to be *solvable* if there exists a chain of subgroups $\{e\} = G_0 \subseteq G_1 \subseteq \cdots \subseteq G_n = G$ so that G_{j-1} is normal in G_j and G_j/G_{j-1} is abelian, $1 \leq j \leq n$.

Obviously abelian groups are solvable. A nonabelian example is furnished by D_n. To see this, write $\mathcal{R}_n$ for the set of rotations in D_n (Exercise 2.144) and recall that $\mathcal{R}_n \cong \mathbb{Z}_n$ is a normal subgroup of D_n with $D_n/\mathcal{R}_n \cong \mathbb{Z}_2$. Then the chain of subgroups $\{e\} \subseteq \mathcal{R}_n \subseteq D_n$ shows that D_n is solvable.

For the next example (mostly a nonexample), recall that simple groups have no proper normal subgroups. Looking at the chain $\{e\} \subseteq G$, it follows that simple groups are solvable if and only if they are abelian, i.e., if and only if they are isomorphic to $\mathbb{Z}_p$, p prime (Exercise 2.283).

Of related special importance to us is S_n for $n \geq 5$. In this case, the only proper normal subgroup is A_n (Exercise 2.287). Since $S_n/A_n \cong \mathbb{Z}_2$, it follows that S_n is solvable if and only if A_n is solvable. However, since A_n is simple (Theorem 2.80) and nonabelian, A_n is not solvable. Thus S_n is not solvable either.

There are a number of famous theorems relating the order of a finite group to solvability, some very difficult to prove. One is a theorem of Burnside that says every group of order $p^n q^m$ is solvable when p and q are prime. A second is a theorem of Frobenius stating that every group with square free order is solvable. A third is a conjecture of Burnside (proved by Feit and Thompson) that says every group of odd order is solvable.

As far as general properties go, it is straightforward to see that subgroups and quotients of solvable groups remain solvable (Exercise 4.150). These two facts and the Correspondence Theorem easily show that if N is a normal subgroup of G, then G is solvable if and only if N and G/N are solvable (Exercise 4.151).

DEFINITION 4.52. Let F be a field, let $f \in F[x]$, and let E_f be a splitting field for f over F. The *Galois group* of f is $\operatorname{Gal} f = \operatorname{Gal}_F E_f$.

Since E_f is only unique up to an F-isomorphism, note that $\operatorname{Gal} f$ is only defined up to isomorphism.

LEMMA 4.53. *Let $F \subseteq E \subseteq K$ be fields.*

(1) *If K is a splitting field for a polynomial over either F or E, then $\operatorname{Gal}_F E$ is isomorphic to a quotient of a subgroup of $\operatorname{Gal}_F K$. If E is also a splitting field for a polynomial over F, then $\operatorname{Gal}_E K$ is a normal subgroup of $\operatorname{Gal}_F K$ and*

$$\operatorname{Gal}_F E \cong (\operatorname{Gal}_F K) / (\operatorname{Gal}_E K).$$

(2) *If $b \in F$ and $f = x^n - b$ for $n \in \mathbb{N}$, then $\operatorname{Gal} f$ is solvable.*

PROOF. For part (1), let $H = \{\sigma \in \operatorname{Gal}_F K \mid \sigma(E) = E\}$, clearly a subgroup of $\operatorname{Gal}_F K$. Of course $\rho : H \to \operatorname{Gal}_F E$ given by mapping σ to $\sigma|_E$ is a homomorphism with $\ker \rho = \operatorname{Gal}_E K$. Note that if K is a splitting field over F, then it is also a splitting field over E. In either case, as in the proof of Corollary 4.45, Theorem 4.18, part (2), shows ρ is onto, which finishes the first part of (1). For the second part of (1), simply use Theorem 4.35, part (1), to see that, in that case, $\sigma(E) = E$ for all $\sigma \in \operatorname{Gal}_F K$ so that $H = \operatorname{Gal}_F K$.

For part (2), the case of $b = 0$ is trivial so suppose $b \neq 0$. Let E_f be a splitting field for f over F and write the distinct roots of f as $\{\beta_1, \ldots, \beta_m\}$. If $\operatorname{char} F = 0$, then β_j is clearly not a root of $f' = nx^{n-1}$. Therefore $m = n$ by Theorem 3.30. If $\operatorname{char} F = p > 0$, write $n = p^d q$ with $(q, p) = 1$. Using the Frobenius endomorphism (Exercise 3.43), $f = (x^q - \beta_1^q)^{p^d}$. Since $x^q - \beta_1^q$ is separable (Theorem 4.38), $m = q$. If $\operatorname{char} F = 0$, we unify the notation by setting $d = 0$.

Next we claim that $x^n - 1$ also splits in E_f. To see this, suppose that u is in some extension of E_f and that it satisfies $u^n = 1$. Then $u\beta_1$ is still a root of f and therefore $u\beta_1 = \beta_j$ for some j. Since $\beta_1 \neq 0$ ($b \neq 0$), it follows that $u = \beta_1^{-1}\beta_j \in E_f$ as desired. Writing $x^n - 1 = (x^m - 1)^{p^d}$, it follows that there are m distinct roots of $x^n - 1$. Since these roots clearly form a subgroup of $E_f^{\times}$, Corollary 3.31 shows they are cyclic. Therefore the roots of $x^n - 1$ can be written as $\{\zeta, \zeta^2, \ldots, \zeta^m\}$ for some root $\zeta \in E_f$.

Now consider the fields $F \subseteq F[\zeta] \subseteq E_f$ and the groups $\operatorname{Gal}_{F[\zeta]} E_f \subseteq \operatorname{Gal} f = \operatorname{Gal}_F E_f$. Using properties of solvable groups and the second part of (1), it suffices to show $\operatorname{Gal}_{F[\zeta]} E_f$ and $(\operatorname{Gal}_F E_f) \,/\, (\operatorname{Gal}_{F[\zeta]} E_f) \cong \operatorname{Gal}_F F[\zeta]$ are solvable. In fact, we show they are abelian.

Start with $\operatorname{Gal}_F F[\zeta]$. By Theorem 4.35, part (1), the elements of $\operatorname{Gal}_{F[\zeta]} E_f$ are of the form σ_j for certain (but definitely not all) $[j] \in \mathbb{Z}_m$ where σ_j is determined by $\sigma_j(\zeta) = \zeta^j$. Since these automorphisms clearly satisfy $\sigma_i\sigma_j = \sigma_{ij}$ (apply both sides to ζ), it follows that $\operatorname{Gal}_{F[\zeta]} E_f$ is isomorphic to a subgroup of U_m.

For $\operatorname{Gal}_{F[\zeta]} E_f$, observe that the roots of f may be written as $\{\beta_1, \beta_1\zeta, \beta_1\zeta^2, \ldots, \beta_1\zeta^{m-1}\}$. Therefore the elements of $\operatorname{Gal}_{F[\zeta]} E_f$ are of the form τ_j for certain $[j] \in \mathbb{Z}_m$ where τ_j is determined by $\tau_j(\beta_1) = \beta_1\zeta^j$. Since these automorphisms satisfy $\tau_i\tau_j = \tau_{i+j}$ (apply both sides to β_1), it follows that $\operatorname{Gal}_{F[\zeta]} E_f$ is isomorphic to a subgroup of $\mathbb{Z}_m$ and we are done. □

THEOREM 4.54. *If $F \subseteq E$ is a radical extension, then $\operatorname{Gal}_F E$ is solvable.*

PROOF. Use Lemma 4.50 to find a field K with $F \subseteq E \subseteq K$ so that K is a radical extension and a splitting field for some polynomial over F. Using properties of solvable groups and Lemma 4.53, the first part of (1), it suffices to show that $\operatorname{Gal}_F K$ is solvable. Write $K = F[\alpha_1, \ldots, \alpha_n]$ with $\alpha_j^{k_j} \in F[\alpha_1, \ldots, \alpha_{j-1}]$ and let $f_n = x^{k_n} - \alpha_n^{k_n} \in F[x]$. We will use induction on n to show that $\operatorname{Gal}_F K$ is solvable.

First choose a splitting field L for f_n over K. Note that L is also a splitting field for some polynomial over F. Looking at $F \subseteq K \subseteq L$, Lemma 4.53, the first part of (1), again shows that it suffices to verify that $\operatorname{Gal}_F L$ is solvable. Also note that by attaching the distinct roots of f_n one at a time, it follows that L is a radical extension of K and therefore a radical extension of F.

Next let M be the subfield of L generated by F and all the roots of f_n so that $F \subseteq M \subseteq L$. Observe that M is a splitting field for f_n over F. Therefore Lemma 4.53, the second part of (1), shows that it suffices to verify that $(\operatorname{Gal}_F L) \,/\, (\operatorname{Gal}_M L) \cong \operatorname{Gal}_F M$ and $\operatorname{Gal}_M L$ are solvable. Of course $\operatorname{Gal}_F M$ is solvable by Lemma 4.53, part (2). Since $L = M[\alpha_1, \ldots, \alpha_{n-1}]$, the induction hypothesis can be used to show that $\operatorname{Gal}_M L$ is solvable and the proof is complete. □

With Theorem 4.54 in hand, it is now possible to state the relationship between solvable extensions and solvable groups. We show below that if $f \in F[x]$ is solvable, then $\operatorname{Gal} f$ is solvable. Though we do not prove it here, the converse is nearly true, though an additional hypothesis on $\operatorname{char} F$ is required. Precisely, it turns out that if $\operatorname{char} F = 0$ or if $\operatorname{char} F = p > 0$ with $p \nmid [E_f : F]$, then $\operatorname{Gal} f$ solvable implies f is solvable.

THEOREM 4.55. *If $f \in F[x]$ is solvable, then* $\operatorname{Gal} f$ *is solvable.*

PROOF. By definition and Lemma 4.50, there is a chain of fields $F \subseteq E_f \subseteq K$ where E_f is a splitting field for f over F and K is a radical extension of F and a splitting field for some polynomial over F. By Lemma 4.53, the first part of (1), and properties of solvable groups, it suffices to show $\operatorname{Gal}_F K$ is solvable. However, this is established in Theorem 4.54. □

COROLLARY 4.56. *There exist nonsolvable 5th degree polynomials in $\mathbb{Q}[x]$. In particular, no formula exists for the general solution to a 5th degree polynomial equation using only radical and arithmetic operations.*

PROOF. Consider $f = 2x^5 - 10x + 5$ which is irreducible by Eisenstein's Criterion. Since $f' = 10(x-1)(x+1)(x^2+1)$ (of course this example really started with f' and worked backwards), f has exactly two critical points, $x = \pm 1$. Since $f(-1) = 13$ and $f(1) = -3$, the graph of f has the general shape

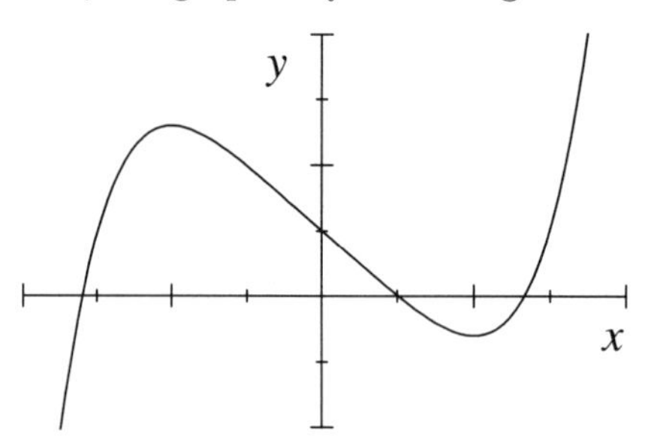

and it has exactly 3 real roots.

Let E_f be the splitting field of f in $\mathbb{C}$ with real roots $\{\alpha_1, \alpha_2, \alpha_3\}$ and imaginary roots $\{\alpha_4, \alpha_5 = \overline{\alpha}_4\}$. Identifying $\operatorname{Gal} f$ with a subgroup of S_5, clearly complex conjugation corresponds to the 2-cycle $\sigma = (45)$. On the other hand, $[\mathbb{Q}[\alpha_1] : \mathbb{Q}] = 5$ so that $5 \mid [E_f : \mathbb{Q}] = |\operatorname{Gal} f|$. Cauchy's Theorem shows $\operatorname{Gal} f$ contains an element τ of order 5 which must be a 5-cycle by the Cyclic Decomposition Theorem (cf. Exercise 2.197). Since τ is a 5-cycle, some power of τ maps 4 to 5. Replacing τ by this power and possibly relabeling $\{\alpha_1, \alpha_2, \alpha_3\}$, we may assume $\tau = (12345)$. It is now a simple exercise to show $\operatorname{Gal} f$ contains all 2-cycles. For instance applying c_{τ^k} to (45) generates the transpositions $S = \{(45), (51), (12), (23), (34)\}$. The rest of the transpositions can clearly be obtained by conjugating elements of S amongst themselves to produce larger "gaps" (cf. Exercise 2.200). Thus $\operatorname{Gal} f = S_5$ by the Parity Theorem. As already noted, S_5 is not a solvable group so Theorem 4.55 finishes the argument. □

In fact, it is possible to find a nonsolvable rational polynomial of degree n for any $n \geq 5$, though we do not prove it here. Exercise 4.119 handles the prime cases and Exercise 4.131, part (d), addresses the situation of a general polynomial.

5.3. Exercises 4.136–4.162.

5.3.1. *Compass and Straightedge Constructions.*

EXERCISE 4.136. Show how to construct an equilateral triangle with base at $(0,0)$ and $(1,0)$ using a compass and straightedge.

EXERCISE 4.137. **(a)** Given an angle, show how to bisect it using a compass and straightedge.

(b) Show that $\cos\frac{\pi}{2^n}$ and $\sin\frac{\pi}{2^n}$ are constructible for $n \in \mathbb{N}$.

EXERCISE 4.138. **(a)** Show that $\sin\theta$ is constructible if and only if $\cos\theta$ is constructible.

(b) Show that $\cos\theta$ is constructible if and only if $\cos 2\theta$ is constructible.

EXERCISE 4.139. Given a square, show how to construct a square with twice the area using a compass and straightedge.

EXERCISE 4.140. Given a circle, show it is possible to construct a circle with twice the area using a compass and straightedge.

EXERCISE 4.141. Show that $\frac{\pi}{4}$ can be trisected with a compass and ruler.

EXERCISE 4.142. Using only the compass and straightedge, show there exist spheres for which a cube with the same volume cannot be constructed.

EXERCISE 4.143 (Alternate Definition of Constructible Numbers). Show $\alpha \in \mathbb{R}$ is constructible if and only if there exist fields $\mathbb{Q} = F_0 \subseteq F_1 \subseteq \cdots \subseteq F_n \subseteq \mathbb{R}$ so that $\alpha \in F_n$ and $[F_j : F_{j-1}] = 2$, $1 \le j \le n$.

5.3.2. *Solvability of Polynomials.*

EXERCISE 4.144. Find a radical extension of $\mathbb{Q}$ containing:

(a) $\sqrt[5]{2+\sqrt[3]{5}}$,

(b) $\left(8+4\sqrt[6]{1+\sqrt[3]{7+\sqrt{13}}}\right)(\sqrt{2}+11)$.

EXERCISE 4.145. Show that radical extensions are finite extensions.

EXERCISE 4.146. If F is a finite field, show that every finite extension is radical.

EXERCISE 4.147. Let F be a field and let $f, g \in F[x]$. Show fg is solvable if and only if f and g are solvable.

EXERCISE 4.148. Let $U_{n\times n}$ be the set of upper triangular $n \times n$ matrices with 1's on the main diagonal and entries in a field F. Show $U_{n\times n}$ is solvable.

EXERCISE 4.149. Show S_2, S_3, and S_4 are solvable.

EXERCISE 4.150. **(a)** If G is a solvable group and if H is a subgroup of G, show H is solvable.

(b) If G is a solvable group and if N is a normal subgroup of G, show G/N is solvable.

(c) Show that the image under a homomorphism of a solvable group is solvable.

EXERCISE 4.151. If N is a normal subgroup of G, show G is solvable if and only if N and G/N are solvable.

EXERCISE 4.152. Recall that the commutator subgroup G' of G is the subgroup generated by $\{g_1g_2g_1^{-1}g_2^{-1} \mid g_i \in G\}$ (cf. Exercise 2.73). Set $G^{(0)} = G$, $G^{(1)} = G'$, and $G^{(k)} = \left(G^{(k-1)}\right)'$ for $k \in \mathbb{N}$, called the *derived series* of G. Show G is solvable if and only if $G^{(k)} = \{e\}$ for some k. *Hint:* Exercise 2.155.

EXERCISE 4.153. If G is a group with $|G| = pq$ for p, q prime, show G is solvable.

EXERCISE 4.154. Show that every p-group is solvable. *Hint:* Exercise 2.254.

EXERCISE 4.155. Show that the following polynomials over $\mathbb{Q}$ are not solvable:
(a) $x^5 - 6x + 3$,
(b) $x^5 - x - 1$,
(c) $x^5 - 6x^2 + 2$.

EXERCISE 4.156. Let $F \subseteq E$ be fields with E normal and finite over F. Show that there is an $f \in F[x]$ so that $\operatorname{Gal}_F E \cong \operatorname{Gal} f$.

EXERCISE 4.157. Let F be a field and let $f, g \in F[x]$. Show $\operatorname{Gal}(fg)$ is isomorphic to a subgroup of $\operatorname{Gal} f \times \operatorname{Gal} g$.

EXERCISE 4.158. **(a)** If $f \in \mathbb{Q}[x]$ is irreducible and $\deg f = 2$, show $\operatorname{Gal} f \cong \mathbb{Z}_2$.
(b) If $f \in \mathbb{Q}[x]$ is irreducible and $\deg f = 3$, show $\operatorname{Gal} f$ is isomorphic to $\mathbb{Z}_3$ or S_3 and so is solvable.
(c) If $f \in \mathbb{Q}[x]$ is irreducible and $\deg f = 4$, show $\operatorname{Gal} f$ is isomorphic to a subgroup of S_4 and so is solvable.

EXERCISE 4.159. Find $f \in \mathbb{Q}[x]$ with $\deg f = 7$ and $\operatorname{Gal} f \cong S_7$.

EXERCISE 4.160 (Primitive Roots of Unity). Let F be a field and let $n \in \mathbb{N}$. We say ζ in some extension of F is an nth *root of unity* if $\zeta^n = 1$. If $|\zeta| = n$ (order with respect to the multiplicative structure), we say ζ is a *primitive* n^{th} *root of unity*.
(a) Show that the nth roots of unity in $\mathbb{C}$ are $\{e^{\frac{2\pi ik}{n}} \mid 0 \le k < n\}$ and that $e^{\frac{2\pi ik}{n}}$ is primitive if and only if $(n, k) = 1$.
(b) Given a field F, let E be the splitting field of $x^n - 1 \in F[x]$. Show E has a primitive n^{th} root of unity if and only if $\operatorname{char} F = 0$ or $\operatorname{char} F = p > 0$ and $p \nmid n$. *Hint:* See the proof of Lemma 4.53, part (2).
(c) Let $m = n$ if $\operatorname{char} F = 0$ or $\operatorname{char} F = p > 0$ and $p \nmid n$. Otherwise, write $n = p^d m$ with $(m, p) = 1$. Show $\operatorname{Gal}_F E$ is isomorphic to a subgroup of U_m.

EXERCISE 4.161. Let $F \subseteq K$ be fields. If K is Galois over F and if $\operatorname{Gal}_F K$ is a cyclic group, we say K is a *cyclic extension*. Assume that F contains a primitive n^{th} root of unity (Exercise 4.160).
(a) If $K = F[\alpha]$ with $\alpha^n \in F$, show K is a cyclic extension with $|\operatorname{Gal}_F K| \mid n$. *Hint:* See the proof of Lemma 4.53, part (2), and Exercise 2.63.
(b) If E is a cyclic extension of F with $[E : F] = n$, write σ for a generator of $\operatorname{Gal}_F E$ and ζ for an nth root of unity in F. The *Lagrange resolvent* of $\beta \in E$ is $(\beta, \zeta) = \sum_{j=1}^{n} \zeta^j \sigma^j(\beta)$. Show $\sigma^k((\beta, \zeta)) = \zeta^{-k}(\beta, \zeta)$ and $\sigma((\beta, \zeta)^n) = (\beta, \zeta)^n$.
(c) Show there is a $\beta \in E$ so that $\gamma = (\beta, \zeta) \ne 0$. *Hint:* Exercise 4.91.
(d) If ζ is also a primitive n^{th} root of unity, show that the only intermediate field between F and E containing γ is E. Conclude $E = F[\gamma]$ with $\gamma^n \in F$.

EXERCISE 4.162 (Kummer Extension). If G is a group, the smallest $m \in \mathbb{N}$ so that $g^m = e$ for all $g \in G$ is called the *exponent* of G, when it exists. Let F be a field containing a primitive n^{th} root of unity (Exercise 4.160). A Galois extension K of F is called n-*Kummer* if $\mathrm{Gal}_F K$ is abelian with exponent dividing n. Show that a finite extension K of F is n-Kummer if and only if $K = F[\alpha_1, \ldots, \alpha_k]$ where $\alpha_j \in K$ satisfies $\alpha_j^n \in F$. *Hint:* Exercise 4.161 and the Fundamental Theorem of Finite Abelian Groups.

6. Cyclotomic Fields

If F is a field, a splitting field for $x^n - 1 \in F[x]$ over F is called a *cyclotomic extension* of F (cf. Exercise 4.160). The most important case occurs when $F = \mathbb{Q}$. In this setting, it is therefore important to study the set of nth *roots of unity* in $\mathbb{C}$. As is well known, this set is given by

$$\mu_n = \{e^{\frac{2\pi i k}{n}} \mid 0 \leq k < n\}.$$

As a group under multiplication, clearly $\mu_n \cong \mathbb{Z}_n$ by mapping $e^{\frac{2\pi i k}{n}} \to [k]$. We say $\zeta \in \mu_n$ is a *primitive* n^{th} *root of unity* if $|\zeta| = n$ (using the order with respect to the multiplicative structure). In particular, $e^{\frac{2\pi i k}{n}}$ is primitive if and only if $(n, k) = 1$ by Theorem 2.11 and there are exactly $\varphi(n)$ primitive n^{th} roots of unity. If ζ_n is a primitive n^{th} root of unity, we get another isomorphism $\mu_n \cong \mathbb{Z}_n$ by mapping $\zeta_n^k \to [k]$. Again by Theorem 2.11, note that ζ_n^k is primitive if and only if $(n, k) = 1$.

Definition 4.57. For $n \in \mathbb{N}$, the nth *cyclotomic polynomial*, initially defined as an element of $\mathbb{C}[x]$, is

$$\Phi_n = \prod_{\text{primitive } \zeta \in \mu_n} (x - \zeta) = \prod_{\substack{1 \leq k < n \\ (k,n)=1}} (x - \zeta_n^k)$$

where ζ_n is a primitive n^{th} root of unity.

For example,

$$\begin{aligned}
\Phi_1 &= x - 1, \\
\Phi_2 &= x + 1, \\
\Phi_3 &= \left(x - \frac{-1 + i\sqrt{3}}{2}\right)\left(x - \frac{-1 - i\sqrt{3}}{2}\right) = x^2 + x + 1, \\
\Phi_4 &= (x - i)(x + i) = x^2 + 1.
\end{aligned}$$

Observe that all of the above polynomials are actually in $\mathbb{Z}[x]$, that they are irreducible in $\mathbb{Q}[x]$, and that they have coefficients in $\{0, \pm 1\}$. We will see that the first two observations hold in general. However, the third observation eventually fails, though the first counterexample does not appear until Φ_{105}! In fact, it is known that every integer appears as a coefficient of some cyclotomic polynomial.

Theorem 4.58. *Let* $n \in \mathbb{N}$.

(1) Φ_n *is a monic polynomial of degree* $\varphi(n)$ *satisfying*

$$x^n - 1 = \prod_{\substack{1 \leq d \leq n \\ d \mid n}} \Phi_d.$$

(2) $\Phi_n \in \mathbb{Z}[x]$.

(3) Φ_n *is irreducible in* $\mathbb{Q}[x]$.

PROOF. Start with part (1). By definition, Φ_n is monic and has degree $\varphi(n)$ since there are $\varphi(n)$ primitive n^{th} roots of unity. Next observe that each $\zeta \in \mu$ is a root of $x^n - 1$. Therefore

$$x^n - 1 = \prod_{\zeta \in \mu_n} (x - \zeta).$$

If $\zeta \in \mu_n$, write $d = |\zeta|$. It follows that $d \mid n$ and that ζ is a primitive d^{th} root of unity. Conversely, if $d \mid n$ and if ζ is a (primitive) d^{th} root of unity, then it is also an nth root of unity. It follows that μ_n is the disjoint union of the primitive d^{th} roots of unity over all d dividing n. Therefore

$$x^n - 1 = \prod_{d \mid n} \prod_{\text{primitive } \zeta \in \mu_d} (x - \zeta) = \prod_{\substack{1 \le d \le n \\ d \mid n}} \Phi_d.$$

For part (2), note that $\Phi_1 = x - 1 \in \mathbb{Z}[x]$ and induct on n. Using part (1) and the inductive hypothesis, write $x^n - 1 = \Phi_n f$ for $f = \prod_{1 \le d < n,\ d \mid n} \Phi_d$ with $f \in \mathbb{Z}[x]$. Since $x^n - 1$ and f lie in $\mathbb{Q}[x]$, the uniqueness part of the Division Algorithm shows $\Phi_n \in \mathbb{Q}[x]$ (cf. Exercise 3.97). By Gauss's Lemma, there are $r, s \in \mathbb{Q}^\times$ so that $r\Phi_n, sf \in \mathbb{Z}[x]$ and $x^n - 1 = (r\Phi_n)(sf)$. Since Φ_n and f are monic, this forces $r, s \in \mathbb{Z}$ with $rs = 1$ so that $r = s = \pm 1$. In either case, $\Phi_n \in \mathbb{Z}[x]$ as desired.

For part (3), fix a primitive n^{th} root of unity ζ_n and let g be its minimal polynomial over $\mathbb{Q}$. Clearly $g \mid \Phi_n$ in $\mathbb{Q}[x]$. It suffices to show ζ_n^k is a root of g when $(k, n) = 1$ since then $g = \Phi_n$. Writing k as its prime factorization and iterating the argument, it actually suffices to show ζ^p is a root of g whenever ζ is a primitive n^{th} root of unity satisfying $g(\zeta) = 0$ and p is prime with $p \nmid n$.

Since g is an irreducible factor of Φ_n, write $\Phi_n = gh$. By Gauss's Lemma and the fact that Φ_n is monic, we may assume $g, h \in \mathbb{Z}[x]$ with g, h monic. Let ζ be a (primitive) n^{th} root of unity satisfying $g(\zeta) = 0$ and let p be a prime with $p \nmid n$. Since ζ^p is a root of Φ_n, it is a root of g or a root of h. Argue by contradiction and suppose ζ^p is not a root of g. Then ζ^p is a root of h and ζ is a root of $h(x^p)$.

Since g is also the minimal polynomial of ζ, we can write $h(x^p) = g(x)u(x)$ for some monic $u \in \mathbb{Z}[x]$. Applying the reduction homomorphism from $\mathbb{Z}[x]$ to $\mathbb{Z}_p[x]$ and using the Frobenius automorphism, we get

$$\overline{g}(x)\overline{u}(x) = \overline{h}(x^p) = \left(\overline{h}(x)\right)^p.$$

Let $\overline{v} \in \mathbb{Z}_p[x]$ be an irreducible factor of $\overline{g}$. Since $\mathbb{Z}_p[x]$ is a UFD, it follows that $\overline{v} \mid \overline{h}$. As $x^n - [1] = \overline{g}\overline{h} = \overline{\Phi}_n$, we see that $\overline{v}^2 \mid x^n - [1]$. This would force $x^n - [1] \in \mathbb{Z}_p[x]$ to have roots with multiplicity at least two, which is a contradiction to the fact that it is separable (Theorem 3.30). □

Note that part (1) of Theorem 4.58 gives a recursive procedure for calculating Φ_n via long division. For instance, $x^5 - 1 = \Phi_1\Phi_5$ so that

$$\Phi_5 = \frac{x^5 - 1}{x - 1} = x^4 + x^3 + x^2 + x + 1.$$

In general $\Phi_p = x^{p-1} + x^{p-2} + \cdots + x + 1$ for p prime (Exercise 4.163).

COROLLARY 4.59. *Let $n \in \mathbb{N}$ and let ζ_n be a primitive n^{th} root of unity. The splitting field for $x^n - 1$ over $\mathbb{Q}$ is $\mathbb{Q}[\zeta_n]$ and satisfies*

$$[\mathbb{Q}[\zeta_n] : \mathbb{Q}] = \varphi(n) \text{ with } \operatorname{Gal}_{\mathbb{Q}} \mathbb{Q}[\zeta_n] \cong U_n.$$

PROOF. Clearly $\mathbb{Q}[\zeta_n]$ is the splitting field for $x^n - 1$. Part (3) of Theorem 4.58 shows that Φ_n is actually the minimal polynomial over $\mathbb{Q}$ for ζ_n so that $[\mathbb{Q}[\zeta_n] : \mathbb{Q}] = \varphi(n)$. Finally, write the set of primitive n^{th} roots as $\{\zeta_n^k \mid (n,k) = 1\}$. By Theorem 4.35, part (3), the elements of $\text{Gal}_{\mathbb{Q}} \mathbb{Q}[\zeta_n]$ are precisely those automorphisms σ_k, $[k] \in U_n$, where σ_k is determined by $\sigma_k(\zeta_n) = \zeta_n^k$. Since the automorphisms satisfy $\sigma_i \sigma_j = \sigma_{ij}$, it follows that $\text{Gal}_{\mathbb{Q}} \mathbb{Q}[\zeta_n]$ is isomorphic to U_n under the map $\sigma_k \to [k]$. $\square$

In general, an extension $F \subseteq K$ is called *abelian* if it is Galois and $\text{Gal}_F K$ is abelian. For example, $\mathbb{Q}[\zeta_n]$ is an abelian extension of $\mathbb{Q}$. Moreover, using Corollary 4.59, it is not too hard to see that any finite abelian group can be realized as the Galois group over $\mathbb{Q}$ of a subfield of some cyclotomic field $\mathbb{Q}[\zeta_n]$ (Exercise 4.177). It is a famous theorem of *Kronecker-Weber* that this is the only way to get finite abelian extensions over $\mathbb{Q}$: every abelian extension of $\mathbb{Q}$ is a subfield of some cyclotomic extension of $\mathbb{Q}$.

6.1. Exercises 4.163–4.179.

EXERCISE 4.163. Demonstrate the following.

(a) $\Phi_p = x^{p-1} + x^{p-2} + \cdots + x + 1$ for p prime (cf. Exercise 3.145).

(b) $\Phi_6 = x^2 - x + 1$.

(c) $\Phi_8 = x^4 + 1$.

(d) $\Phi_9 = x^6 + x^3 + 1$.

(e) $\Phi_{10} = x^4 - x^3 + x^2 - x + 1$.

(f) $\Phi_{12} = x^4 - x^2 + 1$.

(g) $\Phi_{p^n} = 1 + x^{p^{n-1}} + x^{2p^{n-1}} + \cdots + x^{(p-1)p^{n-1}}$ for p prime (cf. Exercises 3.146 or 4.169).

(h) $\Phi_{2p} = x^{p-1} - x^{p-2} + x^{p-3} - \cdots - x + 1$ for p odd and prime.

EXERCISE 4.164. Use Theorem 4.58 to show $n = \sum_{d|n} \varphi(d)$. (N.B.: Compare this to Exercise 2.47.)

EXERCISE 4.165. Show $\Phi_1(0) = -1$ and $\Phi_n(0) = 1$ for $n \geq 2$.

EXERCISE 4.166. If $n \in \mathbb{Z}$ is odd and $n \geq 3$, show $\Phi_n(-1) = 1$.

EXERCISE 4.167. If $n \in \mathbb{Z}$ is odd and $n \geq 3$, show $\Phi_{2n} = \Phi_n(-x)$. *Hint:* Show that the negative of the set of primitive $2n^{\text{th}}$ roots of unity is the set of primitive n^{th} roots of unity.

EXERCISE 4.168. For $n \in \mathbb{N}$, show $x^n + 1 = \prod_{d|(2n) \text{ but } d\nmid n} \Phi_d$. *Hint:* $\frac{x^{2n}-1}{x^n-1}$.

EXERCISE 4.169. **(a)** Let p be a positive prime. Show $(x^k - 1)\Phi_p(x^k) = x^{pk} - 1$. *Hint:* $(x^k)^p - 1$.

(b) Show $\Phi_p(x^k) = \prod_{d|(pk) \text{ but } d\nmid k} \Phi_d$.

(c) Show $\Phi_p(x^k)$ is irreducible if and only if $k = p^m$ for some $m \in \mathbb{Z}_{\geq 0}$.

(d) Conclude $\Phi_{p^n}(x) = \Phi_p(x^{p^{n-1}})$.

(e) Show $\Phi_{2^n} = x^{2^{n-1}} + 1$ and factor $x^{2^n} - 1$ into irreducibles over $\mathbb{Q}[x]$.

EXERCISE 4.170. Let p be an odd positive prime and let ζ_p be a primitive p^{th} root of unity.

(a) For each $k \in \mathbb{N}$ with $k \mid (p-1)$, write $p-1 = kr$. Show there exists a unique intermediate field K_r between $\mathbb{Q}$ and $\mathbb{Q}[\zeta_p]$ with $[\mathbb{Q}[\zeta_p] : K_r] = k$ and $[K_r : \mathbb{Q}] = r$. Show that these are the only intermediate fields.

(b) Let $a_p = \operatorname{Re}\zeta_p = \frac{\zeta_p + \zeta_p^{-1}}{2}$. Show that $K_{\frac{p-1}{2}} = \mathbb{Q}[a_p]$ and that $K_{\frac{p-1}{2}} = \mathbb{R} \cap \mathbb{Q}[\zeta_p]$. *Hint:* Show $\zeta_p^2 - 2a_p\zeta_n + 1 = 0$.

(c) Show $K_2 = \mathbb{Q}\left[\sqrt{(-1)^{\frac{p-1}{2}}p}\right]$. *Hint:* Differentiate $x^p - 1 = \prod_{j \neq 0}(x - \zeta_p^j)$, set $x = \zeta_p^k$, and take the product over $k = 0, \ldots, p-1$.

EXERCISE 4.171. Let $n, m \in \mathbb{N}$ and let l be the least common multiple, $l = [n, m]$. Show $\mathbb{Q}[\zeta_n, \zeta_m] = \mathbb{Q}[\zeta_l]$ where ζ_k is a primitive k^{th} root of unity.

EXERCISE 4.172 (Möbius Inversion Formula). For $n \in \mathbb{N}$, the *Möbius μ-function* $\mu : \mathbb{N} \to \{0, \pm 1\}$ is defined by $\mu(1) = 1$, $\mu(n) = 0$ if $n = p^2 m$ for some prime p, and $\mu(n) = (-1)^k$ if $n = p_1 \cdots p_k$ for distinct primes.

(a) Show $\sum_{d|n} \mu(d) = \begin{cases} 1 & \text{if } n = 1, \\ 0 & \text{if } n \neq 1. \end{cases}$

Hint: If $n > 1$ has r distinct prime divisors, show $\sum_{d|n} \mu(d) = \sum_{k=0}^{r} \binom{r}{k}(-1)^k$ and then use the Binomial Theorem.

(b) Given $f : \mathbb{N} \to \mathbb{N}$, let $F : \mathbb{N} \to \mathbb{N}$ be defined by $F(n) = \sum_{d|n} f(d)$. Show that

$$f(n) = \sum_{d|n} \mu(d) F\left(\frac{n}{d}\right) = \sum_{d|n} \mu\left(\frac{n}{d}\right) F(d),$$

called the *Möbius inversion formula. Hint:* For the first formula, substitute in F, show $d \mid n$ and $c \mid \frac{n}{d}$ if and only if $c \mid n$ and $d \mid \frac{n}{c}$, and use part (a).

(c) Given $g : \mathbb{N} \to \mathbb{N}$, let $G : \mathbb{N} \to \mathbb{N}$ be defined by $G(n) = \prod_{d|n} g(d)$. Show that $g(n) = \prod_{d|n} G\left(\frac{n}{d}\right)^{\mu(d)}$ and $g(n) = \prod_{d|n} G(d)^{\mu\left(\frac{n}{d}\right)}$. *Hint:* Essentially the same proof but switch sums to products and products to powers.

(d) Show

$$\Phi_n = \prod_{d|n} (x^d - 1)^{\mu\left(\frac{n}{d}\right)} = \prod_{d|n} (x^{\frac{n}{d}} - 1)^{\mu(d)}.$$

Hint: Show that it is enough to check this equality for all large integer values of x. Note that $\Phi_n(x) \to \infty$ as $x \to \infty$ and use $g(n) = \Phi_n(x)$.

EXERCISE 4.173. **(a)** For $n \in \mathbb{Z}$ with $n > 1$, let $p_1, \ldots, p_r$ be the distinct positive primes dividing n and $m = p \cdots p_r$. Show $\Phi_n(x) = \Phi_m\left(x^{\frac{n}{m}}\right)$. *Hint:* Exercise 4.172, part (d).

(b) If $p \mid n$ for a prime p, show $\Phi_{pn}(x) = \Phi_n(x^p)$.

(c) If $(p, n) = 1$ for a prime p, show $\Phi_{pn}(x) = \frac{\Phi_n(x^p)}{\Phi_n(x)}$. *Hint:* Expand $\Phi_n(x^p)$ with Exercise 4.172, part (d), and note that $d \mid (pm)$ is of the form c or pc for $c \mid m$.

EXERCISE 4.174. For $n, p \in \mathbb{N}$ with p prime, let $\psi(n)$ be the number of irreducible polynomials of degree n in $\mathbb{Z}_p[x]$. Show $\psi(n) = \frac{1}{n}\sum_{d|n} \mu(d) p^{\frac{n}{d}}$. *Hint:* Show $p^n = \sum_{d|n} d\psi(d)$ (cf. Exercise 4.80) and then apply Möbius inversion.

EXERCISE 4.175. Let $n, k, p \in \mathbb{N}$ with p prime with $p \mid \Phi_n(k)$. This exercise shows that either $p \mid n$ or $p \equiv 1 \bmod(n)$.

(a) Applying the reduction homomorphism, show $x - [k]$ divides $\overline{\Phi}_n$ in $\mathbb{Z}_p[x]$.

(b) Show $[k] \in U_p$ with $d_0 = |[k]|$ satisfying $d_0 \mid n$ and $d_0 \mid (p-1)$. *Hint:* Use Theorem 4.58 to show $p \mid (k^n - 1)$.

(c) If $p \nmid n$, show $d_0 = n$. *Hint:* Use Theorem 4.58 on $x^{d_0} - 1$. If $d_0 < n$, write $x^n - [1] = \overline{\Phi}_n \left(x^{d_0} - [1]\right) \overline{f}$ for some $f \in \mathbb{Z}_p[x]$ and consider separability at $x = [k]$ to get a contradiction.

(d) If $p \nmid n$, conclude that $p \equiv 1 \bmod(n)$.

EXERCISE 4.176 (Dirichlet's Theorem, $k = 1$). Let $k, n \in \mathbb{N}$. It is a difficult and famous theorem of *Dirichlet* that there are an infinite number of primes in the arithmetic progression k, $k+n$, $k+2n$, $k+3n$, $\ldots$ if and only if $(n,k) = 1$. This exercise proves Dirichlet's Theorem in the special case of $k = 1$.

(a) If $(n,k) > 1$, show that there are at most a finite number of primes in $\{k + nj \mid j \in \mathbb{Z}_{\geq 0}\}$.

(b) For $l \in \mathbb{N}$, show $(\Phi_n(nl), n) = 1$. *Hint:* If q is a positive prime dividing $\Phi_n(nl)$ and n, show $[nl] = [0]$ is a root of $\overline{\Phi}_n$ in $\mathbb{Z}_q[x]$ and therefore, impossibly, a root of $x^n - [1]$.

(c) Show $|\Phi_n(nl_1)| > 1$ for all sufficiently large $l_1 \in \mathbb{N}$. *Hint:* If not, show $\Phi_n(nx)$, $\Phi_n(nx) + 1$, or $\Phi_n(nx) - 1$ has an infinite number of zeros. Alternately, use analysis and observe that $\Phi_n(x) \to \infty$ as $x \to \infty$.

(d) Choose l_1 as in part (c) and let p_1 be a positive prime dividing $\Phi_n(nl_1)$. Conclude $p_1 \equiv 1 \bmod(n)$. *Hint:* Exercise 4.175.

(e) Show there is an $l_2 \in \mathbb{N}$ so that $|\Phi_n(np_1l_2)| > 1$ and choose a positive prime p_2 dividing $\Phi_n(np_1l_2)$. Show that $p_2 \neq p_1$ and that $p_2 \equiv 1 \bmod(n)$.

(f) Show there are an infinite number of primes in $\{1 + nj \mid j \in \mathbb{Z}_{\geq 0}\}$.

EXERCISE 4.177 (Finite Abelian Groups Are Galois Groups over $\mathbb{Q}$). **(a)** Let G be a finite abelian group and write $G \cong \mathbb{Z}_{n_1} \times \cdots \times \mathbb{Z}_{n_k}$. Show there exist distinct primes p_j so that $p_j \equiv 1 \bmod(n_j)$. *Hint:* Exercise 4.176.

(b) Let $n = \prod_j p_j$. If ζ_n is a primitive n^{th} root of unity, show $\operatorname{Gal}_{\mathbb{Q}} \mathbb{Q}[\zeta_n] \cong \prod_j \mathbb{Z}_{p_j - 1}$.

(c) Write $p_j - 1 = m_j n_j$ and let $H_j = \langle [n_j] \rangle \subseteq \mathbb{Z}_{p_j-1}$. Show $H_j \cong \mathbb{Z}_{m_j}$ and $\mathbb{Z}_{p_j-1}/H_j \cong \mathbb{Z}_{n_j}$.

(d) Let $H = H_1 \times \cdots \times H_k$ and let E be the intermediate field between $\mathbb{Q}$ and $\mathbb{Q}[\zeta_n]$ corresponding to H in the Galois correspondence. Show $\operatorname{Gal}_{\mathbb{Q}} E \cong G$.

EXERCISE 4.178 (Wedderburn's Theorem). Let R be a finite division ring. This exercise shows R is a field.

(a) Write Z for the center of a division ring R. Show that Z is a finite field with $q = |Z| \geq 2$ and that $|R| = q^n$ for some $n \in \mathbb{N}$. *Hint:* Pick a basis over Z.

(b) Choose representatives $u_1, \ldots, u_m$ for the distinct nonsingleton conjugacy classes in $R^\times$. Show $|R^\times| = |Z^\times| + \sum_{i=1}^m |R^\times| / |Z_{R^\times}(u_i)|$. *Hint:* Apply the Class Equation to $R^\times$.

(c) If $m > 0$, show $|Z_{R^\times}(u_i)| = q^{d_i} - 1$ for some $d_i \in \mathbb{N}$ and conclude that $d_i \mid n$. *Hint:* Note that $Z \subseteq Z_R(u_i)$ and then recall Exercise 4.74.

(d) If $m > 0$, show $d_i < n$ to conclude $\Phi_n(q) \mid (|R^\times| / |Z_{R^\times}(u_i)|)$. Use this to show $\Phi_n(q) \mid (q-1)$. *Hint:* If $d_i < n$, write $x^n - 1 = \Phi_n \left(x^{d_i} - 1\right) f$ for some $f \in \mathbb{Z}[x]$. Observe also that $\Phi_n(q) \mid |R^\times|$ and then use part (b).

(e) Show R is a field. *Hint:* If $m > 0$, note that $n > 1$ so that $|q - \zeta_n| > |q - 1|$ for ζ_n a primitive n^{th} root of unity. Use this to see that $|\Phi_n(q)| > (q-1)$ and get a contradiction.

EXERCISE 4.179 (Constructing Regular n-gons). A *Fermat prime* is a prime of the form $2^{2^n}+1$. This exercise shows that a regular n-gon in the plane can be constructed with a compass and straightedge if and only if n is of the form $n = 2^k p_1 \cdots p_r$ where $p_1, \ldots, p_r$ are distinct Fermat primes.

(a) Show that a regular n-gon is constructible if and only if $a_n = \operatorname{Re} \zeta_n$ is constructible where ζ_n is a primitive n^{th} root of unity.

(b) If a regular n-gon can be constructed, show $\varphi(n)$ is a power of 2. *Hint:* Show $\zeta_n^2 - 2x\zeta_n + 1 = 0$, look at $\mathbb{Q} \subseteq \mathbb{Q}[a_n] \subseteq \mathbb{Q}[\zeta_n]$, show $[\mathbb{Q}[a_n] : \mathbb{Q}] = \frac{\varphi(n)}{2}$, and use Exercise 4.143.

(c) If $\varphi(n)$ is a power of 2, show that a regular n-gon is constructible. *Hint:* Start with the structure of $\operatorname{Gal}_{\mathbb{Q}} \mathbb{Q}[\zeta_n]$ to get at $\operatorname{Gal}_{\mathbb{Q}} \mathbb{Q}[a_n]$ and use the Fundamental Theorem of Abelian Groups to find a chain of successive index two subgroups.

(d) Show $\varphi(n)$ is a power of 2 if and only if $n = 2^k p_1 \cdots p_r$ where $p_1, \ldots, p_r$ are distinct Fermat primes.

(e) Show that the first five Fermat primes are 3, 5, 17, 257, and 65537 (currently these are the only known Fermat primes).

(f) Show that a regular n-gon is constructible if $n = 3$, 4, 5, 6, 8, 10, 12, 15, 16, 17, 20, 24, ... but not if $n = 7$, 9, 11, 13, 14, 18, 19, 21, 22, 23, 25, ...

Index

Titles in This Series

Volume

This series was founded by the highly respected mathematician and educator, Paul J. Sally, Jr.

11 **Mark R. Sepanski**
Algebra

10 **Sue E. Goodman**
Beginning Topology

9 **Ronald Solomon**
Abstract Algebra

8 **I. Martin Isaacs**
Geometry for College Students

7 **Victor Goodman and Joseph Stampfli**
The Mathematics of Finance: Modeling and Hedging

6 **Michael A. Bean**
Probability: The Science of Uncertainty with Applications to Investments, Insurance, and Engineering

5 **Patrick M. Fitzpatrick**
Advanced Calculus, Second Edition

4 **Gerald B. Folland**
Fourier Analysis and Its Applications

3 **Bettina Richmond and Thomas Richmond**
A Discrete Transition to Advanced Mathematics

2 **David Kincaid and Ward Cheney**
Numerical Analysis: Mathematics of Scientific Computing, Third Edition

1 **Edward D. Gaughan**
Introduction to Analysis, Fifth Edition